Optimal Measurement Methods for Distributed Parameter System Identification

Systems and Control Series
Edited by Eric Rogers

Optimal Measurement Methods for Distributed Parameter System Identification

Dariusz Uciński

CRC Press
Taylor & Francis Group
Boca Raton London New York

CRC Press is an imprint of the
Taylor & Francis Group, an **informa** business

CRC Press
Taylor & Francis Group
6000 Broken Sound Parkway NW, Suite 300
Boca Raton, FL 33487-2742

First issued in paperback 2019

© 2005 by Taylor & Francis Group, LLC
CRC Press is an imprint of Taylor & Francis Group, an Informa business

No claim to original U.S. Government works

ISBN-13: 978-0-8493-2313-3 (hbk)
ISBN-13: 978-0-367-39398-4 (pbk)

Library of Congress Cataloging-in-Publication Data

Catalog record is available from the Library of Congress

**Visit the Taylor & Francis Web site at
http://www.taylorandfrancis.com**

**and the CRC Press Web site at
http://www.crcpress.com**

To Ania, Ola and Piotr

About the Author

Dariusz Uciński was born in Gliwice, Poland, in 1965. He studied electrical engineering at the Technical University of Zielona Góra, Poland, from which he graduated in 1989. He received Ph.D. (1992) and D.Sc. (2000) degrees in automatic control and robotics from the Technical University of Wrocław, Poland. He is currently an associate professor at the University of Zielona Góra, Poland.

He has authored and co-authored numerous journal and conference papers. He also co-authored two textbooks in Polish: *Artificial Neural Networks — Foundations and Applications* and *Selected Numerical Methods of Solving Partial Differential Equations.* For fifteen years his major activities have been concentrated on measurement optimization for parameter estimation in distributed systems. In 2001 his habilitation thesis on this subject was granted an award from the Minister of National Education. In his career, he has been both a leader and a member of several national and international research projects. Other areas of expertise include optimum experimental design, algorithmic optimal control, robotics and cellular automata. Since 1992 he has been the scientific secretary of the editorial board of the *International Journal of Applied Mathematics and Computer Science.*

Contents

Preface

It is well understood that the choice of experimental conditions for distributed systems has a significant bearing upon the accuracy achievable in parameter-estimation experiments. Since for such systems it is impossible to observe their states over the entire spatial domain, close attention has been paid to the problem of optimally locating discrete sensors to estimate the unknown parameters as accurately as possible. Such an optimal sensor-location problem has been widely investigated by many authors since the beginning of the 1970s (for surveys, see [135, 145, 237, 298, 307]), but the existing methods are either restricted to one-dimensional spatial domains for which some theoretical results can be obtained for idealized linear models, or onerous, not only discouraging interactive use but also requiring a large investment in software development. The potential systematic approaches could be of significance, e.g., for environmental monitoring, meteorology, surveillance, hydrology and some industrial experiments, which are typical exemplary areas where we are faced with the sensor-location problem, especially owing to serious limitations on the number of costly sensors.

This was originally the main motivation to pursue the laborious research detailed in this monograph. My efforts on optimum experimental design for distributed-parameter systems began some fifteen years ago at a time when rapid advances in computing capabilities and availability held promise for significant progress in the development of a practically useful as well as theoretically sound methodology for such problems. At present, even low-cost personal computers allow us to solve routinely certain computational problems which would have been out of the question several years ago.

The aim of this monograph is to give an account of both classical and recent work on sensor placement for parameter estimation in dynamic distributed systems modelled by partial differential equations. We focus our attention on using real-valued functions of the Fisher information matrix of parameters as the performance index to be minimized with respect to the positions of point-wise sensors. Particular emphasis is placed on determining the 'best' way to guide scanning and moving sensors and making the solutions independent of the parameters to be identified. The bulk of the material in the corresponding chapters is taken from a collection of my original research papers. My main objective has been to produce useful results which can be easily translated into computer programmes. Apart from the excellent monograph by Werner Müller [181], which does not concern dynamic systems and has been written from a statistician's point of view, it is the first up-to-date and comprehensive monograph which systematizes characteristic features of the problem, analyses the existing approaches and proposes a wide range of original solutions. It brings together a large body of information on the topic, and presents it within a unified and relatively simple framework. As a result, it should provide researchers and engineers with a sound understanding of sensor-location techniques, or more generally, modern optimum experimental design, by offering a step-by-step guide to both theoretical aspects and practical design methods of sensor location, backed by many numerical examples.

The study of this subject is at the interface of several fields: optimum experimental design, partial differential equations, nonlinear programming, optimal control, stochastic processes, and numerical methods. Consequently, in order to give the reader a clear image of the proposed approach, the adopted strategy is to indicate direct arguments in relevant cases which preserve the essential features of the general situation, but avoid many technicalities.

This book is organized as follows. In Chapter 1, a brief summary of concrete applications involving the sensor-location problem is given. Some of these examples are used throughout the monograph to motivate and illustrate the demonstrated developments. A concise general review of the existing literature and a classification of methods for optimal sensor location are presented. Chapter 2 provides a detailed exposition of the measurement problem to be discussed in the remainder of the book and expounds the main complications which make this problem difficult. In Chapter 3 our main results for stationary sensors are stated and proved. Their extensions to the case of moving internal observations are reported in Chapter 4. Efficient original policies of activating scanning sensors are first proposed and examined. Then optimal design measures are treated in the context of moving sensors. A more realistic situation with nonnegligible dynamics of the vehicles conveying the sensors and various restrictions imposed on their motions is also studied therein and the whole problem is then formulated as a state-constrained optimal-control problem. Chapter 5 deals with vital extensions towards sensor location with alternative design objectives, such as prediction or model discrimination. Then Chapter 6 establishes some methods to overcome the difficulties related to the dependence of the optimal solutions on the parameters to be identified. Some indications of possible modifications which can serve to attack problems generally considered 'hard' are contained in Chapter 7. Chapter 8 attempts to treat in detail some case studies which are close to practical engineering problems. Finally, some concluding remarks are made in Chapter 9. The core chapters of the book are accompanied by nine appendices which collect accessory material, ranging from proofs of theoretical results to MATLAB implementations of the proposed algorithms.

The book may serve as a reference source for researchers and practitioners working in various fields (applied mathematics, electrical engineering, civil/geotechnical engineering, mechanical engineering, chemical/environmental engineering) who are interested in optimum experimental design, spatial statistics, distributed parameter systems, inverse problems, numerical analysis, optimization and applied optimal control. It may also be a textbook for graduate and postgraduate students in science and engineering disciplines (e.g., applied mathematics, engineering, physics/geology). As regards prerequisites, it is assumed that the reader is familiar with the basics of partial differential equations, vector spaces, probability and statistics. Appendices constitute an essential collection of mathematical results basic to the understanding of the material in this book.

Dariusz Uciński

Acknowledgments

It is a pleasure to express my sincere thanks to Professor Ewaryst Rafajłowicz, whose works introduced me to the field of experimental design for distributed-parameter systems, for many valuable suggestions and continuous support. In addition, I wish to express my gratitude to Professor Józef Korbicz for his encouragement in this project and suggesting the problem many years ago. Particular thanks are due to Dr. Maciej Patan, my talented student, who invested many hours of his time in developing numerical examples for Chapter 8. I wish to thank the reviewers from Taylor & Francis for their valuable suggestions. My appreciation is also extended to the editorial and production staff at CRC Press for their help in improving the manuscript and bringing it to production. Finally, I would like to thank the Polish State Committee for Scientific Research for supporting the writing of this book through a grant (contract 7 T11A 023 20).

1

Introduction

1.1 The optimum experimental design problem in context

Distributed-parameter systems (DPSs) are dynamical systems whose state depends not only on time but also on spatial coordinates. They are frequently encountered in practical engineering problems. Examples of a thermal nature are furnaces for heating metal slabs or heat exchangers; examples of a mechanical nature are large flexible antennas, aircrafts and robot arms; examples of an electrical nature are energy transmission lines.

Appropriate mathematical modelling of DPSs most often yields partial differential equations (PDEs), but descriptions by integral equations or integro-differential equations can sometimes be considered. Clearly, such models involve using very sophisticated mathematical methods, but in recompense for this effort we are in a position to describe the process more accurately and to implement more effective control strategies. Early lumping, which means approximation of a PDE by ordinary differential equations of possibly high order, may completely mask the distributed nature of the system and therefore is not always satisfactory.

For the past forty years DPSs have occupied an important place in control and systems theory. This position has grown in relevance due to the ever-expanding classes of engineering systems which are distributed in nature, and for which estimation and control are desired. DPSs, or more generally, infinite-dimensional systems are now an established area of research with a long list of journal articles, conference proceedings and several textbooks to its credit [59, 74, 76, 103, 133, 138, 140, 152, 167, 179, 200, 273, 357], so the field of potential applications could hardly be considered complete [17, 151, 317–319].

One of the basic and most important questions in DPSs is parameter estimation, which refers to the determination from observed data of unknown parameters in the system model such that the predicted response of the model is close, in some well-defined sense, to the process observations [200]. The parameter-estimation problem is also referred to as parameter identification or simply the inverse problem [118]. There are many areas of technological importance in which identification problems are of crucial significance. The importance of inverse problems in the petroleum industry, for example, is well

1

documented [80, 138]. One class of such problems involves determination of the porosity (the ratio of pore volume to total volume) and permeability (a parameter measuring the ease with which the fluids flow through the porous medium) of a petroleum reservoir based on field production data. Another class of inverse problems of interest in a variety of areas is to determine the elastic properties of an inhomogeneous medium from observations of reflections of waves travelling through the medium. The literature on the subject of DPS identification is considerable. Kubrusly [144] and Polis [216] have surveyed the field by systematically classifying the various techniques. A more recent book by Banks and Kunisch [16] is an attempt to present a thorough and unifying account of a broad class of identification techniques for DPS models, also see [14, 320].

In order to identify the unknown parameters (in other words, to calibrate the considered model), the system's behaviour or response is observed with the aid of some suitable collection of sensors termed the measurement or observation system. In many industrial processes the nature of state variables does not allow much flexibility as to which they can be measured. For variables which can be measured online, it is usually possible to make the measurements continuously in time. However, it is generally impossible to measure process states over the entire spatial domain. For example [211], the temperature of molten glass flowing slowly in a forehearth is described by a linear parabolic PDE, whereas the displacements occasioned by dynamic loading on a slender airframe can be described by linear second-order hyperbolic PDEs. In the former example, temperature measurements are available at selected points along the spatial domain (obtained by a pyrometer or some other device), whereas, in the latter case, strain gauge measurements at selected points on the airframe are reduced to yield the deflection data. In both cases the measurements are incomplete in the sense that the entire spatial profile is not available. Moreover, the measurements are inexact by virtue of inherent errors of measurement associated with transducing elements and also because of the measurement environment.

The inability to take distributed measurements of process states leads to the question of where to locate sensors so that the information content of the resulting signals with respect to the distributed state and PDE model be as high as possible. This is an appealing problem since in most applications these locations are not prespecified and therefore provide design parameters. The location of sensors is not necessarily dictated by physical considerations or by intuition and, therefore, some systematic approaches should still be developed in order to reduce the cost of instrumentation and to increase the efficiency of identifiers.

As already mentioned, the motivations to study the sensor-location problem stem from practical engineering issues. Optimization of air quality-monitoring networks is among the most interesting. As is well known, due to traffic emissions, residential combustion and industry emissions, air pollution has become a big social problem. One of the tasks of environmental protection systems is to provide expected levels of pollutant concentrations. In case smog is expected, a local community can be warned and some measures can be taken to prevent or minimize the release of prescribed substances and to render such substances harmless. But to produce such a forecast, a smog-prediction model is necessary [115, 284, 285, 331], which is usually chosen in the form of an advection-diffusion PDE. Its calibration requires parameter estimation (e.g., the unknown spatially varying turbulent diffusivity tensor should be identfied based on the measurements from monitoring stations [198, 199]). Since measurement transducers are usually rather costly and their number is limited, we are inevitably faced with the problem of how to optimize their locations in order to obtain the most precise model. A need for the appropriate strategies of optimally allocating monitoring stations is constantly indicated in the works which report the implementations of systems to perform air-quality management [3, 181, 194, 281, 332]. Of course, some approaches have already been advanced [84, 181]. Due to both the complexity of urban and industrial areas and the influence of meteorological quantities, the suggested techniques are not easy to apply and further research effort is required.

Another stimulating application concerns groundwater modelling employed in the study of groundwater-resources management, seawater intrusion, aquifer remediation, etc. To build a model for a real groundwater system, some observations of state variables such as the head and concentration are needed. But the cost of experiments is usually very high, which results in many efforts regarding, e.g., optimizing the decisions on the state variables to be observed, the number and location of observation wells and the observation frequency (see [282] and references given therein). Besides, it is easy to imagine that similar problems appear, e.g., in the recovery of valuable minerals and hydrocarbon from underground permeable reservoirs [80], in gathering measurement data for calibration of mathematical models used in meteorology and oceanography [20, 60, 114, 166, 190], in automated inspection in static and active environments, or in hazardous environments where trial-and-error sensor planning cannot be used (e.g., in nuclear power plants [136, 138]), or, in recent years, in emerging smart material systems [17, 151].

1.2 A general review of the literature

In general, the following main strategies of taking measurements can be distunguished:

- Locating a given number of stationary sensors

- Using moving sensors

- Scanning, i.e., only a part of a given total number of stationary sensors take measurements at a given moment in time

As a matter of fact, every real measuring transducer averages the measured quantity over some portion of the spatial domain. In most applicatons, however, this averaging is approximated by assuming that pointwise measurements are available at a number of spatial locations. Otherwise, the problem of finding an optimal geometry of the sensor support can be formulated [72].

Trying to implement the above-mentioned techniques, we are faced with the corresponding problems:

- How to determine an optimal sensor placement in a given admissible spatial domain?

- How to design optimal sensor trajectories?

- How to select the best subset of all available sensors to take measurements at a given time?

Additionally, in all cases we should also address the question of a minimal number of sensors which will guarantee sufficient accuracy of the estimates.

The literature on optimal sensor location in DPSs is plentiful, but most works deal with state estimation, see [2, 73, 75, 145] for surveys of the state of the art in the mid-1980s or more recent overviews [72, 135, 307]. There have been considerably fewer works on this subject related to parameter identification. This is mainly because a direct extension of the appropriate results from state estimation is not straightforward and has not been pursued. That problem is essentially different from the optimal measurement problem for parameter identification, since in the latter case the current state usually depends strongly nonlinearly on unknown parameters [135], even though the PDE is linear in these parameters, while the dependence of the current state on the initial one is linear for linear systems, which makes state estimation easier.

The existing methods of sensor location for parameter identification can be gathered in three major groups:

1. Methods leading to state estimation

2. Methods employing random fields analysis

3. Methods using optimum experimental design theory

Group 1 brings together some attempts to transform the problem into a state-estimation one (by augmenting the state vector) and then to use well-developed methods of optimal sensor location for state estimation. However, since the state and parameter estimation are to be carried out simultaneously, the whole problem becomes strongly nonlinear. To overcome this difficulty, a sequence of linearizations at consecutive state trajectories was performed by Malebranche [168] and a special suboptimal filtering algorithm was used by Korbicz *et al.* [137]. Nevertheless, the viability of this approach is rather questionable owing to the well-known severe difficulties inherent in nonlinear state estimation.

The methods of Group 2 are based on random fields theory. Since DPSs are described by PDEs, direct application of that theory is impossible, and therefore this description should be replaced by characteristics of a random field, e.g., mean and covariance functions. Such a method for a beam vibrating due to the action of stochastic loading was considered by Kazimierczyk [128] who made extensive use of optimum experimental design for random fields [31]. Although the flexibility of this approach seems rather limited, it can be useful in some case studies, see, e.g., [282].

The methods belonging to major Group 3 originate from the classical theory of optimum experimental design [77,82,85,132,208,224,245,349] and its extensions to models for dynamic systems, especially in the context of the optimal choice of sampling instants and input signals [101,123,142,174,175,292,349]. Consequently, the adopted optimization criteria are essentially the same, i.e., various scalar measures of performance based on the Fisher information matrix (FIM) associated with the parameters to be identified are minimized. The underlying idea is to express the goodness of parameter estimates in terms of the covariance matrix of the estimates. For sensor-location purposes, one assumes that an unbiased and efficient (or minimum-variance) estimator is employed so that the optimal sensor placement can be determined independently of the estimator used. This leads to a great simplification since the Cramér-Rao lower bound for the aforementioned covariance matrix is merely the inverse of the FIM, which can be computed with relative ease, even though the exact covariance matrix of a particular estimator is very difficult to obtain.

There is a fundamental complication in the application of the resulting optimal location strategies and that is the dependence of the optimal solutions on the parameters to be identified. It seems that we have to know their true values in order to calculate an optimal sensor configuration for estimating them. As a result, practically all the works in context are based on the assumption of *a priori* estimates for the true parameters.

As regards dynamic DPSs, the first treatment of this type for the sensor-location problem was proposed by Quereshi *et al.* [231]. Their approach was based on maximization of the determinant of the FIM and examples regarding a damped vibrating string and a heat-diffusion process were used to illustrate the advantages and peculiarities of the method. Besides sensor location, the determination of boundary perturbations was also considered. In order to avoid computational difficulties, sinusoidal excitations were assumed and the position of a single sensor was optimized.

The same optimality criterion was used by Rafajłowicz [232] in order to optimize both sensor positions and a distributed control for parameter estimation of a static linear DPS. Reduction of the problem to a form where results of the classical theory of optimum experimental design can be applied, was accomplished after eigenfunction expansion of the solution to the PDE considered and subsequent truncation of the resulting infinite series. Consequently, the FIM was associated with system eigenvalues, rather than with the system parameters. A separation principle was proved which allows the possibility of finding an optimal control and an optimal sensor configuration independently of each other. The delineated approach was generalized in [233] to a class of DPSs described by linear hyperbolic equations with known eigenfunctions and unknown eigenvalues. The aim was to find conditions for optimality of measurement design and of optimal spectral density of the stochastic input. It was indicated that common numerical procedures from classical experimental design for linear regression models could be adopted to find optimal sensor location. Moreover, the demonstrated optimality conditions imply that the optimal input comprises a finite number of sinusoidal signals and that optimal sensor positions are not difficult to find in some cases. A similar problem was studied in [234] in a more general framework of DPSs which can be described in terms of Green's functions.

The idea of generalizing methods of optimum experimental design for parameter identification of lumped systems was also applied to solve the optimal measurement problem for moving sensors [238]. The approach was based on looking for a time-dependent measure, rather than for the trajectories themselves. Various sufficient optimality conditions were presented, among others the so-called *quasi-maximum principle*. In spite of their somewhat abstract forms, they made it possible to solve relatively easily a number of nontrivial examples. The problem of moving sensors in DPSs was also revisited in [240, 241], but without direct reference to parameter estimation.

As regards other works by the same author which pertain to the optimal measurement problem for DPSs, let us also mention [239] where the notion of equi-informative sensor-actuator configurations in the sense of the coincidence of the corresponding FIMs was introduced and studied for a class of static DPSs, and [244] where a sensor allocation was sought so as to maximize the identification accuracy for the mean values of random pointwise inputs to a static DPS described by Green's function. In turn, optimization of input signals for a fixed sensor configuration was exhaustively treated by Rafajłowicz [235–237, 242, 243] and Rafajłowicz and Myszka [247].

The approach based on maximization of the determinant of the appropriate FIM is not restricted to theoretical considerations and there are examples which do confirm its effectiveness in practical applications. Thus, in [185] a given number of stationary sensors were optimally located using nonlinear programming techniques for a biotechnological system consisting of a bubble column loop fermenter. On the other hand, Sun [282] advocates using optimum experimental design techniques to solve inverse problems in groundwater modelling. How to monitor the water quality around a landfill place is an example of such a network design. Sun's monograph constitutes an excellent introductory text to applied experimental design for DPSs, as it covers a broad range of issues motivated by engineering problems. Nonlinear programming techniques are also used there to find numerical approximations to the respective exact solutions.

A similar approach was used by Kammer [124, 125] for on-orbit modal identification of large space structures. Although the respective models are not PDEs, but their discretized versions obtained through the finite-element method, the proposed solutions can still be of interest owing to the striking similitude of both the formulations. A fast and efficient approach was delineated for reducing a relatively large initial candidate sensor-location set to a much smaller optimum set which retains the linear independence of the target modes and maintains the determinant of the FIM resulting in improved modal-response estimates.

A related optimality criterion was given in [213] by the maximization of the Gram determinant, which is a measure of the independence of the sensitivity functions evaluated at sensor locations. The authors argue that such a procedure guarantees that the parameters are identifiable and the correlation between the sensor outputs is minimized. The form of the criterion itself resembles the D-optimality criterion proposed by Quereshi *et al.* and Rafajłowicz, but the counterpart of the FIM takes on much larger dimensions, which suggests that the approach involves more cumbersome calculations. The delineated technique was successfully applied to a laboratory-scale, catalytic fixed-bed reactor [333].

At this juncture, it should be noted that spatial design methods related to the design of monitoring networks are also of great interest to statisticians and a vast amount of literature on the subject already exists [181, 194, 195] contributing to the research field of spatial statistics [57] motivated by practical problems in agriculture, geology, meteorology, environmental sciences and economics. However, the models considered in the statistical literature are quite different from the dynamic models described by PDEs discussed here. Spatiotemporal data are not considered in this context (hence techniques discussed here such as moving sensors and scanning are not applicable within that framework) and the main purpose is to model the spatial process by a spatial random field, incorporate prior knowledge and select the best subset of points of a desired cardinality to best represent the field in question. The motivation is a need to interpolate the observed behaviour of a process at unobserved spatial locations, as well as to design a network of optimal observation locations which allows an accurate representation of the process. The field itself is modelled by some multivariate distribution, usually Gaussian [5]. Designs for spatial trend and variogram estimation can be considered. The basic theory of optimal design for spatial random fields is outlined in the excellent monograph by Müller [181] which bridges the gap between spatial statistics and classical optimum experimental design theory. The optimal design problem can also be formulated in terms of information-based criteria whose application amounts to maximizing the amount of information (of the Kullback-Leibler type) to be gained from an experiment [41, 42]. However, the applicability of all those fine statistical results in the engineering context discussed here is not clear for now and more detailed research into this direction should be pursued in the near future (specifically, generalizations regarding time dynamics are not obvious).

In summary, this brief review of the state of the art in the sensor-location problem indicates that, as was already emphasized by Kubrusly and Malebranche, more attention should be paid to the problem of sensor allocation for parameter identification of DPSs, as from an engineering point of view the use of the existing scarce methods is restricted owing to computational and/or realizing difficulties. Few works have appeared about the results regarding two- or three-dimensional spatial domains and spatially varying parameters. Thus, some generalizations are still expected in this connection. Furthermore, most of the contributions deal with stationary sensors. On the other hand, the optimal measurement problem for scanning or spatially movable sensors seems to be very attractive from the viewpoint of the degree of optimality and should receive more attention. Similarly, the dependence of the solutions on the values assumed for the unknown parameters to be identified should be addressed with greater care, as this limitation of the existing methods seems to be one of the main impediments to persuading engineers to apply these methods in practice.

This monograph constitutes an attempt to systematize the existing approaches to the sensor-location problem and to meet the above mentioned needs created by practical applications through the development of new techniques and algorithms or adopting methods which have been successful in related fields of optimum experimental design. It is an outgrowth of original research papers and some results which have not been published yet. We believe that the approach outlined here has significant practical and theoretical advantages which will make it, with sufficient development, a versatile tool in numerous sensor-location problems encountered in engineering practice.

2

Key ideas of identification and experimental design

The construction of a DPS model is a procedure which involves three basic steps [46]:

Step 1: A careful analysis of the governing physical laws, which results in a set of PDEs accompanied by the appropriate boundary conditions such that the well posedness of such a description is guaranteed.

Step 2: Reduction of the problem to a finite-dimensional representation which can be treated on a computer. Finite-element or finite-difference techniques are most often used at this stage.

Step 3: Model calibration, which refers to the determination from observed data of unknown parameters in the system model obtained in the previous step such that some (error, loss or objective) function of the differences between the predicted response of the model and the observed output of the DPS is minimized.

The last step is called parameter identification and the present chapter establishes the relation between the accuracy of the parameter estimates produced by this process and the choice of experimental conditions.

2.1 System description

In this section, we introduce the class of systems to be considered. Let Ω be a bounded, simply connected open domain of a d-dimensional Euclidean space $\mathbb{R}^d$ with sufficiently smooth boundary $\partial\Omega$. Since the results outlined in this monograph are strongly motivated by two-dimensional situations, the explicit formulae will be written for $d = 2$, but bear in mind that they can easily be generalized to $d = 3$ or limited to $d = 1$. Accordingly, the spatial coordinate vector will be denoted by $x = (x_1, x_2) \in \bar{\Omega} = \Omega \cup \partial\Omega$. As our fundamental state system we consider the scalar (possibly nonlinear) distributed-parameter system described by a partial differential equation of the form

$$\frac{\partial y}{\partial t} = \mathcal{F}\left(x, t, y, \frac{\partial y}{\partial x_1}, \frac{\partial y}{\partial x_2}, \frac{\partial^2 y}{\partial x_1^2}, \frac{\partial^2 y}{\partial x_2^2}, \theta\right), \quad x \in \Omega, \quad t \in T, \tag{2.1}$$

where t stands for time, $T = (0, t_f)$, $y = y(x, t)$ denotes the state variable with values in $\mathbb{R}$ and $\mathcal{F}$ is some known function which may include terms accounting for given *a priori* forcing inputs. The system evolves from $t = 0$ to $t = t_f < \infty$, the period over which observations are available.

Equation (2.1) is accompanied by the boundary condition of the general form

$$\mathcal{E}\left(x, t, y, \frac{\partial y}{\partial x_1}, \frac{\partial y}{\partial x_2}, \theta\right) = 0, \quad x \in \partial\Omega, \quad t \in T \tag{2.2}$$

and the initial condition

$$y(x, 0) = y_0(x), \quad x \in \Omega, \tag{2.3}$$

where $\mathcal{E}$ and y_0 denote some known functions.

We assume the existence of a unique solution to (2.1)–(2.3), which is sufficiently regular. The system model above contains an unknown parameter vector denoted by θ (note that it may also appear in the boundary conditions), which is assumed to belong to a parameter space Θ. The possible forms of Θ are [200]:

1. Constant parameters

$$\Theta_1 = \left\{\theta = (\theta_1, \theta_2, \ldots, \theta_m) \in \mathbb{R}^m\right\}.$$

2. Assumed functional form, i.e., some or all components of θ are assumed to have known functional forms which contain unknown parameters. Thus,

$$\theta = \theta(x, t, y) = g(x, t, y, \theta_1, \theta_2, \ldots, \theta_m),$$

where θ_i, $i = 1, \ldots, m$ are unknown constants and the functional form of each component of g is assumed to be known:

$$\Theta_2 = \left\{\theta = (\theta_1, \theta_2, \ldots, \theta_m) \in \mathbb{R}^m\right\}.$$

3. General functions of space x, and/or time t, and/or the state y, i.e.,

$$\Theta_3 = \{\theta = (\theta_1(x, t, y), \ldots, \theta_m(x, t, y))\},$$

where the parameter space in this case is infinite dimensional.

From a practical point of view, Case 2 does not differ in anything from Case 1, and Case 3 must be approximated eventually by a finite-dimensional space to obtain numerical results (within certain limitations discussed on p. 14), so in what follows we will focus our attention on Case 1, which is also a common procedure in the literature.

Example 2.1

A chief aim of the intensive studies on the atmospheric aspects of air pollution is to be able to describe mathematically the spatiotemporal distribution of contaminants released into the atmosphere. The phenomenon of pollutant advection and diffusion over a spatial domain Ω containing an urban region is governed by the equation [115]

$$\frac{\partial y}{\partial t} = -v_1 \frac{\partial y}{\partial x_1} - v_2 \frac{\partial y}{\partial x_2} + \kappa \left(\frac{\partial^2 y}{\partial x_1^2} + \frac{\partial^2 y}{\partial x_2^2} \right) - \gamma y + \frac{1}{H}(E - \sigma y) + Q$$

$$\text{in } \Omega \times T,$$

subject to the boundary condition

$$\frac{\partial y}{\partial n}(x,t) = 0 \quad \text{on } \partial\Omega \times T \text{ for } \langle v, n \rangle \geq 0,$$
$$y(x,t) = 0 \quad \text{on } \partial\Omega \times T \text{ for } \langle v, n \rangle < 0$$

and the initial condition

$$y(x,0) = y_0 \quad \text{in } \Omega,$$

where $y = y(x,t)$ denotes the pollutant concentration, $v_i = v_i(x,t)$, $i = 1,2$ stand, respectively, for the x_1 and x_2 directional wind velocities, $v = (v_1, v_2)$, $Q = Q(x,t)$ signifies the intensity of the pointwise emission sources, all these quantities being averaged in the vertical direction over the mixing layer whose height is H. The term $E - \sigma y$ represents the ground-level stream of the pollutant, where $E = E(x,t)$ is the intensity of the area emission and σ denotes the dry deposition coefficient. Furthermore, κ is the horizontal diffusion coefficient and γ is the wet deposition factor, which usually depends on the precipitation intensity. Here the notation $\partial y/\partial n$ means the partial derivative of y with respect to the outward normal of $\partial\Omega$, n, and $\langle \cdot, \cdot \rangle$ is used to denote the scalar product of ordinary vectors.

Some parameters which appear as coefficients in this model are known or can be obtained from direct observations (e.g., the wind field v is usually available from weather forecasting centres), but the others cannot be measured (e.g., the diffusivity κ, cf. [198, 199]) and consequently they constitute components of the vector θ introduced above. Due to spatial changes in environmental conditions, it is highly likely that the unknown parameters will also be spatially varying. Their precise determination is essential to the process of accurately simulating and predicting the spatial distribution of air-pollutant concentrations. ▯

The objective of parameter estimation is to choose a parameter θ^* in Θ so that the solution y to (2.1)–(2.3) corresponding to $\theta = \theta^*$ agrees with the 'true' observed state $\tilde{y}$. In general, however, measurements of the state may not be possible, rather only measurements for some observable part of the actual state $\tilde{y}$ may be available [16]. This is related to the fact that there are several possible manners in which the measurements themselves are made. For example, measurements may be carried out [49]

(M1) over the entire spatial domain continuously in time,

(M2) over the entire spatial domain at discrete points in time,

(M3) at discrete spatial locations continuously in time, or

(M4) at discrete spatial locations at discrete points in time.

Manners (M1) and (M2) are considered to be of little significance, since it is generally not possible to carry out measurements over the entire domain of the system. Some authors adopt them notwithstanding and obtain 'distributed' measurements by an appropriate interpolation of the pointwise data [146, 149]. Note, however, that recent technological advances in measuring instrumentation are very promising for this type of observation (e.g., thermography cameras used to continuously monitor manufacturing processes, scanning tunnelling microscopes for imaging solid surfaces, or scanning thermal conductivity microscopes capable of producing thermal conductivity maps of specimen surfaces with submicron resolution).

In contrast, measurements at discrete spatial locations are commonly encountered in engineering applications and, as a consequence, they dominate in the literature on parameter identification. Manner (M3) causes the problem of choosing the spatial locations and Manner (M4) involves a choice of both spatial locations and measurement timing. If measurements can only be made intermittently, the time interval between measurements is usually determined by the time requirements of the analytical procedure. If the meaurements are not costly, then one will take data as frequently as possible. As long as the timing of measurements is not a decision variable, Manners (M3) and (M4) are basically equivalent. Thus we focus here on Manner (M3) only. Based on the developed ideas, it is a simple matter to procure the corresponding results for Manner (M4).

Consequently, it is further assumed that the observation process is described by the equation of the form

$$z(t) = y_{\mathrm{m}}(t) + \varepsilon_{\mathrm{m}}(t), \quad t \in T, \tag{2.4}$$

where

$$y_{\mathrm{m}}(t) = \mathrm{col}[y(x^1, t), \dots, y(x^N, t)],$$
$$\varepsilon_{\mathrm{m}}(t) = \mathrm{col}[\varepsilon(x^1, t), \dots, \varepsilon(x^N, t)],$$

$z(t)$ is the N-dimensional observation vector, $x^j \in \bar{\Omega}$, $j = 1, \ldots, N$ denote the pointwise and stationary sensor locations, and $\varepsilon = \varepsilon(x, t)$ is a white Gaussian noise process (a formal time derivative of a Wiener process) whose statistics are

$$E\{\varepsilon(x, t)\} = 0, \quad E\{\varepsilon(x, t)\varepsilon(x', t')\} = q(x, x', t)\delta(t - t'), \tag{2.5}$$

δ being the Dirac delta function concentrated at the origin.

Note that pointwise sensors require a corresponding smoothness of solutions to (2.1)–(2.3). In fact, each solution y may be thought of as a collection of functions of space parameterized by time, $\{y(\,\cdot\,, t)\}_{t \in T}$. But a 'snapshot' $y(\,\cdot\,, t) \in L^2(\Omega)$ need not be continuous and it is therefore not meaningful to talk about its values at particular points. On the other hand, explicitly requiring $y(\,\cdot\,, t)$ to be continuous is not convenient, since $C(\bar{\Omega})$ is not a Hilbert space. Consequently, we must be careful in specifying the functional framework for our problem. For the rest of this book, we shall consider pointwise observations generally assuming that $\mathcal{F}$, $\mathcal{E}$ and the boundary $\partial\Omega$ are sufficiently regular to ensure the enclosure of outputs into $L^2(T; \mathbb{R}^N)$. Let us observe that this condition is met, e.g., if $y \in L^2(T; H^2(\Omega))$, where $H^2(\Omega)$ is the second-order Sobolev space. In fact, from the Sobolev embedding theorem [58, Th. 8.5, p. 141] it follows that $H^2(\Omega) \subset C(\bar{\Omega})$ with continuous injection provided that the dimension of Ω (i.e., d) is less than or equal to three. This means that there exists a constant $K \in (0, \infty)$ such that

$$\|v\|_{C(\bar{\Omega})} \le K\|v\|_{H^2(\Omega)}, \quad \forall v \in H^2(\Omega). \tag{2.6}$$

The result is

$$\|y_{\mathrm{m}}\|^2_{L^2(T; \mathbb{R}^N)} = \int_T \|y_{\mathrm{m}}(t)\|^2_{\mathbb{R}^N} \, \mathrm{d}t \le N \int_T \|y(\,\cdot\,, t)\|^2_{C(\bar{\Omega})} \, \mathrm{d}t$$
$$\le NK^2 \int_T \|y(\,\cdot\,, t)\|^2_{H^2(\Omega)} \, \mathrm{d}t = NK^2\|y\|^2_{L^2(T; H^2(\Omega))}. \tag{2.7}$$

2.2   Parameter identification

Based on a collection of data $\{z(t)\}_{t \in T}$ which have been observed from our physical process, we wish to calibrate the model (2.1)–(2.3), i.e., to determine a parameter vector $\hat{\theta}$ such that the predicted response of the model is close, in some well-defined sense, to the process observations. Such calibration is generally performed for one of the following objectives: 1) accurate determination of parameter values which may have some physical significance, such as specific heat and thermal conductivity of materials; 2) response prediction and forecasting; and 3) control-system design. Mathematically, the problem is usually cast as an optimization one, which leads to the so-called *weighted least-squares* approach to parameter estimation in which we seek to minimize the output-error criterion (also called the fit-to-data criterion)

$$\mathcal{J}(\theta) = \frac{1}{2} \int_T \|z(t) - \hat{y}_m(t; \theta)\|^2_{Q^{-1}(t)} \, dt, \qquad (2.8)$$

where $Q(t) = \left[q(x^i, x^j, t)\right]^N_{i,j=1} \in \mathbb{R}^{N \times N}$ is assumed to be positive definite,

$$\|a\|^2_{Q^{-1}(t)} = a^\mathsf{T} Q^{-1}(t) a, \quad \forall a \in \mathbb{R}^N,$$
$$\hat{y}_m(t; \theta) = \mathrm{col}[\hat{y}(x^1, t; \theta), \ldots, \hat{y}(x^N, t; \theta)],$$

and $\hat{y}(\cdot, \cdot; \theta)$ stands for the solution to (2.1)–(2.3) corresponding to a given parameter θ.

A decided advantage of using (2.8) stems from the fact that the system model within which the unknown parameters are imbedded is largely irrelevant and hence techniques for the estimation of parameters in lumped-parameter systems can be readily extended to the distributed-parameter case. Moreover, it can often be used effectively even if only a minimal data set is available. In turn, a disadvantage is that this criterion is almost never quadratic in the parameters and therefore the addressed optimization problem may be poorly conditioned (e.g., the surface plot of the criterion may be very flat). Another drawback to this approach manifests itself when the infinite-dimensional parameter space is replaced by a space of finite dimension. If the number of parameters is kept small, a well-behaved solution results. However, the modelling error introduced is significant, since the corresponding subspace of θ's is too restricted to provide a good approximation of an arbitrary θ. As the number of parameters is increased, on the other hand, numerical instabilities appear, manifested by spatial oscillations in the estimated θ, the frequency and amplitude of which are inconsistent with the expected smoothness of the true θ. Restoring a type of problem stability necessitates employing some regularization approaches [14, 16, 48, 141, 146, 149], which leads to an increased technical complexity. But in spite of these inconveniences, in actual fact (2.8) is commonly adopted as the first choice in parameter-estimation problems. For some alternatives, the interested reader is referred to the comprehensive monographs [16, 17].

2.3 Measurement-location problem

Clearly, the parameter estimate $\hat{\theta}$ resulting from minimization of the fit-to-data criterion depends on the sensor positions since one can observe the quantity z in the integrand on the right-hand side of (2.8). This fact suggests that we may attempt to select sensor locations which lead to best estimates of the system parameters. To form a basis for the comparison of different locations, a quantitative measure of the 'goodness' of particular locations is required. A logical approach is to choose a measure related to the expected accuracy of the parameter estimates to be obtained from the data collected. Such a measure is usually based on the concept of the *Fisher Information Matrix* (FIM) [238,282], which is widely used in optimum experimental design theory for lumped systems [85,349]. When the time horizon is large, the nonlinearity of the model with respect to its parameters is mild, and the measurement errors are independently distributed and have small magnitudes, the inverse of the FIM constitutes an approximation of the covariance matrix for the estimate of θ [85, 349]. Derivation of this fundamental property is centred on the use of the Cramér-Rao inequality* [101]

$$\operatorname{cov} \hat{\theta} \succeq M^{-1} \tag{2.9}$$

as the starting point (here M is just the FIM), which requires the additional qualification that the estimator $\hat{\theta}$ is unbiased. Accordingly, it is sensible to assume that the estimator is *efficient* (minimum-variance) in the sense that the parameter covariance matrix achieves the lower bound, i.e., (2.9) becomes an equality, which is justified in many situations [237]. This leads to a great simplification since the minimum variance given by the Cramér-Rao lower bound can be easily computed in a number of estimation problems, even though the exact covariance matrix of a particular estimator is very difficult to obtain.

It is customary to restrict investigations to spatial uncorrelated observations, i.e.,

$$\mathrm{E}\{\varepsilon(x^i, t)\varepsilon(x^j, t')\} = \sigma^2 \delta_{ij}\delta(t - t'), \tag{2.10}$$

where δ_{ij} denotes the Kronecker delta function and $\sigma > 0$ is the standard deviation of the measurement noise. The reason obviously lies in the simplicity of a subsequent analysis. Such an assumption yields the following explicit form of the FIM [231]:

$$M = \frac{1}{\sigma^2} \sum_{j=1}^{N} \int_0^{t_f} \left(\frac{\partial y(x^j, t)}{\partial \theta}\right)^{\mathsf{T}} \left(\frac{\partial y(x^j, t)}{\partial \theta}\right) \mathrm{dt}, \tag{2.11}$$

which is encountered in the bulk of the literature on sensor location. We shall also adopt this approach in the remainder of this chapter.

Clearly, the elements of M depend on the corresponding sensor positions. We shall emphasize this dependence setting:

$$s = (x^1, \ldots, x^N) \in \mathbb{R}^{2N} \tag{2.12}$$

---

*Recall that (2.9) should be interpreted in terms of the Löwner ordering of symmetric matrices, i.e., the matrix $\operatorname{cov} \hat{\theta} - M^{-1}$ is required to be nonnegative definite.

and writing here and subsequently $M = M(s)$.

It is our chief aim here to determine a sensor configuration s such that θ is estimated with as much precision as possible. From (2.9) we see that the FIM is a good candidate for quantifying the accuracy of the θ estimate. It reflects the variability in the estimate that is induced from the stochastic variability of the output $z(\cdot)$. A 'larger' value of M reflects more precision — lower variability — in the estimate. Given the close connection of the inverse of the covariance matrix to the precision matrix M, a natural goal in selecting the sensor configuration is to find sensor locations which make the matrix M 'large.' We might first ask if it is possible, in general, to find such a solution. That is [274], does there exist a configuration $s^\star$ such that $M(s^\star) \succeq M(s)$ for all possible s? The answer is no, as only in exceptional circumstances is such an $s^\star$ likely to exist and it is not difficult to find examples to destroy our vain hope, cf. [269, p. 5].

Since there is little point in seeking a configuration $s^\star$ which is optimal in the very strong sense outlined above, we have to content ourselves with introducing a weaker metric that is similar in spirit [274]. Accordingly, we attempt to find an $s^\star$ which minimizes some real-valued function Ψ defined on all possible FIMs. Various choices exist for such a function [85, 208, 224, 349], including, e.g., the following:

- The D-optimality (determinant) criterion

$$\Psi(M) = -\ln\det(M). \tag{2.13}$$

- The E-optimality criterion (smallest eigenvalue; $\lambda_{\max}(\cdot)$ denotes the maximum eigenvalue of its argument)

$$\Psi(M) = \lambda_{\max}(M^{-1}). \tag{2.14}$$

- The A-optimality (trace) criterion

$$\Psi(M) = \operatorname{tr}(M^{-1}). \tag{2.15}$$

- The sensitivity criterion

$$\Psi(M) = -\operatorname{tr}(M). \tag{2.16}$$

A D-optimum design minimizes the volume of the uncertainty ellipsoid for the estimates. An E-optimum design minimizes the length of the largest axis of the same ellipsoid. An A-optimum design suppresses the average variance of the estimates. An important advantage of D-optimality is that it is invariant under scale changes in the parameters and linear transformations of the output, whereas A-optimality and E-optimality are affected by these transformations. The sensitivity criterion is often used due to its simplicity, but it sometimes leads to serious problems with identifiability as it may result in a singular FIM [356], so in principle it should be used only to obtain startup locations for one of the above criteria. The introduction of an optimality criterion renders it possible to formulate the sensor-location problem as an optimization problem.

The above criteria are no doubt the most popular ones in modern experimental design theory, but they are by no means the only options. In fact, there is a plethora of other criteria which are labelled by other letters (some authors even speak of 'alphabetic' criteria [85]). A rather general class of optimality criteria employs the following family of costs functions [85, 349]:

$$\Psi_\gamma(M) = \begin{cases} \left[\dfrac{1}{m}\operatorname{tr}(PM^{-1}P^\mathsf{T})^\gamma\right]^{1/\gamma} & \text{if } \det M \neq 0, \\ \infty & \text{if } \det M = 0, \end{cases} \tag{2.17}$$

where $P \in \mathbb{R}^{m \times m}$ is a weighting matrix. For example, setting $P = I_m$ (the identity matrix), we obtain $\gamma = 1$, $\gamma \to \infty$ and $\gamma \to 0$ for the A-, E- and D-optimum design criteria, respectively.

Example 2.2
In order to illustrate the introduced ideas, consider a thin rod or wire whose lateral surface is impervious to heat (i.e., insulated). For modelling purposes, we assume the rod coincides with the x-axis from $x = 0$ to $x = 1$, is made of uniform material, and has a uniform cross-section. We know that the initial temperature of the rod is specified as $\sin(\pi x)$. The temperature distribution $y = y(x, t)$ at some later time in the absence of any heat source is then a solution to the one-dimensional heat equation

$$\frac{\partial y}{\partial t}(x, t) = \theta \frac{\partial^2 y}{\partial x^2}(x, t), \quad x \in (0, 1), \ t \in (0, t_f) \tag{2.18}$$

and the prescribed initial condition

$$y(x, 0) = \sin(\pi x), \quad x \in (0, 1), \tag{2.19}$$

where θ stands for the diffusivity of the material forming the rod. The temperature inside the rod will also be affected by how the ends of the rod exchange heat energy with the surrounding medium. Suppose the two ends of the rod are at time $t = 0$ suddenly placed in contact with ice packs at $0°$ and that the temperature at the ends is maintained at all later times. This corresponds to the boundary conditions

$$y(0, t) = y(1, t) = 0, \quad t \in (0, t_f). \tag{2.20}$$

The model is specified up to the value of the diffusivity coefficient θ, which is not known exactly. Assume that the temperature $y(x, t)$ can be measured continuously by one thermocouple and the measurements are described by

$$z(t) = y(x^1, t) + \varepsilon(t), \tag{2.21}$$

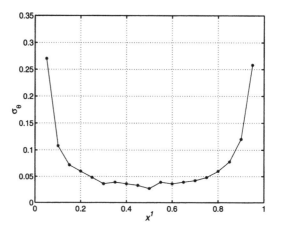

FIGURE 2.1
Empirical mean-squared estimation error for θ of (2.18)–(2.20) versus different sensor locations.

where the measurement noise ε is Gaussian and white with variance σ^2, and x^1 signifies the measurement location. It is desirable to select an optimal sensor location x^1 so as to obtain a best estimate of θ. This is to be accomplished prior to the experiment itself and the subsequent identification process.

In order to settle our problem, first let us note that (2.18)–(2.20) has the solution

$$y(x,t) = \exp(-\theta\pi^2 t)\sin(\pi x). \qquad (2.22)$$

Hence the FIM (which reduces to a scalar, since there is only one parameter to be identified) is of the form

$$
\begin{aligned}
M(x^1) &= \frac{1}{\sigma^2} \int_0^{t_f} \left(\frac{\partial y(x^1,t)}{\partial \theta} \right)^2 dt \\
&= \frac{1}{\sigma^2} \int_0^{t_f} \left\{ -\pi^2 t \exp(-\theta\pi^2 t)\sin(\pi x) \right\}^2 dt \\
&= \underbrace{\frac{1}{\sigma^2}\pi^4 \int_0^{t_f} t^2 \exp(-2\theta\pi^2 t)\,dt}_{\text{a positive constant}} \sin^2(\pi x)
\end{aligned}
\qquad (2.23)
$$

and it is evident that it attains a maximum at $x^1 = 1/2$. Since practically all the design criteria in common use meet the condition of monotonicity (cf. (A4) on p. 41), the centre of the rod corresponds to a point of the minimum of $\Psi[M(x^1)]$ for any Ψ as well. That is where the sensitivity of the output with respect to changes in θ is maximal. Consequently, measurements at the centre convey more information about the dynamics of our heat transfer process.

The benefits of the optimum sensor location can be evaluated via simulations. For that purpose, a computer code was written to simulate the system behaviour for $\theta_{\text{true}} = 1$ and $t_f = 0.2$. An $\mathcal{N}(0, 0.05)$ normal distribution was assumed for the noise and uniform sampling with time period $\Delta t = 0.01$ simulated time-continuous measurements. As potential sensor locations, the points $x_i^1 = 0.05i$, $i = 1, \ldots, 19$ were tested by performing $L = 150$ identification experiments at each of them (to this end, the routine dbrent from [219] was used as the minimizer). Figure 2.1 shows the empirical mean-squared estimation error

$$
\hat{\sigma}_\theta(x^1) = \sqrt{\frac{1}{L-1} \sum_{j=1}^{L} \left(\hat{\theta}_j(x^1) - \frac{1}{L} \sum_{\ell=1}^{L} \hat{\theta}_\ell(x^1) \right)^2}
$$

as a function of the sensor location. As predicted, the best point for taking measurements is just the centre of the rod, since at that point the dispersion of the estimates is the least of all. Also note that the accuracy of the estimates is almost ten times as great for the optimal point as for the outermost allowable locations, which indicates that even sophisticated techniques of sensor placement can be worthwile if we wish to identify the parameters with great precision. ☐

2.4 Main impediments to solving the sensor-placement problem

Transformation of the initial sensor location problem to minimization of a criterion defined on the FIM may lead to the conclusion that the only question remaining is that of selecting an appropriate solver from a library of numerical optimization routines. Unfortunately, the reality turns out to be harsher than the previous perfunctory analysis suggests. This is mainly because of the four problems outlined in what follows, which explain to a certain extent why so few works have been published on this subject so far when compared, e.g., with the sensor-location problem for state estimation.

2.4.1 High dimensionality of the multimodal optimization problem

In practice, the number of sensors to be placed in a given region may be quite large. For example, in the research carried out to find spatial predictions for ozone in the Great Lakes of the United States, measurements made by approximately 160 monitoring stations were used [195]. When trying to treat the task as a constrained nonlinear programming problem, the actual number of variables is even doubled, since the position of each sensor is determined by its two spatial coordinates, so that the resulting problem is rather of large scale. What is more, a desired global extremum is usually hidden among many poorer local extrema. Consequently, to directly find a numerical solution may be extremely difficult. One approach which makes this problem dimensionality substantially lower is delineated in the next chapter.

2.4.2 Loss of the underlying properties of the estimator for finite horizons of observation

As a matter of fact, the approximation of the covariance matrix for the parameter estimates by the inverse of the FIM is justified when $t_f \to \infty$ [238, 349]. In practice, this is a rare case, as the observation horizon is usually limited by imposed technical requirements. But the resulting loss in accuracy is commonly neglected in practical applications.

2.4.3 Sensor clusterization

One of the most serious problems which complicate the selection of measurement points is the sensors' clusterization being a consequence of the assumption that the measurement noise is spatially uncorrelated. This means that in an optimal solution different sensors often tend to take measurements at one point, which is obviously unacceptable from a technical point of view. This phenomenon is illustrated with the following example.

Example 2.3
Let us consider the following one-dimensional heat equation:

$$\frac{\partial y(x,t)}{\partial t} = \theta_1 \frac{\partial^2 y(x,t)}{\partial x^2}, \quad x \in (0,\pi), \quad t \in (0,1)$$

supplemented by the conditions

$$\begin{cases} y(0,t) = y(\pi,t) = 0, & t \in (0,1), \\ y(x,0) = \theta_2 \sin(x), & x \in (0,\pi). \end{cases}$$

We assume that constant coefficients θ_1 and θ_2 are unknown and estimated based on the data gathered by two stationary sensors. Let us try to determine the sensors' locations x^1 and x^2 so as to maximize the determinant of the FIM.

Of course, in this elementary case the solution can be obtained in closed form as

$$y(x,t) = \theta_2 \exp(-\theta_1 t) \sin(x).$$

There is no loss of generality in assuming $\sigma = 1$ as this value has no influence

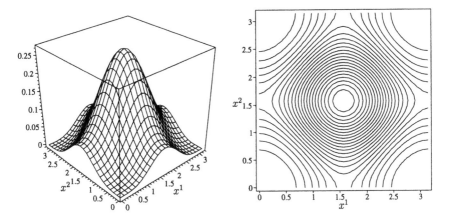

FIGURE 2.2
Surface and contour plots of $\det(M(x^1, x^2))$ of Example 2.3 versus the sensors' locations ($\theta_1 = 0.1$, $\theta_2 = 1$).

on the sensor positions. After some easy calculations, we get

$$\det(M(x^1, x^2))$$
$$= \underbrace{\frac{\theta_2^2}{16\theta_1^4}\left(-4\theta_1^2\exp(-2\theta_1) - 2\exp(-2\theta_1) + \exp(-4\theta_1) + 1\right)}_{\text{constant term}}$$
$$\cdot \underbrace{\left(2 - \cos^2(x^1) - \cos^2(x^2)\right)^2}_{\text{term dependent on } x^1 \text{ and } x^2}.$$

Figure 2.2 shows the corresponding surface and contour plots. Clearly, the maximum of the above criterion is attained for

$$x^{1\star} = x^{2\star} = \frac{\pi}{2},$$

but this means that both the sensors should be placed at the same point at the centre of the interval $(0, \pi)$. ▯

In the literature on stationary sensors, a common remedy for such a predicament is to guess *a priori* a set of N' possible locations, where $N' > N$, and then to seek the best set of N locations from among the N' possible, so that the problem is then reduced to a combinatorial one [300]. (If the system is solved numerically, the maximum value of N' is the number of grid points in the domain $\bar{\Omega}$.)

2.4.4 Dependence of the solution on the parameters to be identified

Another great difficulty encountered while trying to design the sensors' locations is the dependence of the optimal solution on the parameters to be identified, which are unknown before the experiment. In other words, one cannot determine an optimal design setting for estimating θ without having to specify an initial estimate θ^0 of θ. This peculiarity is illustrated by the following example.

Example 2.4

Let us reconsider the heat process of Example 2.3:

$$\frac{\partial y(x,t)}{\partial t} = \theta \frac{\partial^2 y(x,t)}{\partial x^2}, \quad x \in (0,\pi), \quad t \in (0,t_f)$$

with the same homogeneous boundary conditions and a slightly changed initial condition:

$$\begin{cases} y(0,t) = y(\pi,t) = 0, & t \in (0,t_f), \\ y(x,0) = \sin(x) + \frac{1}{2}\sin(2x), & x \in (0,\pi). \end{cases}$$

Its solution can be easily found in explicit form as

$$y(x,t) = \exp(-\theta t)\sin(x) + \frac{1}{2}\exp(-4\theta t)\sin(2x).$$

This time we assume that we have at our disposal only one stationary sensor and we would like it to be placed at the best position x^1 in order to estimate the constant parameter θ as precisely as possible.

Based on the previous considerations and using any reasonable computer-algebra system, we can write the following expression for the FIM (here again we assume $\sigma = 1$):

$$M(x^1) = \int_0^{t_f} \left(\frac{\partial y(x^1,t;\theta)}{\partial \theta} \right)^2 dt$$

$$= \frac{1}{4\theta^3} \Big\{ -\sin^2(x^1)\left(2\theta^2 t_f^2 + 2t_f\theta + 1\right)\exp(-2t_f\theta)$$

$$- \frac{32}{125}\cos(x^1)\sin^2(x^1)\left(10t_f\theta + 2 + 25\theta^2 t_f^2\right)\exp(-5t_f\theta)$$

$$- \frac{1}{4}\cos^2(x^1)\sin^2(x^1)\left(1 + 32\theta^2 t_f^2 + 8t_f\theta\right)\exp(-8t_f\theta)$$

$$+ \frac{1}{500}\sin^2(x^1)\left(500 + 256\cos(x^1) + 125\cos^2(x^1)\right) \Big\}.$$

Since there is only one parameter to be identified, the FIM is actually a scalar value and hence finding a minimum of any design criterion Ψ defined on

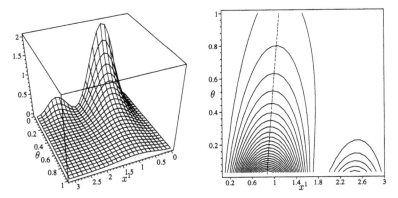

FIGURE 2.3

Surface and contour plots for $M(x^1)$ of Example 2.4 as a function of the parameter θ (the dashed line represents the best measurement points).

it leads to the same solution due to the monotonicity of Ψ. Figure 2.3 shows the corresponding graphs after setting $t_f = 1$, from which two important features can be deduced. First of all, there may exist local minima of Ψ which interfere with numerical minimization of the adopted performance criterion. Apart from that, the optimal sensor position which corresponds to the global minimum of $\Psi(M(x^1))$ depends on the true value of θ (cf. the dashed line joining points $(0.87, 0)$ and $(1.1, 1)$ on the contour plot).   ☐

This dependence on θ is an unappealing characteristic of nonlinear optimum-experimental design and was most appropriately depicted by Cochran [131]: 'You tell me the value of θ and I promise to design the best experiment for estimating θ.' This predicament can be partially circumvented by relying on a nominal value of θ, the results of a preliminary experiment or a *sequential design* which consists of multiple alternation of experimentation and estimation steps. Unfortunately, such strategies are often impractical, because the required experimental time may be too long and the experimental cost may be too high. An altervative is to exploit the so-called *robust-design* strategies [282, 349] which allow us to make optimal solutions independent of the parameters to be identified. The approach (called the *average-optimality* approach) relies on a probabilistic description of the prior uncertainty in θ, characterized by a prior distribution $\pi_p(\theta)$ (this distribution may have been inferred, e.g., from previous observations collected in similar processes). In the same spirit, the *minimax optimality* produces the best sensor positions in the worst circumstances. Both the approaches are treated in detail in Chapter 6.

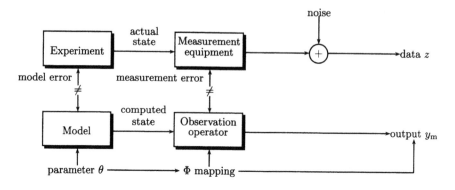

FIGURE 2.4

An interpretation of the identification problem.

2.5   Deterministic interpretation of the FIM

It is worth pointing out that Fisher's information matrix may also be given a deterministic interpretation. To this end, set up a somewhat more abstract conceptual framework for the identification problem under consideration. First, we define three function spaces [52, 141]: the parameter space Θ, the state space Y and the observation space Z, to which belong θ, y and z, respectively. The set of physically admissible parameters is denoted by $\Theta_{\mathrm{ad}} \subset \Theta$. Then solving the PDE for a given value of θ is represented by a solution operator $\mathcal{S} : \Theta_{\mathrm{ad}} \to Y$ defined by

$$y = \mathcal{S}(\theta). \tag{2.24}$$

The type of measurement available is characterized by an observation operator $\mathcal{O} : Y \to Z$ defined by

$$y_{\mathrm{m}} = \mathcal{O}(y). \tag{2.25}$$

Combining (2.24) and (2.25), y_{m} is given by

$$y_{\mathrm{m}} = \Phi(\theta), \tag{2.26}$$

where $\Phi = \mathcal{O} \circ \mathcal{S}$ signifies the composite mapping of $\mathcal{S}$ and $\mathcal{O}$. The situation is depicted in Fig. 2.4 [47].

Given an observation $z \in Z$, the inverse problem under consideration consists in finding a model parameter $\hat{\theta}$ which solves the operator equation

$$\Phi(\hat{\theta}) = z. \tag{2.27}$$

However, because of the measurement and model errors, this equation usually has no solution, so we attempt to solve it approximately by minimizing the least-squares functional on Θ_{ad}:

$$J(\theta) = \frac{1}{2}\|\Phi(\theta) - z\|_Z^2, \qquad (2.28)$$

which quantifies the discrepancy between experimentally measured and analytically predicted response data.

Our task now is to study the differentiability of J at a given point $\bar{\theta} \in \Theta_{ad}$. Let us orient Φ by the requirement that it be continuously Fréchet differentiable in a neighbourhood of $\bar{\theta}$ and, additionally, that it have the second-order Gâteaux derivative at $\bar{\theta}$. Making use of the chain rule of differentation, we see at once that J itself is continuously Fréchet differentiable and that

$$J'(\bar{\theta})\delta\theta = \langle \Phi(\bar{\theta}) - z, \Phi'(\bar{\theta})\delta\theta \rangle_Z = \langle \Phi'^*(\bar{\theta})\left[\Phi(\bar{\theta}) - z\right], \delta\theta \rangle_\Theta \qquad (2.29)$$

or, equivalently,

$$\nabla J(\bar{\theta}) = \Phi'^*(\bar{\theta})\left[\Phi(\bar{\theta}) - z\right], \qquad (2.30)$$

where $\nabla J(\bar{\theta})$ stands for the gradient of J at $\bar{\theta}$ and $\Phi'^*(\bar{\theta})$ is the adjoint operator of $\Phi'(\bar{\theta})$.

Furthermore, we deduce that

$$\frac{1}{\lambda}\left\{J'(\bar{\theta} + \lambda\delta\theta')\delta\theta - J'(\bar{\theta})\delta\theta\right\}$$
$$= \frac{1}{\lambda}\left\{\langle \Phi(\bar{\theta} + \lambda\delta\theta') - z, \Phi'(\bar{\theta} + \lambda\delta\theta')\delta\theta \rangle_Z \right.$$
$$\left. - \langle \Phi(\bar{\theta}) - z, \Phi'(\bar{\theta})\delta\theta \rangle_Z \right\} \qquad (2.31)$$
$$= \langle \frac{1}{\lambda}\left\{\Phi(\bar{\theta} + \lambda\delta\theta') - \Phi(\bar{\theta})\right\}, \Phi'(\bar{\theta} + \lambda\delta\theta')\delta\theta \rangle_Z$$
$$+ \langle \Phi(\bar{\theta}) - z, \frac{1}{\lambda}\left\{\Phi'(\bar{\theta} + \lambda\delta\theta')\delta\theta - \Phi'(\bar{\theta})\delta\theta\right\} \rangle_Z.$$

From this, letting $\lambda \to 0$ and using the continuity of Φ', we get

$$\lim_{\lambda \to 0} \frac{1}{\lambda}\left\{J'(\bar{\theta} + \lambda\delta\theta')\delta\theta - J'(\bar{\theta})\delta\theta\right\}$$
$$= \langle \Phi'(\bar{\theta})\delta\theta', \Phi'(\bar{\theta})\delta\theta \rangle_Z + \langle \Phi(\bar{\theta}) - z, \Phi''(\bar{\theta})(\delta\theta, \delta\theta') \rangle_Z, \quad (2.32)$$

the limit being uniform with respect to all $\delta\theta$ such that $\|\delta\theta\|_\Theta = 1$. Consequently, the second-order Gâteaux derivative of J exists and is given by

$$J''(\bar{\theta})(\delta\theta, \delta\theta') = \langle \delta\theta', \Phi'^*(\bar{\theta})\Phi'(\bar{\theta})\delta\theta \rangle_\Theta$$
$$+ \langle \Phi(\bar{\theta}) - z, \Phi''(\bar{\theta})(\delta\theta, \delta\theta') \rangle_Z. \qquad (2.33)$$

If the measurements and model errors are small and $\bar{\theta}$ is in the proximity of a global minimum of J, then the second term on the right-hand side may be neglected using the fact that $\Phi(\bar{\theta}) - z \approx 0$. Hence

$$J''(\bar{\theta})(\delta\theta, \delta\theta') \approx \langle \delta\theta', \Phi'^*(\bar{\theta})\Phi'(\bar{\theta})\delta\theta \rangle_\Theta, \qquad (2.34)$$

which forces

$$H(\bar{\theta}) \approx \Phi'^*(\bar{\theta})\Phi'(\bar{\theta}), \tag{2.35}$$

where $H(\bar{\theta})$ denotes the Hessian of $\mathcal{J}$ at $\bar{\theta}$.

Let us now unravel what (2.35) means in case $\Theta = \mathbb{R}^m$ and $Z = L^2(T; \mathbb{R}^N)$, where $T = (0, t_f)$. As was already noted, this setting corresponds to an N-sensor parameter-output mapping:

$$y_{\mathrm{m}}(\cdot) = \Phi(\theta) = \mathrm{col}[y(x^1, \cdot\,; \theta), \ldots, y(x^N, \cdot\,; \theta)]. \tag{2.36}$$

Clearly, there exist partial Fréchet derivatives Φ'_{θ_i} of Φ with respect to individual parameters $\theta_i \in \mathbb{R}$ and each of them may be identified with an element $g_i(\bar{\theta}) \in Z$ (see Appendix E.4), so that we have

$$\delta y_{\mathrm{m}} = \Phi'(\bar{\theta})\delta\theta = \sum_{i=1}^m \delta\theta_i g_i(\bar{\theta}). \tag{2.37}$$

An easy computation shows that

$$\begin{aligned} \langle h, \Phi'(\bar{\theta})\delta\theta \rangle_Z &= \Big\langle h, \sum_{i=1}^m \delta\theta_i g_i(\bar{\theta}) \Big\rangle_Z \\ &= \sum_{i=1}^m \langle h, g_i(\bar{\theta}) \rangle_Z\, \delta\theta_i = \langle A(\bar{\theta})h, \delta\theta \rangle_{\mathbb{R}^m}, \quad \forall\, h \in Z, \end{aligned} \tag{2.38}$$

where $A(\bar{\theta})h = \mathrm{col}[\langle h, g_1(\bar{\theta}) \rangle_Z, \ldots, \langle h, g_m(\bar{\theta}) \rangle_Z]$, which yields $\Phi'^*(\bar{\theta}) \equiv A(\bar{\theta})$. For the composite mapping in (2.35) we thus get

$$\begin{aligned} H(\bar{\theta}) &= \begin{bmatrix} \langle g_1(\bar{\theta}), g_1(\bar{\theta}) \rangle_Z & \cdots & \langle g_m(\bar{\theta}), g_1(\bar{\theta}) \rangle_Z \\ \vdots & & \vdots \\ \langle g_1(\bar{\theta}), g_m(\bar{\theta}) \rangle_Z & \cdots & \langle g_m(\bar{\theta}), g_m(\bar{\theta}) \rangle_Z \end{bmatrix} \\ &= \sum_{j=1}^N \int_0^{t_f} \gamma_j(t; \bar{\theta}) \gamma_j^{\mathsf{T}}(t; \bar{\theta})\, \mathrm{d}t, \end{aligned} \tag{2.39}$$

where $\gamma_j(\cdot\,; \bar{\theta}) = \mathrm{col}[g_{1j}(\bar{\theta})(\cdot), \ldots, g_{mj}(\bar{\theta})(\cdot)]$, g_{ij} being the j-th component of g_i (viz. the component of the derivative which corresponds to the i-th parameter and the j-th sensor), $i = 1, \ldots, m$ and $j = 1, \ldots, N$. But if the solution to (2.1)–(2.3) is sufficiently regular and $\bar{\theta}$ is merely the actual parameter vector, then (2.39) is nothing but the FIM calculated for $\sigma = 1$. The interpretation of the FIM as the Hessian of the least-squares criterion will be exploited in Section 3.6, p. 92.

2.6   Calculation of sensitivity coefficients

A basic step in the design of optimal sensor positions is to devise an effective numerical procedure for the computation of the so-called sensitivity coefficients, i.e., the derivatives of the states with respect to system parameters, which is necessary to form the elements of the appropriate FIM. The problem is closely related to design sensitivity analysis, which plays a critical role in inverse and identification studies, as well as numerical optimization and reliability analyses [13, 79, 110, 296] for which there exist many comprehensive monographs and surveys. In the context of parameter estimation of DPSs the problem of calculating sensitivities dates back to the early works of Chavent [45]. As for more recent developments, let us cite the monograph [282] where the issue was addressed from an engineering point of view with application to groundwater resources management. At the other extreme, Brewer [29] studied the differentiability with respect to a parameter of the solution to a linear inhomogeneous abstract Cauchy problem by employing the theory of strongly continuous semigroups. Some of his ideas were then used to identify spatially varying unknown coefficients in parabolic PDEs via quasilinearization [108].

   In what follows, three classical methods of calculating sensitivities are briefly delineated and compared. The reader interested in technical formalities regarding the method widely exploited throughout this monograph is referred to Appendices F.2 and G.2. In order to maintain the discussion at a reasonable level of clarity and to avoid tedious calculations, we will consider the system (2.1) with initial conditions (2.3) and Dirichlet boundary conditions

$$y(x,t) = b(x,t) \quad \text{on } \partial\Omega \times T, \tag{2.40}$$

where b is some prescribed function.

2.6.1   Finite-difference method

The finite-difference method is undoubtedly the easiest method to implement as it does not require any analytical or programming effort by the user, but suffers from computational inefficiency and possible errors. We employ the Taylor series expansion to approximate the derivative, which gives

$$y(x,t; \theta + \Delta\theta_i e_i) = y(x,t; \theta) + \frac{\partial y(x,t; \theta)}{\partial \theta_i} \Delta\theta_i + o(\Delta\theta_i), \tag{2.41}$$

where

$$e_i = (0,0,\ldots,0, \overbrace{1}^{i\text{-th component}}, 0, \ldots, 0) \tag{2.42}$$

and $\Delta\theta_i$ represents the parameter perturbation. The above is solved for $\partial y/\partial\theta_i$ to obtain the forward-difference approximation

$$\frac{\partial y(x,t;\theta)}{\partial\theta_i} = \frac{y(x,t;\theta+\Delta\theta_i e_i) - y(x,t;\theta)}{\Delta\theta_i} + O(\Delta\theta_i), \qquad (2.43)$$

where we see that the truncation error of the approximation is of the order $O(\Delta\theta_i)$. Thus, a smaller $\Delta\theta_i$ yields a more accurate approximation. However, if $\Delta\theta_i$ is too small, then a numerical round-off error will erode the accuracy of the computations. The simultaneous choice of the perturbation $\Delta\theta_i$ and of a numerical tolerance used for the numerical simulation of the model equation is critical and may be delicate in some applications [213]. To obtain a second-order accurate approximation, the central-difference approximation

$$\frac{\partial y(x,t;\theta)}{\partial\theta_i} = \frac{y(x,t;\theta+\Delta\theta_i e_i) - y(x,t;\theta-\Delta\theta_i e_i)}{2\Delta\theta_i} + o(\Delta\theta_i) \qquad (2.44)$$

is commonly employed.

Forward and central differences require, respectively, $m+1$ and $2m+1$ simulation runs of the model (2.1), (2.3), (2.40) to get all the sensitivity coefficients for a given $\theta \in \mathbb{R}^m$, i.e., this number is proportional to the number of unknown parameters.

2.6.2 Direct-differentiation method

In the direct-differentation method we differentiate the system equations (2.1), (2.3), (2.40) with respect to the individual parameters, which gives, after some rearrangement, the so-called *sensitivity equations*

$$
\begin{aligned}
\frac{\partial}{\partial t}\left[\frac{\partial y}{\partial\theta_i}\right] &= \frac{\partial\mathcal{F}}{\partial y}\frac{\partial y}{\partial\theta_i} + \frac{\partial\mathcal{F}}{\partial y_{x_1}}\frac{\partial}{\partial x_1}\left[\frac{\partial y}{\partial\theta_i}\right] + \frac{\partial\mathcal{F}}{\partial y_{x_2}}\frac{\partial}{\partial x_2}\left[\frac{\partial y}{\partial\theta_i}\right] \\
&+ \frac{\partial\mathcal{F}}{\partial y_{x_1 x_1}}\frac{\partial^2}{\partial x_1^2}\left[\frac{\partial y}{\partial\theta_i}\right] + \frac{\partial\mathcal{F}}{\partial y_{x_2 x_2}}\frac{\partial^2}{\partial x_2^2}\left[\frac{\partial y}{\partial\theta_i}\right] \\
&+ \frac{\partial\mathcal{F}}{\partial\theta_i} \qquad \text{in } Q = \Omega \times T
\end{aligned} \qquad (2.45)
$$

subject to

$$\frac{\partial y}{\partial\theta_i}(x,0) = 0 \quad \text{in } \Omega, \qquad (2.46)$$

$$\frac{\partial y}{\partial\theta_i}(x,t) = 0 \quad \text{on } \Sigma = \partial\Omega \times T \qquad (2.47)$$

for $i = 1,\ldots,m$, where y_{x_j} and $y_{x_j x_j}$ denote

$$\frac{\partial y}{\partial x_j} \quad \text{and} \quad \frac{\partial^2 y}{\partial x_j^2},$$

respectively, $j = 1, 2$.

Let us note that these equations are linear even if the system equation is nonlinear. All the derivatives of $\mathcal{F}$ are calculated for a solution to (2.1), (2.3), (2.40) corresponding to a given value of θ and therefore this solution must be obtained prior to solving sensitivity equations and then stored, or computed simultaneously with these.

If the system equation is linear, then the form of the sensitivity problem is exactly the same as that of simulating the original system. Consequently, we can use the same computer code to solve both of them. The total computational effort is the same as using the finite-difference method ($m+1$ simulation runs), but this time the results obtained are exact (for the numerical solution) and the difficulty of determining the size of perturbation increments can thus be avoided.

2.6.3 Adjoint method

Suppose that the values of all the sensitivity coefficients are required at a point $x^0 \in \Omega$ for a given time moment $t^0 \in T$. It turns out that this can be accomplished without the necessity of solving sensitivity equations, as this step can be eliminated via the calculus of variations. Indeed, the variation in y due to variations in the parameter vector θ is given by the solution of

$$
\begin{aligned}
\delta y_t &= \frac{\partial \mathcal{F}}{\partial y} \delta y + \frac{\partial \mathcal{F}}{\partial y_{x_1}} \delta y_{x_1} + \frac{\partial \mathcal{F}}{\partial y_{x_2}} \delta y_{x_2} \\
&+ \frac{\partial \mathcal{F}}{\partial y_{x_1 x_1}} \delta y_{x_1 x_1} + \frac{\partial \mathcal{F}}{\partial y_{x_2 x_2}} \delta y_{x_2 x_2} + \frac{\partial \mathcal{F}}{\partial \theta} \delta\theta \quad \text{in } Q = \Omega \times T
\end{aligned} \tag{2.48}
$$

subject to

$$
\delta y(x, 0) = 0 \quad \text{in } \Omega, \tag{2.49}
$$
$$
\delta y(x, t) = 0 \quad \text{on } \Sigma = \partial\Omega \times T. \tag{2.50}
$$

We introduce the adjoint state $\psi = \psi(x, t)$, multiplying (2.48) by it and integrating the result over Q while making use of the Green formulae (cf. Appendix F.1), to give

$$
\begin{aligned}
\int_\Omega \psi \delta y \big|_{t=t_f} \, dx \\
= \int_Q \bigg\{ \frac{\partial \psi}{\partial t} + \psi \frac{\partial \mathcal{F}}{\partial y} - \frac{\partial}{\partial x_1} \left(\psi \frac{\partial \mathcal{F}}{\partial y_{x_1}} \right) - \frac{\partial}{\partial x_2} \left(\psi \frac{\partial \mathcal{F}}{\partial y_{x_2}} \right) \\
+ \frac{\partial^2}{\partial x_1^2} \left(\psi \frac{\partial \mathcal{F}}{\partial y_{x_1 x_1}} \right) + \frac{\partial^2}{\partial x_2^2} \left(\psi \frac{\partial \mathcal{F}}{\partial y_{x_2 x_2}} \right) \bigg\} \delta y \, dx \, dt \\
+ \int_\Sigma \psi \bigg\{ \frac{\partial \mathcal{F}}{\partial y_{x_1 x_1}} \delta y_{x_1} \nu_1 + \frac{\partial \mathcal{F}}{\partial y_{x_2 x_2}} \delta y_{x_2} \nu_2 \bigg\} \, d\sigma \, dt \\
+ \int_Q \psi \frac{\partial \mathcal{F}}{\partial \theta} \delta\theta \, dx \, dt,
\end{aligned} \tag{2.51}
$$

where ν_i's signify the direction cosines of the unit outward normal to $\partial\Omega$. The quantity ψ acts as the Lagrange multiplier and for now it is arbitrary.

Writing the value of y at (x^0, t^0) in the form

$$y(x^0, t^0) = \int_Q y(x,t)\delta(x_1 - x_1^0)\delta(x_2 - x_2^0)\delta(t - t^0)\,dx\,dt, \qquad (2.52)$$

where δ is the Dirac delta function, we see that

$$\delta y(x^0, t^0) = \int_Q \delta y(x,t)\delta(x_1 - x_1^0)\delta(x_2 - x_2^0)\delta(t - t^0)\,dx\,dt. \qquad (2.53)$$

The summation of (2.51) and (2.53) gives

$$\delta y(x^0, t^0)$$
$$= -\int_\Omega \psi\delta y\big|_{t=t_f}\,dx + \int_Q \Bigg\{ \frac{\partial\psi}{\partial t} + \psi\frac{\partial\mathcal{F}}{\partial y} - \frac{\partial}{\partial x_1}\left(\psi\frac{\partial\mathcal{F}}{\partial y_{x_1}}\right)$$
$$- \frac{\partial}{\partial x_2}\left(\psi\frac{\partial\mathcal{F}}{\partial y_{x_2}}\right) + \frac{\partial^2}{\partial x_1^2}\left(\psi\frac{\partial\mathcal{F}}{\partial y_{x_1 x_1}}\right) + \frac{\partial^2}{\partial x_2^2}\left(\psi\frac{\partial\mathcal{F}}{\partial y_{x_2 x_2}}\right)$$
$$+ \delta(x_1 - x_1^0)\delta(x_2 - x_2^0)\delta(t - t^0) \Bigg\}\delta y\,dx\,dt$$
$$+ \int_\Sigma \psi\left\{ \frac{\partial\mathcal{F}}{\partial y_{x_1 x_1}}\delta y_{x_1}\nu_1 + \frac{\partial\mathcal{F}}{\partial y_{x_2 x_2}}\delta y_{x_2}\nu_2 \right\} d\sigma\,dt$$
$$+ \int_Q \psi\frac{\partial\mathcal{F}}{\partial\theta}\delta\theta\,dx\,dt. \qquad (2.54)$$

Since ψ is arbitrary, we may select it to annihilate the terms related to δy. We therefore specify that ψ be governed by

$$\frac{\partial\psi}{\partial t} = -\psi\frac{\partial\mathcal{F}}{\partial y} + \frac{\partial}{\partial x_1}\left(\psi\frac{\partial\mathcal{F}}{\partial y_{x_1}}\right) + \frac{\partial}{\partial x_2}\left(\psi\frac{\partial\mathcal{F}}{\partial y_{x_2}}\right) - \frac{\partial^2}{\partial x_1^2}\left(\psi\frac{\partial\mathcal{F}}{\partial y_{x_1 x_1}}\right)$$
$$- \frac{\partial^2}{\partial x_2^2}\left(\psi\frac{\partial\mathcal{F}}{\partial y_{x_2 x_2}}\right) - \delta(x_1 - x_1^0)\delta(x_2 - x_2^0)\delta(t - t^0) \quad \text{in } Q, \qquad (2.55)$$

subject to the final condition

$$\psi(x, t_f) = 0 \quad \text{in } \Omega \qquad (2.56)$$

and the boundary condition

$$\psi(x, t) = 0 \quad \text{on } \Sigma. \qquad (2.57)$$

Substituting (2.55)–(2.57) into (2.54), we obtain

$$\delta y(x^0, t^0) = \int_Q \psi\frac{\partial\mathcal{F}}{\partial\theta}\delta\theta\,dx\,dt \qquad (2.58)$$

and hence

$$\frac{\partial y}{\partial \theta}(x^0, t^0) = \int_Q \psi \frac{\partial \mathcal{F}}{\partial \theta} \, dx \, dt, \tag{2.59}$$

which is due to the fact that $\theta \in \mathbb{R}^m$.

From what has just been outlined, it follows that the adjoint method is implemented as follows:

1. Solve the state equation (2.1) from $t = 0$ to $t = t_f$ subject to (2.3), (2.40) and store the resulting values of the state y obtained on a spatial grid at selected time instants.

2. Solve the adjoint problem (2.55) backwards in time, i.e., from $t = t_f$ to $t = 0$, subject to (2.56) and (2.57) using interpolated values of y.

3. Compute $\partial y(x^0, t^0)/\partial \theta$ from the reduced equation (2.59).

A decided advantage of the method lies in the fact that the amount of computations is not proportional to the number of unknown parameters, as only solution of one adjoint equation is required in addition to the solution of the state equation (compare this with the necessity of solving $m + 1$ equations in the finite-difference and direct-differentiation methods). Moreover, the adjoint method also yields exact (for the numerical solution) results.

However, the solution of the adjoint problem may be computationally more delicate than that of the original problem. In particular, the adjoint PDE contains point sources, which makes the spatial discretization more complicated [116, 213] and eventually the resulting increase in computation time may outweigh the benefits from the reduction in the number of equations. Moreover, the adjoint equation frequently proves to be numerically unstable, especially for long terminal times. This means that any round-off error which becomes 'mixed into' the calculation at an early stage is successively magnified until it comes to swamp the true answer. But the main drawback to the method is that it is primarily intended to calculate the sensitivity vector at a number of points in the space-time domain Q, which is less than the number of parameters. This makes its usefulness questionable in optimally locating sensors, but it can be valuable in some studies related to identifiability [282].

2.7 A final introductory note

The undeniable relationship between the sensor location and the achievable accuracy in parameter identification for distributed systems motivates the development of systematic methods for finding points which are best suited for taking measurements. This chapter has dealt with the characterization of the optimal sensor-location problem from formulation to discussion of some specific technical issues, such as calculation of the sensitivity coefficients. The problem has been ultimately formalized as minimization of some scalar measure of performance based on the Fisher information matrix whose inverse constitutes the Cramér-Rao bound on the covariance matrix for the estimates. Unfortunately, some severe difficulties encountered while naïvely trying to treat this problem as an ordinary nonlinear programming problem exclude straightforward approaches to finding the corresponding solutions. Owing to the same reasons, the existing techniques are limited to some particular situations and are far from being flexible enough to tackle a broad class of problems facing engineers who deal with applications. Accordingly, the remainder of this monograph is intended as an attempt to fill this gap (at least to a certain extent), to overcome some of the shortcomings of the earlier approaches, and to provide a unified methodology for optimally locating measurement transducers along a given spatial domain.

3

Locally optimal designs for stationary sensors

System identification problems occur in many diverse fields. Although system parameters are usually identified by applying known experimental conditions and observing the system response, in most situations there are a number of variables which can be adjusted, subject to certain constraints, so that the information provided by the experiment is maximized. As indicated by Walter and Pronzato [348], experimental design includes two steps. The first is qualitative and consists in selecting a suitable configuration of the input/output ports so as to make, if possible, all the parameters of interest identifiable. The second step is qualitative and based on the optimization of a suitable criterion with respect to free variables such as input profiles, sampling schedules, etc. Since the estimation accuracy can be considerably enhanced by the use of the settings so found, a comprehensive optimum experimental design methodology has been developed and applied in areas such as statistics, engineering design and quality control, social science, agriculture, process control, and medical clinical trials and pharmaceuticals [274].

This elegant theory, which extensively employs results of convex analysis, is mostly focused on static regression models [8, 56, 77, 78, 82, 85, 208, 224, 269], but successful attempts at extending it to dynamic systems were also made, cf. [101, 174, 175, 274, 349, 355]. First investigations of its applicability in the context of DPSs date back to the pioneering works by Rafajłowicz [232–235, 238–241, 244] who demonstrated several interesting properties of optimum sensor locations and set forth an approach to the optimum input design in frequency domain [236, 237, 242, 243]. The purpose of this chapter is to provide a detailed exposition of the results which can be obtained while following this line of research.

3.1 Optimum experimental design for continuous-time, linear-in-parameters lumped models

As an introduction to the measurement system design for DPSs, we shall first investigate a slightly simpler linear case of a lumped system. The presented results are in principle rather easy alterations of their counterparts from the classical theory of optimum experimental design, the main difference being the assumption of continuous-time observations in lieu of a finite collection of measurements at selected time instants.

3.1.1 Problem statement

Consider the dynamic system

$$z(t) = F^{\mathsf{T}}(t)\theta + \varepsilon(t), \quad t \in T = [0, t_f], \tag{3.1}$$

where t denotes time and t_f is a fixed finite time horizon. Here $\theta \in \mathbb{R}^m$ signifies a vector of constant parameters and F is given by

$$F(\,\cdot\,) = \left[f(x^1, \,\cdot\,) \ldots f(x^N, \,\cdot\,) \right], \tag{3.2}$$

where the quantities $x^1, \ldots, x^N$ are treated as variables whose values belong to a compact set $X \subset \mathbb{R}^n$ (for now, we freeze their values) and $f \in C(X \times T; \mathbb{R}^m)$ is known *a priori*. Furthermore, ε is a zero-mean white (in time) Gaussian process which plays the role of a disturbance. Its covariance meets the condition

$$\mathrm{E}\{\varepsilon(t)\varepsilon^{\mathsf{T}}(\tau)\} = C(t)\delta(t - \tau), \tag{3.3}$$

where δ denotes the Dirac delta function and $C(t) \in \mathbb{R}^{N \times N}$ is symmetric and positive definite for any $t \in T$.

In this setting, the parameter estimation problem is as follows: Given the history of F, i.e., the set $\{F(t)\}_{t \in T}$, and the outcomes of the measurements $\{z(t)\}_{t \in T}$ find, among all possible values of the parameter vector θ, a parameter which minimizes the weighted least-squares criterion

$$\mathcal{J}(\theta) = \frac{1}{2} \int_0^{t_f} \left[z(t) - F^{\mathsf{T}}(t)\theta \right]^{\mathsf{T}} C^{-1}(t) \left[z(t) - F^{\mathsf{T}}(t)\theta \right] \, \mathrm{d}t. \tag{3.4}$$

Differentiating (3.4) with respect to θ shows that the sought value $\hat{\theta}$ satisfies the equation (often called the *normal equation*)

$$\nabla \mathcal{J}(\hat{\theta}) = - \int_0^{t_f} F(t)C^{-1}(t) \left[z(t) - F^{\mathsf{T}}(t)\hat{\theta} \right] \, \mathrm{d}t = 0. \tag{3.5}$$

If the matrix

$$M = \int_0^{t_f} F(t)C^{-1}(t)F^{\mathsf{T}}(t) \, \mathrm{d}t \tag{3.6}$$

is nonsingular, then there is a unique solution which can be expressed as

$$\hat{\theta} = M^{-1} \int_0^{t_f} F(t)C^{-1}(t)z(t) \, \mathrm{d}t. \tag{3.7}$$

The estimator (3.7) is unbiased, since we have

$$\begin{aligned}
\mathrm{E}\{\hat{\theta}\} &= M^{-1} \int_0^{t_f} F(t) C^{-1}(t) \, \mathrm{E}\{z(t)\} \, \mathrm{d}t \\
&= M^{-1} \underbrace{\int_0^{t_f} F(t) C^{-1}(t) F^{\mathsf{T}}(t) \, \mathrm{d}t}_{M} \, \theta = \theta
\end{aligned} \tag{3.8}$$

provided that θ is the true parameter value. Moreover, it can be shown that it is the best linear unbiased estimator of θ, cf. Appendix C.2. Its covariance is given by

$$\begin{aligned}
\mathrm{cov}\{\hat{\theta}\} &= \mathrm{E}\{(\hat{\theta} - \theta)(\hat{\theta} - \theta)^{\mathsf{T}}\} \\
&= M^{-1} \int_0^{t_f} \int_0^{t_f} F(t) C^{-1}(t) \, \mathrm{E}\{\varepsilon(t)\varepsilon^{\mathsf{T}}(\tau)\} C^{-1}(\tau) F^{\mathsf{T}}(\tau) \, \mathrm{d}t \, \mathrm{d}\tau M^{-1} \\
&= M^{-1} M M^{-1} = M^{-1}.
\end{aligned} \tag{3.9}$$

The matrix M above plays the role of the Fisher information matrix, cf. (2.9). Let us observe that it does not depend on the observations z but it does depend on the parameters x^j, $j = 1, \ldots, N$. In practice, this means that we may attempt to adjust x^j's prior to any experiment so that they are (in terms of $\mathrm{cov}\{\hat{\theta}\}$) 'better' than others and the information provided by the experiment is maximized. This constitutes the main topic of the remainder of this section.

As was already mentioned (p. 16), a common procedure is to introduce a scalar cost function (design criterion) Ψ defined on the FIM, which permits the optimal experimental design to be cast as an optimization problem:

$$\Psi\big[M(x^1, \ldots, x^N)\big] \longrightarrow \min. \tag{3.10}$$

This leads to the so-called exact designs which can then be calculated with the use of numerous widely accessible nonlinear programming solvers if N is not too large. Unfortunately, the problem quickly becomes computationally too demanding and intractable for larger N's. This predicament has been addressed in plentiful works on optimum experimental design and the most efficient solution is no doubt the introduction of the so-called continuous designs [77, 82, 85, 101, 208, 224, 349]. Such an approach will also be adopted in what follows.

In order to relax the limitations of exact designs, it is necessary to orient the covariance C by the requirement that

$$C = \sigma^2 I, \tag{3.11}$$

where I is the $N \times N$ identity matrix and σ plays the role of a constant standard deviation of the measurement errors (note that we might also assume that C is a diagonal matrix, but this will not be pursued for the sake of simplicity, since the corresponding changes are rather obvious). Such an assumption amounts to accepting the situation when the measurements are constantly independent of one another. Sometimes this is unrealistic (especially in most sensor-location contexts), but the clear advantage of such a procedure, which outweighs all the shortcomings, is that the form of the FIM is then substantially simpler:

$$M = \frac{1}{\sigma^2} \int_0^{t_f} F(t)F^\mathsf{T}(t)\,\mathrm{d}t = \sum_{j=1}^{N} M_j, \qquad (3.12)$$

where

$$M_j = \frac{1}{\sigma^2} \int_0^{t_f} f(x^j,t)f^\mathsf{T}(x^j,t)\,\mathrm{d}t.$$

As a result, the total FIM is the sum of the information matrices M_j for individual observations, which is crucial for the approach.

One more simplification comes in handy, but this time it involves no loss of generality. Since in practice all the design criteria satisfy the condition

$$\Psi(\beta M) = \gamma(\beta)\Psi(M), \quad \beta > 0, \qquad (3.13)$$

γ being a positive function, we may set $\sigma = 1$. Similarly, operating on the so-called *average* (or *normalized*) FIM

$$\bar{M} = \frac{1}{Nt_f} \sum_{j=1}^{N} \int_0^{t_f} f(x^j,t)f^\mathsf{T}(x^j,t)\,\mathrm{d}t \qquad (3.14)$$

is slightly more convenient, so in the sequel we will constantly use it in lieu of M. For simplicity of notation, we will also drop the bar over M.

Since we admit replicated measurements, i.e., some values x^j may appear several times in the optimal solution (this is an unavoidable consequence of independent measurements), it is sensible to distinguish only the components of the sequence $x^1, \ldots, x^N$, which are different and, if there are ℓ such components, to relabel them as $x^1, \ldots, x^\ell$ while introducing $r_1, \ldots, r_\ell$ as the corresponding numbers of replications. The redefined x^i's are said to be the *design* or *support* points. The collection of variables

$$\xi_N = \left\{ \begin{matrix} x^1, \ x^2, \ \ldots, \ x^\ell \\ p_1, \ p_2, \ \ldots, \ p_\ell \end{matrix} \right\}, \qquad (3.15)$$

where $p_i = r_i/N$, $N = \sum_{i=1}^{\ell} r_i$, is called the *exact design* of the experiment. The proportion p_i of observations performed at x^i can be considered as the percentage of experimental effort spent at that point.

On account of the above remarks, we rewrite the FIM in the form

$$M(\xi_N) = \sum_{i=1}^{\ell} p_i \frac{1}{t_f} \int_0^{t_f} f(x^i,t)f^\mathsf{T}(x^i,t)\,\mathrm{d}t. \qquad (3.16)$$

Here the p_i's are rational numbers, since both r_i's and N are integers. Note that the design (3.15) defines a discrete probability distribution on its distinct support points, i.e., $x^1, \ldots, x^\ell$. As will be shown below, criteria quantifying the goodness of given designs are rather involved functions of support points and weights.

The discrete nature of N-measurement exact designs causes serious difficulties, as the resultant numerical analysis problem is not amenable to solution by standard optimization techniques, particularly when N is large. The situation is analogous to the much simpler one when we wish to minimize a function defined on integers. Because of the discrete domain, calculus techniques cannot be exploited in the solution. A commonly used device for this problem is to extend the definition of the design [8, 56, 85, 224, 269, 347]. In such a relaxed formulation, when N is large, the feasible p_i's can be considered as any real numbers in the interval $[0, 1]$ which sum up to unity, and not necessarily integer multiples of $1/N$. It is thus easier to operate on the so-called *approximate* designs of the form

$$\xi = \left\{ \begin{array}{l} x^1, x^2, \ldots, x^\ell, \\ p_1, p_2, \ldots, p_\ell \end{array}; \quad \sum_{i=1}^{\ell} p_i = 1 \right\}, \tag{3.17}$$

where the number of support points ℓ is not fixed and constitutes an additional parameter to be determined, subject to the condition $\ell < \infty$. The key idea is then to use calculus to find the number of support points ℓ and associated values $x^{i\star}$, together with the proportions of measurements assigned to those support points $p_i^\star$ where the minimum of a suitable performance index occurs, and argue that the minimum of the same function over the integer multiples of $1/N$ will occur at adjacents to $p_1^\star, \ldots, p_\ell^\star$.

As a result, the mathematical formulation becomes much nicer and more tractable. For example, in contrast to the set of all exact designs, the family of all approximate designs is convex. This means that for any $\alpha \in [0, 1]$ and any approximate designs ξ_1 and ξ_2 of the form (3.17) the convex combination of these designs $(1 - \alpha)\xi_1 + \alpha\xi_2$ can be expressed in the form (3.17), i.e., it is an approximate design, too. To be more precise, assume that the support points of the designs ξ_1 and ξ_2 coincide (this situation is easy to achieve, since if a support point is included in only one design, we can formally add it to the other design and assign it the zero weight). Thus we have

$$\xi_1 = \left\{ \begin{array}{ccc} x^1, & \ldots, & x^\ell \\ p_1^{(1)}, & \ldots, & p_\ell^{(1)} \end{array} \right\}, \quad \xi_2 = \left\{ \begin{array}{ccc} x^1, & \ldots, & x^\ell \\ p_1^{(2)}, & \ldots, & p_\ell^{(2)} \end{array} \right\} \tag{3.18}$$

and, in consequence, we define

$$\xi = (1-\alpha)\xi_1 + \alpha\xi_2 = \left\{ \begin{array}{ccc} x^1, & \ldots, & x^\ell \\ (1-\alpha)p_1^{(1)} + \alpha p_1^{(2)}, & \ldots, & (1-\alpha)p_\ell^{(1)} + \alpha p_\ell^{(2)} \end{array} \right\}, \tag{3.19}$$

which means that ξ conforms to (3.17).

The above reinterpretation of the admissible designs as discrete probability distributions on finite subsets of X ameliorates to a great extent the tractability of the measurement-selection problem. Nevertheless, there still remain many technicalities which make the resulting calculus rather cumbersome. Paradoxically, the encountered difficulties can be easily overcome by an apparent complication of the problem statement, i.e., further widening the class of admissible designs to all probability measures ξ over X which are absolutely continuous with respect to the Lebesgue measure and satisfy by definition the condition

$$\int_X \xi(\mathrm{d}x) = 1. \tag{3.20}$$

Note that ξ can be defined on subsets of X including single points, i.e., a particular point $x \in X$ may have a probability mass attached to it.

Such an extension of the design concept allows us to replace (3.16) by

$$M(\xi) = \int_X \Upsilon(x)\,\xi(\mathrm{d}x), \tag{3.21}$$

where

$$\Upsilon(x) = \frac{1}{t_f}\int_0^{t_f} f(x,t)f^{\mathsf{T}}(x,t)\,\mathrm{d}t$$

and the integration in (3.20) and (3.21) is to be understood in the Lebesgue-Stieltjes sense [40], cf. Appendix B.8. This leads to the so-called *continuous* designs which constitute the basis of the modern theory of optimal experiments and originate in seminal works by Kiefer and Wolfowitz, cf., e.g., [132]. It turns out that such an approach drastically simplifies the design and the remainder of this section is devoted to this issue.

For clarity, we adopt the following notational conventions: Here and subsequently, we will use the symbol $\Xi(X)$ to denote the set of all probability measures on X. Let us also introduce the notation $\mathfrak{M}(X)$ for the set of all admissible information matrices, i.e.,

$$\mathfrak{M}(X) = \{M(\xi) : \xi \in \Xi(X)\}. \tag{3.22}$$

Then we may redefine an optimal design as a solution to the optimization problem:

$$\xi^\star = \arg\min_{\xi \in \Xi(X)} \Psi[M(\xi)]. \tag{3.23}$$

3.1.2   Characterization of the solutions

In what follows, two basic assumptions are vital:

(A1) X is compact, and

(A2) $f \in C(X \times T; \mathbb{R}^m)$.

We begin with certain convexity and representation properties of $M(\xi)$.

LEMMA 3.1
For any $\xi \in \Xi(X)$ the information matrix $M(\xi)$ is symmetric and nonnegative definite.

PROOF The first part is a direct consequence of the definition (3.21). The other results from the dependence

$$\forall b \in \mathbb{R}^m, \ b^\mathsf{T} M(\xi)b = \int_X b^\mathsf{T} \Upsilon(x) b \, \xi(x)$$

$$= \frac{1}{t_f} \int_X \left\{ \int_0^{t_f} [b^\mathsf{T} f(x,t)]^2 \, \mathrm{d}t \right\} \xi(\mathrm{d}x) \geq 0. \tag{3.24}$$

∎

LEMMA 3.2
$\mathfrak{M}(X)$ *is compact and convex.*

PROOF Let us notice that by Assumption (A2) the function Υ is continuous in X [134, Th. 22, p. 360]. Prokhorov's theorem, cf. [143, Th. 3.16, p. 241] or [120, Th. 5, p. 188], then implies that $\Xi(X)$ is weakly compact, i.e., from any sequence $\{\xi_i\}_{i=1}^\infty$ of $\Xi(X)$ we can extract a subsequence $\{\xi_{i_j}\}_{j=1}^\infty$, which is weakly convergent to a probability measure $\xi_\star \in \Xi(X)$ in the sense that

$$\lim_{j \to \infty} \int_X g(x)\, \xi_{i_j}(\mathrm{d}x) = \int_X g(x)\, \xi_\star(\mathrm{d}x), \quad \forall g \in C(X). \tag{3.25}$$

Choosing g consecutively as the components of the matrix Υ, we get

$$\lim_{j \to \infty} M(\xi_{i_j}) = M(\xi_\star) \tag{3.26}$$

which establishes the first part of our claim. Another way of achieving the same result is based on the observation that the set $\mathcal{S}(X) = \{\Upsilon(x) : x \in X\}$ is compact, being the image of the compact set X under the continuous mapping Υ from X into the space of all $m \times m$ matrices endowed with the Euclidean matrix scalar product

$$\langle A, B \rangle = \mathrm{tr}(A^\mathsf{T} B), \tag{3.27}$$

which turns $\mathbb{R}^{m \times m}$ into a Euclidean space of dimension m^2. $\mathfrak{M}(X)$ constitutes the convex hull of $\mathcal{S}(X)$, and in Euclidean space the convex hull of a compact set is compact [255], which is the desired conclusion.

The other assertion follows immediately from the implication

$$M[(1-\alpha)\xi_1 + \alpha\xi_2] = (1-\alpha)M(\xi_1) + \alpha M(\xi_2), \quad \forall \xi_1, \xi_2 \in \Xi(X) \tag{3.28}$$

valid for any $\alpha \in [0, 1]$.   ∎

REMARK 3.1   Let us observe that Assumption (A2) may be slightly weakened: For the continuity of Υ it suffices to require only $f(\cdot, t)$ to be continuous and to impose the condition

$$\forall x \in X, \ \|f(x, t)\| \le h(t) \tag{3.29}$$

almost everywhere in T for some $h \in L^2(T)$.   ∎

It turns out that, despite a rather abstract framework for continuous designs, the results obtained through their use are surprisingly closely related to discrete designs (they derive their name from 'discrete probability distributions') which concentrate on a support set consisting of a finite number of x values. In other words, the optimal design can be chosen to be of the form

$$\xi^\star = \left\{ \begin{matrix} x^{1\star}, & x^{2\star}, & \dots, & x^{\ell\star} \\ p_1^\star, & p_2^\star, & \dots, & p_\ell^\star \end{matrix} ; \ \sum_{i=1}^{\ell} p_i^\star = 1 \right\}, \tag{3.30}$$

where $\ell < \infty$, which concentrates $N p_1^\star$ measurements at $x^{1\star}$, $N p_2^\star$ at $x^{2\star}$, and so on. In fact, we have the following assertion:

LEMMA 3.3 Bound on the number of support points
For any $M_0 \in \mathfrak{M}(X)$ there always exists a purely discrete design ξ with no more than $m(m+1)/2 + 1$ support points such that $M(\xi) = M_0$. If M_0 lies on the boundary of $\mathfrak{M}(X)$, then the number of support points is less than or equal to $m(m+1)/2$.

PROOF   We first observe that due to the symmetry of FIMs, $\mathfrak{M}(X)$ can be identified with a closed convex set of $\mathbb{R}^{m(m+1)/2}$ (for a unique representation of any $M \in \mathfrak{M}(X)$, it suffices to use only the elements which are on and above the diagonal). Moreover, $\mathfrak{M}(X)$ constitutes the convex hull of the set $\mathcal{S}(X) = \{\Upsilon(x) : x \in X\}$ denoted by $\mathrm{conv}(X)$, cf. Appendix B.4. Hence, from Carathéodory's theorem (cf. Appendix B.4), $M_0 \in \mathfrak{M}(X)$ can be expressed as a convex combination of at most $m(m+1)/2 + 1$ elements of $\mathcal{S}(X)$, i.e.,

$$M_0 = \sum_{i=1}^{\ell} p_i \Upsilon(x^i), \tag{3.31}$$

where

$$p_i \ge 0, \quad i = 1, \dots, \ell, \quad \sum_{i=1}^{\ell} p_i = 1, \tag{3.32}$$

x^i, $i = 1, \ldots, \ell$ being some points of X, $\ell \leq m(m+1)/2 + 1$. Note, however, that $\Upsilon(x^i)$ constitutes the FIM corresponding to the one-point design measure $\delta_{x^i} = \left\{ \begin{smallmatrix} x^i \\ 1 \end{smallmatrix} \right\}$ putting unit weight at the design point x^i (the Dirac measure concentrated at a single point x^i), i.e., $\Upsilon(x^i) = M(\delta_{x^i})$. Consequently, we have

$$M_0 = \sum_{i=1}^{\ell} p_i M(\delta_{x^i}) = M(\xi_0), \tag{3.33}$$

where $\xi_0 = \sum_{i=1}^{\ell} p_i \delta_{x^i}$.

The second part of the lemma is also established based on Carathéodory's theorem, which states that any boundary point of $\mathfrak{M}(X) = \operatorname{conv} \mathcal{S}(X)$ can be expressed as a convex combination of $m(m+1)/2$ or fewer points of the set $\mathcal{S}(X)$. ∎

The above lemma justifies restricting our attention only to discrete designs with a limited number of supporting points, so the introduction of continuous designs, which may seem at first sight a superfluous complication, leads to very tangible results. What is more, this result implies that even if a design contains more than $m(m+1)/2 + 1$ support points, a design with the same information matrix can be found which has support at no more than $m(m+1)/2 + 1$ points.

To make a step further, the following additional assumptions about the design criterion $\Psi : \mathrm{NND}(m) \to \mathbb{R}$ will be needed[*]:

(A3) Ψ is convex.

(A4) If $M_1 \preceq M_2$, then $\Psi(M_1) \geq \Psi(M_2)$ (monotonicity).

(A5) There exists a finite real q such that

$$\{\xi : \Psi[M(\xi)] \leq q < \infty\} = \widetilde{\Xi}(X) \neq \emptyset.$$

(A6) For any $\xi \in \widetilde{\Xi}(X)$ and $\bar{\xi} \in \Xi(X)$, we have

$$\Psi[M(\xi) + \alpha(M(\bar{\xi}) - M(\xi))]$$
$$= \Psi[M(\xi)] + \alpha \int_X \psi(x, \xi)\, \bar{\xi}(\mathrm{d}x) + o(\alpha; \xi, \bar{\xi}), \tag{3.34}$$

where o is the usual Landau symbol, i.e.,

$$\lim_{\alpha \downarrow 0} \frac{o(\alpha; \xi, \bar{\xi})}{\alpha} = 0.$$

---

[*]Recall that $\mathrm{NND}(m)$ stands for the set of nonnegative definite $m \times m$ matrices, cf. Appendix B.2. Similarly, $\mathrm{PD}(m)$ will denote the set of positive definite $m \times m$ matrices.

TABLE 3.1
Functions which define the directional derivatives of the most popular
optimality criteria

$\Psi[M(\xi)]$	$\phi(x,\xi)$	$c(\xi)$
$-\ln\det(M(\xi))$	$\dfrac{1}{t_f}\displaystyle\int_0^{t_f} f^{\mathsf{T}}(x,t)M^{-1}(\xi)f(x,t)\,\mathrm{d}t$	m
$\mathrm{tr}(M^{-1}(\xi))$	$\dfrac{1}{t_f}\displaystyle\int_0^{t_f} f^{\mathsf{T}}(x,t)M^{-2}(\xi)f(x,t)\,\mathrm{d}t$	$\mathrm{tr}(M^{-1}(\xi))$
$-\mathrm{tr}(M(\xi))$	$\dfrac{1}{t_f}\displaystyle\int_0^{t_f} f^{\mathsf{T}}(x,t)f(x,t)\,\mathrm{d}t$	$\mathrm{tr}(M(\xi))$

Assumption (A3) is quite natural, since it allows us to stay within the framework of convex analysis, which greatly facilitates subsequent considerations. In turn, Assumption (A4) characterizes Ψ as a linear ordering of $\Xi(X)$. As for Assumption (A5), it only states that there exist designs with finite values of Ψ, which constitutes a rather mild and quite logical requirement. At this juncture, only Assumption (A6) calls for an appropriate comment, as at first sight it may seem a bit odd. In practice, however, (A6) simply amounts to the existence of the directional derivative

$$\delta_+\Psi[M(\xi); M(\bar{\xi}) - M(\xi)] = \left.\frac{\partial\Psi[M(\xi) + \alpha(M(\bar{\xi}) - M(\xi))]}{\partial\alpha}\right|_{\alpha=0^+} \tag{3.35}$$

whose form must be on one hand specific, i.e., $\int_X \psi(x,\xi)\,\bar{\xi}(\mathrm{d}x)$, but on the other hand, for most practical criteria such a condition is not particularly restrictive.

In fact, requiring Ψ to be differentiable with respect to individual elements of its matrix argument, we obtain

$$\begin{aligned}
\delta_+&\Psi[M(\xi); M(\bar{\xi}) - M(\xi)]\\
&= \mathrm{tr}\Big[\mathring{\Psi}(\xi)(M(\bar{\xi}) - M(\xi))\Big]\\
&= \int_X \mathrm{tr}\Big[\mathring{\Psi}(\xi)\Upsilon(x)\Big]\,\bar{\xi}(\mathrm{d}x) - \mathrm{tr}\Big[\mathring{\Psi}(\xi)M(\xi)\Big]\\
&= \int_X \left\{\mathrm{tr}\Big[\mathring{\Psi}(\xi)\Upsilon(x)\Big] - \mathrm{tr}\Big[\mathring{\Psi}(\xi)M(\xi)\Big]\right\}\,\bar{\xi}(\mathrm{d}x),
\end{aligned} \tag{3.36}$$

where

$$\mathring{\Psi}(\xi) = \left.\frac{\partial\Psi[M]}{\partial M}\right|_{M=M(\xi)}$$

and therefore

$$\psi(x,\xi) = c(\xi) - \phi(x,\xi), \tag{3.37}$$

the functions c and ϕ being respectively defined as

$$c(\xi) = -\operatorname{tr}\!\left[\overset{\circ}{\Psi}(\xi)M(\xi)\right] \tag{3.38}$$

and

$$\phi(x,\xi) = -\operatorname{tr}\!\left[\overset{\circ}{\Psi}(\xi)\Upsilon(x)\right] = -\frac{1}{t_f}\int_0^{t_f} f^{\mathsf T}(x,t)\overset{\circ}{\Psi}(\xi)f(x,t)\,\mathrm{d}t. \tag{3.39}$$

Table 3.1 lists specific forms of the foregoing mappings for the most popular design criteria.

REMARK 3.2   Let $x \in X$. Substituting the one-point (Dirac) design measure $\delta_x = \{\begin{smallmatrix}x\\1\end{smallmatrix}\}$ for $\bar\xi$ in Assumption (A6) gives

$$\psi(x,\xi) = \delta_+\Psi[M(\xi); M(\delta_x) - M(\xi)] = \delta_+\Psi[M(\xi); \Upsilon(x) - M(\xi)], \tag{3.40}$$

i.e., $\psi(x,\xi)$ is the one-sided directional differential of Ψ at $M(\xi)$ with increment $\Upsilon(x) - M(\xi)$.   ∎

REMARK 3.3   In principle, the optimization problem (3.23) can be handled by first finding an optimal FIM from the class defined by all admissible design measures and then relating it to the corresponding design measure. The corresponding method would involve the following two steps:

Step 1: Compute $M^\star = \arg\min\limits_{M\in\mathfrak{M}(X)} \Psi[M]$.

Step 2: Recover $\xi^\star$ satisfying $M(\xi^\star) = M^\star$.

Step 1 reduces to minimization of the convex function Ψ over a convex subset in the Euclidean space of dimension $m(m+1)/2$, while Step 2 aims to identify a measure corresponding to the FIM obtained in the previous step. In spite of its conceptual simplicity, however, this procedure is rather difficult to implement, mainly due to the impossibility of obtaining a simple representation for the set $\mathfrak{M}(X)$ in the bulk of practical situations. Nevertheless, its understanding greatly facilitates derivation of results regarding the existence and characterization of optimal designs. What is more, this interpretation will come in handy as a justification for the convergence of a numerical algorithm aimed at constructing approximated solutions.   ∎

Here are some important properties of the functions introduced above:

LEMMA 3.4

Given a design $\xi \in \tilde{\Xi}(X)$, assume that the mapping $M \rightarrow \Psi[M]$ is continuously differentiable at $M = M(\xi)$. Then we have

$$\overset{\circ}{\Psi}(\xi) \preceq 0, \tag{3.41}$$

$$c(\xi) \geq 0, \tag{3.42}$$

$$\phi(x, \xi) \geq 0, \quad \forall x \in X. \tag{3.43}$$

If, additionally, $\overset{\circ}{\Psi}(\xi) \prec 0$, then the inequalities (3.42) and (3.43) become strict.

PROOF For any $A \succeq 0$ and $\epsilon > 0$, Taylor's Theorem applied to the function $J(\epsilon) = \Psi[M(\xi) + \epsilon A]$ at $\epsilon^0 = 0$ gives

$$\Psi[M(\xi) + \epsilon A] = \Psi[M(\xi)] + \epsilon \operatorname{tr}\left[\overset{\circ}{\Psi}(\xi)A\right] + o(\epsilon). \tag{3.44}$$

But on account of Assumption (A4), we have $\Psi[M(\xi) + \epsilon A] \leq \Psi[M(\xi)]$, which leads to

$$\operatorname{tr}\left[\overset{\circ}{\Psi}(\xi)A\right] \leq \frac{o(\epsilon)}{\epsilon}. \tag{3.45}$$

Letting $\epsilon \downarrow 0$, we thus obtain

$$\operatorname{tr}\left[\overset{\circ}{\Psi}(\xi)A\right] \leq 0. \tag{3.46}$$

Since A is arbitrary, we must have $\overset{\circ}{\Psi}(\xi) \preceq 0$, which is part of the conclusion of Theorem B.8.

Substituting in turn $M(\xi)$ and $\Upsilon(x)$ for A in (3.46), we get (3.42) and (3.43), respectively.

If $\overset{\circ}{\Psi}(\xi) \prec 0$, then $-\overset{\circ}{\Psi}(\xi) \succ 0$, and Theorem B.9 leads to

$$-\operatorname{tr}\left[\overset{\circ}{\Psi}(\xi)A\right] > 0. \tag{3.47}$$

for all $0 \neq A \succeq 0$. In particular, this must hold for $M(\xi)$ and $\Upsilon(x)$, which implies the positiveness of $c(\xi)$ and $\phi(x, \xi)$. ∎

The next three theorems provide characterizations of the optimal designs. Before proceeding, however, the following lemmas which assist in the proofs are presented.

LEMMA 3.5

Each design measure $\xi \in \tilde{\Xi}(X)$ satisfies

$$\min_{x \in X} \psi(x, \xi) \leq 0. \tag{3.48}$$

PROOF Assumption (A6) shows that

$$\int_X \psi(x,\xi)\,\xi(\mathrm{d}x) = 0 \tag{3.49}$$

for any fixed $\xi \in \widetilde{\Xi}(X)$. But

$$\underbrace{\int_X \psi(x,\xi)\,\xi(\mathrm{d}x)}_{=0} \geq \int_X \min_{x \in X} \psi(x,\xi)\,\xi(\mathrm{d}x) = \min_{x \in X} \psi(x,\xi) \tag{3.50}$$

which gives the desired conclusion. ∎

LEMMA 3.6
For any design $\xi \in \widetilde{\Xi}(X)$, the mapping $x \mapsto \psi(x,\xi)$ is continuous on X.

PROOF According to Proposition 4 in [117, p. 206], the one-sided differential of a convex function is continuous in its increment. Consequently, given $\xi \in \widetilde{\Xi}(X)$, the mapping $\delta_+\Psi[M(\xi); \cdot]$ is continuous. The continuity of $\Upsilon(\cdot)$ thus implies that of $x \mapsto \psi(x,\xi) = \delta_+\Psi[M(\xi); \Upsilon(x) - M(\xi)]$. ∎

The experimental design problem in context has been posed as an optimization one, that of minimizing an optimality criterion $\xi \mapsto \Psi[M(\xi)]$ over $\Xi(X)$. The indicated minimization is accomplished by the appropriate choice of an 'admissible' ξ. The purpose of the following theorems is to indicate mathematically such a choice.

THEOREM 3.1
Let Assumptions (A1)–(A6) hold. Then:

(i) *An optimal design exists comprising no more than $m(m+1)/2$ points (i.e., one less than predicted by Lemma 3.3).*

(ii) *The set of optimal designs is convex. If the design criterion Ψ is strictly convex on $\widetilde{\Xi}(X)$, then all optimal designs have the same information matrix.*

(iii) *A design $\xi^\star$ is optimal iff*

$$\min_{x \in X} \psi(x,\xi^\star) = 0. \tag{3.51}$$

(iv) *For any purely discrete optimal design $\xi^\star$, the function $\psi(\cdot,\xi^\star)$ has value zero at all support points corresponding to nonzero weights.*

PROOF   *Part (i):* Each element M of $\mathfrak{M}(X)$ is a symmetric nonnegative definite $m \times m$ matrix which can be represented by a point in $\mathbb{R}^{m(m+1)/2}$, viz. the point with coordinates $\{m_{ij} : 1 \leq i \leq j \leq m\}$ when $M = [m_{ij}]$. The design criterion Ψ can thus be identified with a real-valued function defined on a subset of $\mathbb{R}^{m(m+1)/2}$. The convexity of Ψ then implies its continuity, cf. Theorem B.22. Since, additionally, $\mathfrak{M}(X)$ is compact and convex, cf. Lemma 3.2, Theorem B.23 shows that there exists a matrix $M^\star \in \mathfrak{M}(X)$ such that

$$\Psi[M^\star] = \inf_{M \in \mathfrak{M}(X)} \Psi[M]. \tag{3.52}$$

The definition of $\mathfrak{M}(X)$ guarantees that $M^\star$ corresponds to a design $\xi^\star \in \Xi(X)$, i.e., $M^\star = M(\xi^\star)$. Clearly, this $\xi^\star$ also satisfies

$$\Psi[M(\xi^\star)] = \inf_{\xi \in \Xi(X)} \Psi[M(\xi)], \tag{3.53}$$

which proves the existence of an optimal design.

The monotonicity of Ψ, cf. Assumption (A4), implies that its minimum must occur at a boundary point of $\mathfrak{M}(X)$. Indeed, for if $M(\xi^\star)$ is an interior point of $\mathfrak{M}(X)$, then there exists a $\beta > 1$ such that $\beta M(\xi^\star) \in \mathfrak{M}(X)$. But then

$$\Psi[\beta M(\xi^\star)] = \Psi[M(\xi^\star) + (\beta - 1)M(\xi^\star)] \leq \Psi[M(\xi^\star)]. \tag{3.54}$$

According to Carathéodory's Theorem, cf. Appendix B.4, any boundary point of $\mathfrak{M}(X) = \mathrm{conv}\{\Upsilon(x) : x \in X\}$ can be expressed as a convex combination of at most $m(m+1)/2$ linearly independent elements of $\{\Upsilon(x) : x \in X\}$, which completes this part of the proof.

Part (ii): Let ξ_1 and ξ_2 be two optimal designs, i.e., $\Psi[M(\xi_1^\star)] = \Psi[M(\xi_2^\star)] = \inf_{\xi \in \Xi(X)} \Psi[M(\xi)]$. For $\alpha \in [0,1]$ and the convex combination $\xi^\star = (1-\alpha)\xi_1^\star + \alpha\xi_2^\star$, the convexity of Ψ yields

$$\begin{aligned} \Psi[M(\xi^\star)] &= \Psi[(1-\alpha)M(\xi_1^\star) + \alpha M(\xi_2^\star)] \\ &\leq (1-\alpha)\Psi[M(\xi_1^\star)] + \alpha\Psi[M(\xi_2^\star)] = \inf_{\xi \in \Xi(X)} \Psi[M(\xi)], \end{aligned} \tag{3.55}$$

which gives

$$\Psi[M(\xi^\star)] = \inf_{\xi \in \Xi(X)} \Psi[M(\xi)]. \tag{3.56}$$

This means that $\xi^\star$ is optimal too.

As for the second claim, we can start looking for an optimal measure by first finding $M^\star = \arg\min_{M \in \mathfrak{M}(X)} \Psi[M]$. The set of all information matrices $\widetilde{\mathfrak{M}}(X) = \{M(\xi) : \xi \in \widetilde{\Xi}(X)\}$ is convex, so that for a strictly convex function Ψ there exists at most one minimum of Ψ over $\widetilde{\mathfrak{M}}(X)$, and thus over $\mathfrak{M}(X)$, which is an elementary result of convex optimization, cf. Theorem B.24. But this is precisely the required assertion.

Part (iii): A necessary and sufficient condition for the optimality of $\xi^\star$ over $\Xi(X)$ amounts to that for $M(\xi^\star)$ to minimize $\Psi[M]$ over $\mathfrak{M}(X)$, i.e.,

$$\delta_+\Psi[M(\xi^\star); M(\xi) - M(\xi^\star)] \geq 0, \quad \forall \xi \in \Xi(X), \tag{3.57}$$

cf. Theorem B.25. By Assumption (A6), this is equivalent to the condition

$$\int_X \psi(x, \xi^\star)\,\xi(\mathrm{d}x) \geq 0, \quad \forall \xi \in \Xi(X). \tag{3.58}$$

Setting ξ as the Dirac measure concentrated at x, we get

$$\psi(x, \xi^\star) \geq 0, \quad \forall x \in X. \tag{3.59}$$

Conversely, it is evident that (3.59) implies (3.58).

Therefore, a necessary and sufficient condition for the optimality of $\xi^\star \in \Xi(X)$ has the form

$$\min_{x \in X} \psi(x, \xi^\star) \geq 0. \tag{3.60}$$

Combining it with (3.48), we establish the desired conclusion.

Part (iv): Assumption (A6) implies the equality

$$\int_X \psi(x, \xi^\star)\,\xi^\star(\mathrm{d}x) = 0. \tag{3.61}$$

If $\xi^\star$ is purely discrete, i.e., it is of the form (3.30), then from (3.61) it follows that

$$\sum_{i=1}^{\ell} p_i^\star \psi(x^{i\star}, \xi^\star) = 0. \tag{3.62}$$

But, on account of (3.51), we have

$$\psi(x^{i\star}, \xi^\star) \geq 0, \quad i = 1, \ldots, \ell. \tag{3.63}$$

Since the weights $p_i^\star$, $i = 1, \ldots, \ell$ are nonnegative, too, our claim follows, as otherwise the left-hand side of (3.62) would be positive. ∎

It is now clear that the function ψ is of paramount importance in our considerations, as it determines the location of the support points in the optimal design $\xi^\star$ (they are situated among its points of global minimum). Moreover, given any design ξ, it indicates points at which a new observation contributes to the greatest extent. Indeed, adding a new observation atomized at a single point x^+, which corresponds to the design δ_{x^+} giving unit mass to x^+, amounts to constructing a new design

$$\xi^+ = (1 - \alpha)\xi + \alpha\delta_{x^+} \tag{3.64}$$

for some $\alpha \in (0,1)$. If α is sufficiently small, then from (3.34) it may be concluded that

$$\Psi[M(\xi^+)] - \Psi[M(\xi)] \approx \alpha\psi(x^+,\xi), \qquad (3.65)$$

i.e., the resulting decrease in the criterion value is approximately equal to $-\alpha\psi(x^+,\xi)$. This fact also clarifies why the function $\phi(x,\xi) = -\psi(x,\xi) + c(\xi)$ is usually called the *sensitivity function* (this terminology is somewhat reminiscent of the sensitivity coefficients introduced in Section 2.6, but we hope that it will cause no confusion).

Analytical determination of optimal designs is possible only in simple situations and for general systems it is usually the case that some iterative design procedure will be required. The next theorem is useful in the checking for optimality of designs.

THEOREM 3.2 General Equivalence Theorem

For a criterion Ψ which is differentiable with respect its matrix argument, the following characterizations of an optimal design $\xi^\star$ are equivalent in the sense that each implies the other two:

(i) the design $\xi^\star$ minimizes $\Psi[M(\xi)]$,

(ii) the design $\xi^\star$ minimizes $\max\limits_{x \in X} \phi(x,\xi) - c(\xi)$, and

(iii) $\max\limits_{x \in X} \phi(x,\xi^\star) = c(\xi^\star)$.

All the designs satisfying (i)–(iii) and their convex combinations have the same information matrix $M(\xi^\star)$ provided that the criterion Ψ is strictly convex on $\widetilde{\mathfrak{M}}(X) = \{M(\xi) : \xi \in \widetilde{\Xi}(X)\}$.

PROOF $(i) \Rightarrow (ii)$: From Lemma 3.5 and (3.37) we conclude that

$$\max_{x \in X} \phi(x,\xi) - c(\xi) \geq 0, \quad \forall \xi \in \widetilde{\Xi}(X). \qquad (3.66)$$

In turn, the third part of Theorem 3.1, taken in conjunction with (3.37), shows that $\xi^\star$ satisfies

$$\max_{x \in X} \phi(x,\xi^\star) - c(\xi^\star) = 0, \qquad (3.67)$$

which establishes (ii).

$(ii) \Rightarrow (iii)$: Lemma 3.5 implies that $\max_{x \in X} \phi(x,\xi) - c(\xi)$ is bounded from below by zero. From Theorem 3.1(iii) it follows that this zero bound is achieved at any design minimizing $\Psi[M(\cdot)]$ (the existence of such a design is guaranteed by Theorem 3.1(i). This means that if $\xi^\star$ is a design characterized in (ii), then necessarily $\max_{x \in X} \phi(x,\xi^\star) - c(\xi^\star) = 0$, which is exactly (iii).

$(iii) \Rightarrow (i)$: If $\xi^\star$ satisfies $\max_{x \in X} \phi(x,\xi^\star) = c(\xi^\star)$, then it is immediate that $\min_{x \in X} \psi(x,\xi^\star) = 0$, which automatically forces the optimality of $\xi^\star$, cf. Theorem 3.1(iii).

The uniqueness of the information matrix for each optimal design follows from the convexity of the set $\widetilde{\mathfrak{M}}(X)$ and the strict convexity of the function Ψ on it, cf. Theorem B.24. ∎

When formulated for a particular design criterion, Theorem 3.2 is usually called an *equivalence theorem* and the most famous is the Kiefer-Wolfowitz equivalence theorem corresponding to D-optimum designs. In our framework, this specializes to our next assertion.

THEOREM 3.3 Equivalence Theorem for D-Optimum Designs
The following conditions are equivalent:

(i) the design $\xi^\star$ maximizes $\det M(\xi)$,

(ii) the design $\xi^\star$ minimizes $\max\limits_{x \in X} \dfrac{1}{t_f} \displaystyle\int_0^{t_f} f^{\mathsf{T}}(x,t) M^{-1}(\xi) f(x,t)\,\mathrm{d}t$, *and*

(iii) $\max\limits_{x \in X} \dfrac{1}{t_f} \displaystyle\int_0^{t_f} f^{\mathsf{T}}(x,t) M^{-1}(\xi^\star) f(x,t)\,\mathrm{d}t = m$.

Let us give a thought to the integral expression which appears in the above formulation. Based on the calculated estimate $\hat{\theta}$ we may predict the response $\hat{y}(x_0, t) = f^{\mathsf{T}}(x_0, t)\hat{\theta}$ given a design point $x_0 \in X$ and a $t \in T$. We can then evaluate the prediction error variance according to the formula

$$\mathrm{var}\{\hat{y}(x_0, t)\} = \frac{1}{Nt_f} f^{\mathsf{T}}(x_0, t) M^{-1}(\xi^\star) f(x_0, t), \qquad (3.68)$$

which implies the average variance per unit time in the form

$$\overline{\mathrm{var}}\{\hat{y}(x_0, \cdot)\} = \frac{1}{Nt_f} \left\{ \frac{1}{t_f} \int_0^{t_f} f^{\mathsf{T}}(x_0, t) M^{-1}(\xi^\star) f(x_0, t)\,\mathrm{d}t \right\}, \qquad (3.69)$$

where the expression in the braces is exactly the one which is used in Theorem 3.3. Hence we conclude that in a D-optimum design the observations must be taken at points where the average variance of the predicted response is the largest.

3.1.3 Algorithms

In addition to revealing striking minimax properties of optimal designs, the results of Section 3.1.2 provide us with tests for the optimality of intuitively sensible designs. In particular,

1. If the sensitivity function $\phi(x, \xi)$ is less than or equal to $c(\xi)$ for all $x \in X$, then ξ is optimal.

2. If the sensitivity function $\phi(x, \xi)$ exceeds $c(\xi)$, then ξ is not optimal.

However, to exploit their full potential, we require more than this, namely, we need efficient numerical algorithms which would enable us to construct Ψ-optimum design measures. Note that the problem is not that standard if we cannot initially identify a finite set which includes the support points of an optimal measure. In fact, the General Equivalence Theorem says nothing about the number of the support points of an optimal design, and only a bound of $m(m + 1)/2$ on this number can be obtained from the nature of $M(\xi)$. In principle, we could thus fix the number of support points as $\ell = m(m + 1)/2$ and then treat the minimization of $\Psi[M(\cdot)]$ directly as the following constrained nonlinear programming problem in $(n + 1)\ell$ dimensions: Find $x^1, \dots, x^\ell \in \mathbb{R}^n$ and $p_1, \dots, p_\ell \in \mathbb{R}$ which minimize the performance index

$$J(x^1, \dots, x^\ell, p_1, \dots, p_\ell) = \Psi\left[\sum_{i=1}^{\ell} p_i \Upsilon(x^i)\right] \tag{3.70}$$

subject to (possibly nonlinear) constraints

$$x^i \in X, \quad i = 1, \dots, \ell \tag{3.71}$$

and simple linear constraints

$$\sum_{i=1}^{\ell} p_i = 1, \quad p_i \geq 0, \quad i = 1, \dots, \ell. \tag{3.72}$$

Since ℓ so selected may be excessively high, we could alternatively start with a relatively small, fixed ℓ chosen arbitrarily, solve the optimization problem (3.70)–(3.72), check the solution for optimality based on Theorem 3.2, and possibly repeat this procedure for an incremented ℓ if the optimality conditions for $\xi = \left\{ \begin{smallmatrix} x^1, & \cdots & x^\ell \\ p_1, & \cdots, & p_\ell \end{smallmatrix} \right\}$ are not met. However, such a conceptually simple procedure often fails so far as applications are concerned. Indeed, the criterion Ψ is most often strictly convex on PD(m), and this guarantees that the optimal FIM is unique, but this does not necessarily mean that $\xi \mapsto \Psi[M(\xi)]$ is strictly convex in ξ.

Hence, there is no guarantee that $\xi^\star$ is unique. In particular, multiple global solutions $\xi^\star$ may yield the same minimizing value of $M(\xi)$. Furthermore, there may be multiple local minima to $J(\cdot)$ in (3.70) which highly interferes with the optimization process. For many years, similar complications have stimulated researchers concerned with the classical optimum experimental design and led to the development of special algorithms for this purpose. As will be shown below, they can be easily generalized to the framework discussed in this chapter.

3.1.3.1 Sequential numerical design algorithms with selection of support points

The above-mentioned difficulties related to the large scale of the design problems encountered in practice, the lack of a simple description of the set $\mathfrak{M}(X)$ and the inherent presence of many local minima still stimulate researchers to devise specialized numerical design algorithms which would exploit to a greatest extent specific features of the resulting optimization task. Here we limit ourselves to a sufficiently general class of iterative methods whose structure is as follows [208]:

1. Guess an initial design $\xi^{(0)}$ such that $\Psi[M(\xi^{(0)})] < \infty$.

2. Iteratively compute a sequence of designs $\{\xi^{(k)}\}$ converging to ξ^* as $k \to \infty$, the design $\xi^{(k+1)}$ being obtained by a small perturbation of the design $\xi^{(k)}$, while still requiring $\Psi[M(\xi^{(k+1)})] < \infty$.

3. Stop computing $\{\xi^{(k)}\}$ after a finite number of steps if $\xi^{(k)}$ can be qualified as almost optimal, e.g., in accordance with Theorem 3.2.

The numerical algorithms of this class can be viewed as constrained versions of unconstrained first-order descent algorithms which take account of the fact that the extreme points of the convex set $\mathfrak{M}(X)$ are included in the set $\{\Upsilon(x) : x \in X\}$. Note that the number of the support points is not fixed *a priori* and is subjected to an additional choice during the computations.

The underlying idea is quite simple. Suppose that we have an arbitrary (nonoptimal) design $\xi^{(k)}$ obtained after k iteration steps. Further, let the function $\phi(\cdot, \xi^{(k)})$ attain its maximum (necessarily $> c(\xi^{(k)})$) at $x = x_0^{(k)}$. Then the design

$$\xi^{(k+1)} = (1 - \alpha_k)\xi^{(k)} + \alpha_k \delta_{x_0^{(k)}} \qquad (3.73)$$

(recall that $\delta_{x_0^{(k)}}$ stands for the unit-weight design concentrated at $x_0^{(k)}$) leads to a decrease in the value of $\Psi[M(\xi^{(k+1)})]$ for a suitably small α_k. This follows since the derivative with respect to α_k is negative, i.e.,

$$\frac{\partial}{\partial \alpha_k} \Psi[M(\xi^{(k+1)})]\Big|_{\alpha_k=0+} = c(\xi^{(k)}) - \phi(x_0^{(k)}, \xi^{(k)}) < 0. \qquad (3.74)$$

The steps in using the outlined gradient method can be briefly summarized as follows [77, 85, 245, 349]:

ALGORITHM 3.1 *Iterative construction of Ψ-optimum designs*

Step 1. *Guess a discrete nondegenerate starting design measure $\xi^{(0)}$ (we must have $\Psi[M(\xi^{(0)})] < \infty$). Choose some positive tolerance $\eta \ll 1$. Set $k = 0$.*

Step 2. *Determine* $x_0^{(k)} = \arg\max\limits_{x \in X} \phi(x, \xi^{(k)})$. *If* $\phi(x_0^{(k)}, \xi^{(k)}) < c(\xi^{(k)}) + \eta$, *then* STOP.

Step 3. *For an appropriate value of* $0 < \alpha_k < 1$, *set*

$$\xi^{(k+1)} = (1 - \alpha_k)\xi^{(k)} + \alpha_k \delta_{x_0^{(k)}},$$

increment k *by one and go to Step 2.*

In the same way as for the classical first-order algorithms in common use for many years, it can be shown that the above algorithm converges to an optimal design provided that the sequence $\{\alpha_k\}$ is suitably chosen. For example, the choices which satisfy one of the conditions below will yield the convergence:

(i) Diminishing stepsize (Wynn's algorithm):

$$\lim_{k \to \infty} \alpha_k = 0, \quad \sum_{k=0}^{\infty} \alpha_k = \infty, \tag{3.75}$$

(ii) Limited minimization rule (Fedorov's algorithm):

$$\alpha_k = \arg\min_{\alpha} \Psi[(1 - \alpha)M(\xi^{(k)}) + \alpha\Upsilon(x_0^{(k)})], \tag{3.76}$$

(iii) Armijo rule: Given fixed scalars $\beta, \sigma, \rho \in (0, 1)$, we set

$$\alpha_k = \beta^{r_k}\rho, \tag{3.77}$$

where r_k is the first nonnegative integer r for which

$$\Psi[M(\xi^{(k)})] - \Psi[(1 - \beta^r\rho)M(\xi^{(k)}) + \beta^r\rho\Upsilon(x_0^{(k)})]$$
$$\geq -\sigma\beta^r\rho[c(\xi^{(k)}) - \phi(x_0^{(k)}, \xi^{(k)})]. \tag{3.78}$$

Usually σ is chosen close to zero (e.g., $\sigma \in [10^{-5}, 10^{-1}]$) and the reduction factor β is usually chosen from $1/2$ to $1/10$, cf. [22, p. 29] for details.

We can state these steps simply as follows [278]:

1. Select any nondegenerate starting design.

2. Compute the information matrix.

3. Find the point of maximum sensitivity.

4. Add the point of maximum sensitivity to the design, with measure proportional to its sensitivity.

5. Update the design measure.

In spite of its apparent highly specialized form, the basic version of the computational algorithm presented above constitutes a variety of the general feasible-direction method which is commonly used in nonlinear programming, see, e.g., [22, p. 209]. To see this, recall that the optimization problem in context can equivalently be formulated as follows:

$$\text{Minimize } \Psi[M]$$
$$\text{subject to } M \in \mathfrak{M}(X),$$

where Ψ is convex and $\mathfrak{M}(X)$ is a compact and convex subset of $\text{Sym}(m)$, which can be identified with a subset of $\mathbb{R}^{m(m+1)/2}$. Given a feasible matrix $M \in \mathfrak{M}(X)$, *feasible directions* at M can be interpreted here as the matrices of the form

$$D = \gamma(\bar{M} - M), \quad \gamma > 0, \tag{3.79}$$

where $\bar{M} \in \mathfrak{M}(X)$ is some matrix different from M. The method starts with a feasible matrix $M^{(0)}$ and generates a sequence of feasible matrices $\{M^{(k)}\}$ according to

$$M^{(k+1)} = M^{(k)} + \alpha_k(\bar{M}^{(k)} - M^{(k)}), \tag{3.80}$$

where $\alpha_k \in (0, 1]$, such that if $M^{(k)}$ is nonstationary, then

$$\frac{\mathrm{d}}{\mathrm{d}\alpha}\Psi[M^{(k)} + \alpha(\bar{M}^{(k)} - M^{(k)})]\Big|_{\alpha=0^+} = \text{tr}\left[M^{(k)}(\bar{M}^{(k)} - M^{(k)})\right] < 0, \tag{3.81}$$

i.e., $\bar{M}^{(k)} - M^{(k)}$ is also a descent direction.

Clearly, since the constraint set $\mathfrak{M}(X)$ is convex, we have

$$M^{(k)} + \alpha_k(\bar{M}^{(k)} - M^{(k)}) = (1 - \alpha_k)M^{(k)} + \alpha_k\bar{M}^{(k)} \in \mathfrak{M}(X) \tag{3.82}$$

for all $\alpha_k \in [0, 1]$ when $\bar{M}^{(k)} \in \mathfrak{M}(X)$, so that the generated sequence of iterates $\{M^{(k)}\}$ is feasible. Furthermore, if $M^{(k)}$ is nonstationary, there always exists a feasible direction $\bar{M}^{(k)} - M^{(k)}$ with the descent property (3.81), since otherwise we would have $\delta_+\Psi[M^{(k)}; \bar{M}^{(k)} - M^{(k)}] \geq 0$ for all $M \in \mathfrak{M}(X)$, implying that $M^{(k)}$ is stationary.

A convergence analysis can be made for various rules of choosing the stepsize [22, p. 212], including the limited minimization rule where α_k is selected so that

$$\Psi[M^{(k)} + \alpha_k(\bar{M}^{(k)} - M^{(k)})] = \min_{\alpha \in [0,1]} \Psi[M^{(k)} + \alpha(\bar{M}^{(k)} - M^{(k)})]. \tag{3.83}$$

At this juncture, we see at once striking similarities to our algorithm of finding the best design measures. They are even more visible if we observe that the difficulty in explicitly defining the set $\mathfrak{M}(X)$ while implementing the feasible-direction method can be circumvented by operating directly on the set of feasible designs $\Xi(X)$ whose image is just $\mathfrak{M}(X)$ and making use of the linearity of the FIMs with respect to the designs. Then both the algorithms become identical if we limit the search for feasible directions in (3.80) to the set $\{\Upsilon(x) - M^{(k)} : x \in X\}$, i.e., we consider only $\bar{M}^{(k)}$ corresponding to one-point designs. Then (3.80) becomes

$$\frac{d}{d\alpha}\Psi[M^{(k)} + \alpha(\Upsilon(\bar{x}^{(k)}) - M^{(k)})]\Big|_{\alpha=0^+} = c(\xi^{(k)}) - \phi(\bar{x}^{(k)}, \xi^{(k)}) < 0 \quad (3.84)$$

and the steepest descent regarding the value of Ψ corresponds to $\bar{x}^{(k)} = \arg\max_{x \in X} \phi(x, \xi^{(k)})$. Furthermore, if $\xi^{(k)}$ and thus $M^{(k)} = M(\xi^{(k)})$ is nonoptimal, there always exists a feasible direction $\Upsilon(\bar{x}^{(k)}) - M^{(k)}$ with the descent property (3.84), since otherwise we would have $c(\xi^{(k)}) \geq \phi(x, \xi^{(k)})$ for all $x \in X$, yielding that $\xi^{(k)}$ is optimal.

At this very moment, we should emphasize that Algorithm 3.1 inherits all the drawbacks of its gradient counterparts from mathematical programming. In particular, it usually shows substantial improvements in the first few iterations, but has poor convergence characteristics as the optimal solution is approached.

Computationally, Step 2 is of crucial significance but at the same time it is the most time-consuming step in the algorithm. Complications arise, among other things, due to the necessity of calculating a global maximum of $\phi(\cdot, \xi^{(k)})$ which is usually multimodal (getting stuck in one of local maxima leads to premature termination of the algorithm). Therefore, while implementing this part of the computational procedure an effective global optimizer is essential. Based on numerous computer experiments it was found that the extremely simple adaptive random search (ARS) strategy from [338,349, p. 216] is especially suited for that purpose if the design region X is a hypercube, i.e., the admissible range for x_i is in the form

$$x_{i\,\min} \leq x_i \leq x_{i\,\max}, \quad (3.85)$$

$i = 1, \ldots, n$. The routine choses the initial point x^0 at the centre of X. After q iterations, given the current best point x^q, a random displacement vector Δx is generated and the trial point

$$x^+ = \Pi_X(x^q + \Delta x) \quad (3.86)$$

is checked, where Δx follows a multinormal distribution with zero mean and covariance matrix

$$\text{cov}\{\Delta x\} = \text{diag}[\sigma_1, \ldots, \sigma_n], \quad (3.87)$$

Π_X being the projection onto X.

If $\phi(x^+, \xi^k) < \phi(x^q, \xi^k)$, then x^+ is rejected and consequently we set $x^{q+1} = x^q$, otherwise x^+ is taken as x^{q+1}. The adaptive strategy consists in repeatedly alternating two phases. During the first one (variance selection) $\text{cov}\{\Delta x\}$ is selected from among the sequence $\text{diag}(^1\sigma), \text{diag}(^2\sigma), \ldots, \text{diag}(^5\sigma)$, where

$$^1\sigma = x_{\max} - x_{\min} \quad (3.88)$$

and

$$^i\sigma = {}^{(i-1)}\sigma/10, \quad i = 2, \ldots, 5. \qquad (3.89)$$

With such a choice, $^1\sigma$ is large enough to allow for an easy exploration of X, whereas $^5\sigma$ is small enough for a precise localization of an optimal point. In order to allow a comparison to be drawn, all the possible $^i\sigma$'s are used $100/i$ times, starting from the same initial value of x. The largest $^i\sigma$'s, designed to escape from local maxima, are therefore used more often than the smaller ones.

During the second (exploration) phase, the most successful $^i\sigma$ in terms of the criterion value reached during the variance selection phase is used for 100 random trials started from the best x obtained so far. The variance-selection phase then resumes, unless the decision to stop is taken.

As regards the choice of an optimal α_k in Fedorov's variant of Step 3, it should be emphasized that the situation is a bit different from the well-known case of linear regression considered in classic textbooks for which it is possible to determine a closed-form solution (simplifications resulting in similar circumstances will be discussed in Section 3.7.3). Since in our case the application of the matrix-inversion lemma by no means simplifies the problem, an optimal α_k has to be determined numerically, e.g., with the use of the golden-section search.

Furthermore, while implementing the algorithms, numerous additional problems should be addressed. For instance, it may be possible to achieve a markedly greater decrease in the value of Ψ by removing a measure from a point already in the design $\xi^{(k)}$ and distributing the measure removed among the most promising points of support (e.g., those for which $\phi(x^i, \xi^{(k)}) > c(\xi^{(k)})$), and additionally, distributing the removed measure among those points in a manner proportional to $\phi(x^i, \xi^{(k)}) - c(\xi^{(k)})$. In this way, undesirable and noninformative points of support which were included in the initial design can be eliminated.

Another complication is the tendency of the points produced in Step 2 to cluster around support points of the optimal design. This can be avoided by checking whether the newly generated point is close enough to a point of the current support so as to qualify them as coinciding points. If so, the latter is replaced by the former with a simultaneous update of the weights of all the points according to the rule of Step 3. In more detail, we could check whether $x_0^{(k)}$ is close enough to a point already in the design, e.g., by determining

$$\tilde{x}^{(k)} = \arg \min_{x \in \text{supp}\,\xi^{(k)}} \varrho(x), \qquad (3.90)$$

where

$$\varrho(x) = \|\Upsilon(x) - \Upsilon(x_0^{(k)})\|. \qquad (3.91)$$

If $\varrho(\tilde{x}^{(k)}) < \epsilon$, where ϵ is sufficiently small, then $\tilde{x}^{(k)}$ is to be deleted and its weight to be added to that of $x_0^{(k)}$.

As a supplementary improvement, a cyclic removal of points with negligible weights is also suggested by Rafajłowicz [245] in order to maintain a relatively small number of support points.

A detailed description of all the troubles and corresponding tricks which were invented to alleviate them so as to create efficient codes for constructing optimal experiments can no doubt constitute a subject for a separate monograph. Notwithstanding the fact that the problem outlined in this section is slightly different from the classical formulation, the problems encountered remain in principle the same and hence this topic will not be further discussed owing to a limited volume of this monograph. For a more comprehensive discussion, we refer the reader to the excellent specialized literature [8, 77, 85, 245, 246, 272, 294, 349].

Example 3.1

To get a feel for Theorems 3.2 and 3.3, consider $X = [-1, 1]$ and the following vector of basis functions:

$$f(x, t) = \text{col}[1, t \sin(\pi x), t^2 \exp(-x)]$$

for which a D-optimal design was to be found. To generate a solution, Fedorov's version of Algorithm 3.1 was implemented and the design

$$\xi^0 = \left\{ \begin{matrix} -1/2, & 0, & 1/2 \\ 1/3, & 1/3, & 1/3 \end{matrix} \right\}$$

such that $\det[M(\xi^0)] = 0.023928$, was adopted to launch the computational procedure. After nine cycles of the algorithm the following approximation of the optimal design was obtained:

$$\xi^* = \left\{ \begin{matrix} -1.000000, & -0.512301, & 0.497145 \\ 0.370455, & 0.336952, & 0.292593 \end{matrix} \right\}$$

which corresponds to $\det[M(\xi^*)] = 0.080463$. This result is illustrated in Fig. 3.1, where a solid line represents the optimal sensitivity function. As can be seen, in the interval $[-1, 1]$ it attains the maximal value of $m = 3$ at the points supporting the design. For comparison, the dashed line represents an exemplary nonoptimal design. Clearly, the maximal value of ϕ in the latter case exceeds the number of parameters.   ▯

3.1.3.2 Sequential numerical design algorithms with support points given *a priori*

Suppose we know *a priori* a finite but sufficiently rich collection of points $X_d = \{x^1, \dots, x^\ell\} \subset X$ among which the support points of a Ψ-optimum design measure lie. Then the only problem is to find the appropriate weights

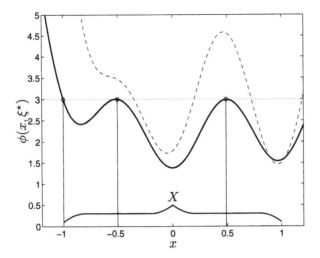

FIGURE 3.1
Sensitivity function for the D-optimal design of Example 3.1 (solid line). The
same function for an exemplary nonoptimal design is also shown (dashed line)
for the sake of comparison.

$p_1, \ldots, p_\ell$ at those points, as they uniquely determine the designs. Such a
formulation turns out to be extremely convenient and it gives rise to a number
of rapidly convergent algorithms which can be used for a fast identification of
points which are not in the support of an optimal design $\xi^\star$.

Given X_d, we thus focus our attention on the designs of the form

$$\xi = \begin{Bmatrix} x^1, & x^2, & \ldots, & x^\ell \\ p_1, & p_2, & \ldots, & p_\ell \end{Bmatrix} \tag{3.92}$$

and consider the problem of finding

$$p^\star = \arg\min_p \Psi[M(\xi)] \tag{3.93}$$

subject to

$$M(\xi) = \sum_{i=1}^{\ell} p_i \Upsilon(x^i) \tag{3.94}$$

and

$$p \in \mathbb{S} = \left\{ p = (p_1, \ldots, p_\ell) : \sum_{i=1}^{\ell} p_i = 1, \; p_i \geq 0, \; i = 1, \ldots, \ell \right\}. \tag{3.95}$$

Clearly, this is a finite-dimensional optimization problem over the canonical
simplex $\mathbb{S}$, which can be tackled by standard numerical methods. A number
of possible choices are indicated below.

Gradient projection method

Owing to a relatively simple form of the constraints (3.95), a straightforward procedure can be proposed, which reduces to using a gradient projection method [22, p. 223]. Generation of a new point in the $(k+1)$-th iteration can be formalized as follows:

$$p^{(k+1)} = p^{(k)} + \alpha_k(\bar{p}^{(k)} - p^{(k)}), \tag{3.96}$$

where

$$\bar{p}^{(k)} = \Pi_{\mathbb{S}}\big[p^{(k)} - \beta_k \nabla_p \Psi[M(\xi^{(k)})]\big], \tag{3.97}$$

$\alpha_k \in (0,1]$ and β_k are a stepsize and a positive scalar, respectively. Here $\Pi_{\mathbb{S}}[\,\cdot\,]$ stands for orthogonal projection into the convex set of admissible weights $\mathbb{S}$. The gradient $\nabla_p \Psi[M(\xi^{(k)})]$ consists of the derivatives

$$\frac{\partial \Psi[M(\xi^{(k)})]}{\partial p_i} = \phi(x^i, \xi^{(k)}), \quad i = 1, \dots, \ell \tag{3.98}$$

which are easy to calculate. The idea is as follows: in order to obtain $\bar{p}^{(k)}$, we make a step along the negative gradient, as in steepest descent. We then project the result $p^{(k)} - \beta_k \nabla_p \Psi[M(\xi^{(k)})]$ onto $\mathbb{S}$, thereby obtaining the feasible weights $\bar{p}^{(k)}$. At last, we make a step along the feasible direction $\bar{p}^{(k)} - p^{(k)}$ using the stepsize α_k.

Note that there exist many possible choices of determining the parameters α_k and β_k steering convergence. For example, we can use the *limited minimization rule* in which we set

$$\beta_k = \beta, \quad k = 0, 1, \dots, \tag{3.99}$$

where β is a constant, and α_k is chosen such that

$$\Psi[M(\xi^{(k)}) + \alpha_k(M(\bar{\xi}^{(k)}) - M(\xi^{(k)}))]$$
$$= \min_{\alpha \in [0,1]} \Psi[M(\xi^{(k)}) + \alpha(M(\bar{\xi}^{(k)}) - M(\xi^{(k)}))], \quad (3.100)$$

$\bar{\xi}^{(k)}$ being the design corresponding to the weights $\bar{p}^{(k)}$. The properties of this rule and other choices (e.g., the Armijo rule along the feasible direction, the Armijo rule along the projection arc, the constant stepsize rule, etc.) are discussed in detail in [22, p. 225].

Clearly, in order for the method to make practical sense, the projection operation onto $\mathbb{S}$ should be fairly simple. In fact, it is, as an algorithm can be developed for that purpose, which is almost as simple as a closed-form solution, cf. Appendix B.9.

Application of semidefinite programming

An alternative approach to determining optimal weights is proposed here, as it makes it possible to employ extremely powerful algorithms for convex optimization based on Linear Matrix Inequalities (LMIs) or, more generally, on Semidefinite Programming (SDP) which has recently become a dynamically expanding research area. The SDP problem can be regarded as an extension of linear programming where the component-wise linear inequalities between vectors are replaced by matrix inequalities, or equivalently, the first orthant is replaced by the cone of nonnegative definite matrices. Most interior-point methods for linear programming have been generalized to semidefinite programmes [28, 102, 334]. As in linear programming, these methods have polynomial worst-case complexity and perform well in practice, rapidly computing the global optima with nonheuristic stopping criteria. Numerical experience shows that in specific problems those algorithms can solve LMI-based problems with a computational effort that is comparable to that required to evaluate the analytic solutions of similar problems (e.g., solving multiple simultaneous Lyapunov inequalities). What is more, numerous efficient public-domain codes are available [280]. SDP has been successfully applied in engineering (from control theory to structural design [27, 37, 69, 112, 336]) and combinatorial optimization [99].

A linear matrix inequality is a matrix inequality of the form

$$G(v) := G_0 + \sum_{i=1}^{\ell} v_i G_i \succeq 0, \qquad (3.101)$$

where $v \in \mathbb{R}^{\ell}$ is the variable and the matrices $G_i \in \text{Sym}(k)$, $i = 1, \ldots, \ell$ are given. The LMI is a convex constraint in v, i.e., the set $\{v \in \mathbb{R}^{\ell} : G(v) \succeq 0\}$ is convex.

In a basic primal SDP problem we minimize a linear function of a variable $v \in \mathbb{R}^{\ell}$ subject to conventional LP constraints and a finite number of LMIs. In more detail, we wish to find the decision vector u that minimizes

$$J(v) = c^{\mathsf{T}} v \qquad (3.102)$$

subject to

$$Hv \geq e, \qquad (3.103)$$
$$Ax = b, \qquad (3.104)$$

$$G_0^{(j)} + \sum_{i=1}^{\ell} v_i G_i^{(j)} \succeq 0, \quad j = 1, \ldots, r, \qquad (3.105)$$

$c \in \mathbb{R}^{\ell}$, $H \in \mathbb{R}^{q \times \ell}$, $e \in \mathbb{R}^q$, $A \in \mathbb{R}^{\kappa \times \ell}$, $b \in \mathbb{R}^{\kappa}$, $G_i^{(j)} \in \text{Sym}(k_j)$. Although the LMI constraints so formulated look rather specialized, they are much more general than linear inequality constraints. For example, when the matrices $G_i^{(j)}$ are diagonal, the j-th LMI is just a system of linear inequalities.

In spite of the fact that potential applications of SDP in optimum experimental design were already indicated by Boyd and Vandenberghe, e.g., in [335], cf. also [28, p. 384], the idea has not been pursued in the optimum experimental design community. In what follows, we present how to adapt Boyd and Vandenberghe's concept in the context of the weight optimization discussed in this section. As there is no method converting the general optimum experimental design problem to the form of (3.102)–(3.105), we consider the three most popular design criteria and show how the corresponding design problems can be cast as convex optimization problems over LMIs.

A-optimality criterion
In A-optimum design we minimize $\text{tr}(M^{-1}(\xi))$ over all weights belonging to the canonical simplex $\mathbb{S}$. Observe, however, that

$$\text{tr}(M^{-1}(\xi)) = \sum_{j=1}^{m} e_j^\mathsf{T} M^{-1}(\xi) e_j, \tag{3.106}$$

where e_j is the usual unit vector along the j-th coordinate of $\mathbb{R}^m$. Introducing additional variables q_j, $j = 1, \ldots, m$ represented in a concise form as the vector $q = \text{col}[q_1, \ldots, q_m]$, the A-optimum design problem can be cast as follows: Find q and p which minimize

$$J(p, q) = \sum_{j=1}^{n} q_j \tag{3.107}$$

subject to

$$e_j^\mathsf{T} M^{-1}(\xi) e_j \le q_j \tag{3.108}$$

and

$$\sum_{i=1}^{\ell} p_i = 1, \quad p_i \ge 0, \quad i = 1, \ldots, \ell. \tag{3.109}$$

Defining the vector $\mathbf{1} \in \mathbb{R}^m$ whose all components are equal to unity and making use of the Schur complement, cf. Theorem B.15, we obtain the equivalent SDP formulation: Find p and q to minimize

$$J(p, q) = \mathbf{1}^\mathsf{T} q \tag{3.110}$$

subject to

$$\mathbf{1}^\mathsf{T} p = 1, \tag{3.111}$$

$$p \ge 0 \tag{3.112}$$

and LMIs

$$\left[\begin{array}{c|c} M(\xi) & e_j \\ \hline e_j^{\mathsf{T}} & q_j \end{array}\right] \succeq 0, \quad j = 1, \dots, m. \tag{3.113}$$

E-optimality criterion

In E-optimum design we minimize the largest eigenvalue of $M^{-1}(\xi)$, which is equivalent to maximizing the smallest eigenvalue $\lambda_{\min}$ of $M(\xi)$ over all weights p from the canonical simplex $\mathbb{S}$. Note, however, that we have

$$\lambda_{\min}[M(\xi)] = \inf_{0 \neq v \in \mathbb{R}^m} \frac{v^{\mathsf{T}} M(\xi) v}{v^{\mathsf{T}} v}, \tag{3.114}$$

cf. Theorem B.4.

Introducing a new real variable q, the E-optimum design problem can be cast as follows: Find p and q, which maximize

$$J(p, q) = q \tag{3.115}$$

subject to

$$q \leq \frac{v^{\mathsf{T}} M(\xi) v}{v^{\mathsf{T}} v} \quad \text{for any } 0 \neq v \in \mathbb{R}^m \tag{3.116}$$

and

$$\sum_{i=1}^{\ell} p_i = 1, \quad p_i \geq 0, \quad i = 1, \dots, \ell. \tag{3.117}$$

But the constraint (3.116) can be written as

$$v^{\mathsf{T}}\big(M(\xi) - qI\big)v \geq 0 \quad \text{for any } 0 \neq v \in \mathbb{R}^m, \tag{3.118}$$

where I is the identity matrix. Since it also holds trivially for $v = 0$, finally we obtain the following equivalent formulation of the E-optimum design problem: Find p and q to minimize

$$\tilde{J}(p, q) = -q \tag{3.119}$$

subject to

$$\mathbf{1}^{\mathsf{T}} p = 1, \quad p \geq 0 \tag{3.120}$$

and the LMI

$$M(\xi) \succeq qI. \tag{3.121}$$

The above reduction to an SDP problem makes it possible to employ efficient interior-point algorithms of polynomial-time complexity. This result can hardly be overestimated in the light of the fact that there appears to be no known efficient algorithms for generating E-optimum designs, cf. Section 3.2.

D-optimality criterion
In D-optimality design we maximize the determinant of the FIM, thereby obtaining the so-called maxdet problem [337], which is equivalent to the minimization of

$$J(p) = \ln \det(M^{-1}(\xi)) \qquad (3.122)$$

subject to the linear constraints

$$\sum_{i=1}^{\ell} p_i = 1, \quad p_i \geq 0, \quad i = 1, \dots, \ell. \qquad (3.123)$$

Although this is not a classical SDP problem, it remains a convex optimization one which can still be solved by several general algorithms of convex programming, e.g., the ellipsoid method. However, they are often efficient only in theory where account of the worst-case complexity is usually taken (the worst-case predictions for the required number of iterations are frequently off by many orders of magnitude from the practically observed number of iterations). Consequently, an interior-point-like method was proposed by Vandenberghe *et al.* [337] for solving the maxdet problem subject to general LMI constraints. The authors treated the maxdet problem as a quite specific extension of the SDP formulaton and thus invented the method which is efficient both in worst-case complexity theory and in practice, as the worst-case complexity of $O(\sqrt{\ell})$ Newton iterations was proved and numerical experiments indicate that the behaviour may be much better.

Although the LMI constraints in our case reduce to a simple system of classical linear constraints, the approach can be of practical importance when the number of support points is pretty high or when efficient codes for solving the maxdet problem are accessible. As an alternative, we outline below a specialized algorithm which originates in an efficient algorithm that has been used in classical optimum experimental design for many years.

Multiplicative algorithm for D-optimum designs

In the case of the D-optimality criterion, the following very simple and numerically effective computational procedure was devised and analysed in [208, 270, 294, 295] for the case of static systems. Below we extend its applicability to systems with continuous-time measurements. An interesting feature of this scheme is that it is globally convergent regardless of the choice of initial weights (they must only correspond to a nonsingular FIM) and that the values of the determinants of successive FIMs form a nondecreasing sequence.

ALGORITHM 3.2 Finding a D-optimum experimental effort

Step 1. *Select weights $p_i^{(0)}$, $i = 1, \ldots, \ell$ (initial guess) which determine the initial design $\xi^{(0)}$ for which we must have $\det(M(\xi^{(0)})) \neq 0$. Choose $0 < \eta \ll 1$, a parameter used in the stopping rule. Set $k = 0$.*

Step 2. *If*

$$\frac{\phi(x^i, \xi^{(k)})}{m} < 1 + \eta, \quad i = 1, \ldots, \ell, \tag{3.124}$$

then STOP.

Step 3. *Evaluate*

$$p_i^{(k+1)} = p_i^{(k)} \frac{\phi(x^i, \xi^{(k)})}{m}, \quad i = 1, \ldots, \ell. \tag{3.125}$$

Form the corresponding design $\xi^{(k+1)}$, increment k by one and go to Step 2.

The idea is reminiscent of the EM algorithm used for maximum likelihood estimation [150]. The properties of this computational scheme are considered here in detail. We start with the following generalization of Proposition V.8 of [208, p. 140], which results from specific properties of the determinant.

LEMMA 3.7
Given any discrete design ξ of the form (3.92), suppose that $\det(M(\xi)) \neq 0$. Then

$$\det(M(\xi)) = \sum_{i_1=1}^{\ell} \cdots \sum_{i_m=1}^{\ell} h(x^{i_1}, \ldots, x^{i_m}) p_{i_1} \cdots p_{i_m}, \tag{3.126}$$

where the function h is nonnegative, independent of the weights $p_1, \ldots, p_\ell$ and symmetric (i.e., invariant under coordinate permutations, cf. Appendix B.6).

PROOF As the matrix $\Upsilon(x^i)$ is nonnegative definite, the eigenvalue decomposition, cf. (B.26), allows us to express it as

$$\Upsilon(x^i) = \sum_{j=1}^{m} \lambda_{ij} v_{ij} v_{ij}^{\mathsf{T}}, \tag{3.127}$$

where the nonnegative numbers λ_{ij}, $j = 1, \ldots, m$ are the eigenvalues of $\Upsilon(x^i)$ counted with their respective multiplicities, and the m-dimensional vectors v_{ij}, $j = 1, \ldots, m$ form an orthonormal system of the corresponding eigenvectors.

Therefore

$$M(\xi) = \sum_{i=1}^{\ell} p_i \Upsilon(x^i) = \sum_{i=1}^{\ell} \sum_{j=1}^{m} p_i \lambda_{ij} v_{ij} v_{ij}^{\mathsf{T}}. \tag{3.128}$$

For notational convenience, we introduce one index k instead of two indices i and j by defining

$$k \equiv (i-1)\ell + j \quad \text{for } i = 1, \ldots, \ell, \quad j = 1, \ldots, m, \tag{3.129}$$

and then replace $p_i \lambda_{ij}$ and v_{ij} by μ_k and w_k, respectively. Hence

$$M(\xi) = \sum_{k=1}^{\ell m} \mu_k w_k w_k^{\mathsf{T}}. \tag{3.130}$$

Denoting by $w_{k(r)}$ the r-th component of w_k, by the multilinearity of the determinant, cf. Appendix B.1, we have

$$\det(M(\xi))$$

$$= \det \left[\sum_{k_1=1}^{\ell m} \mu_{k_1} w_{k_1} w_{k_1(1)} \quad \cdots \quad \sum_{k_m=1}^{\ell m} \mu_{k_m} w_{k_m} w_{k_m(m)} \right] \tag{3.131}$$

$$= \sum_{k_1=1}^{\ell m} \cdots \sum_{k_m=1}^{\ell m} \mu_{k_1} \cdots \mu_{k_m} w_{k_1(1)} \cdots w_{k_m(m)} \det \left[w_{k_1} \; \vdots \; \cdots \; \vdots \; w_{k_m} \right].$$

Note, however, that in the last sum we can only take account of the terms for which $k_1 \neq k_2 \neq \cdots \neq k_m$, as otherwise at least two columns of the matrix whose determinant is to be calculated are the same and hence the determinant becomes zero. Therefore

$$\det(M(\xi))$$

$$= \sum_{k_1 \neq k_2 \neq \cdots \neq k_m} \mu_{k_1} \cdots \mu_{k_m} w_{k_1(1)} \cdots w_{k_m(m)} \det \left[w_{k_1} \; \vdots \; \cdots \; \vdots \; w_{k_m} \right]. \tag{3.132}$$

Summing up terms which have the same subscripts $k_1, \ldots, k_m$ but in different orderings, we obtain

$$\det(M(\xi)) = \sum_{1 \leq k_1 < k_2 < \cdots < k_m \leq \ell m} \mu_{k_1} \cdots \mu_{k_m} \det \left[w_{k_1} \; \vdots \; \cdots \; \vdots \; w_{k_m} \right]$$

$$\cdot \sum_{\sigma} w_{\sigma(1),(1)} \cdots w_{\sigma(m),(m)} \operatorname{sgn}(\sigma), \tag{3.133}$$

where σ in the inner summation runs through the set of all permutations on the set $\{k_1, \ldots, k_m\}$, $\sigma(i)$ stands for the number occupying position i in σ, and $\operatorname{sgn}(\sigma)$ equals $+1$ or -1 depending on whether or not σ is even, cf. Appendix B.1.

But on account of Appendix B.1, the inner summation gives nothing but $\det\left[\,w_{k_1}\,\vdots\,\cdots\,\vdots\,w_{k_m}\,\right]$, which leads to

$$\det(M(\xi)) = \sum_{1 \leq k_1 < k_2 < \cdots < k_m \leq \ell m} \mu_{k_1} \cdots \mu_{k_m} \det^2\left[\,w_{k_1}\,\vdots\,\cdots\,\vdots\,w_{k_m}\,\right]$$

$$= \frac{1}{m!} \sum_{k_1=1}^{\ell m} \cdots \sum_{k_m=1}^{\ell m} \mu_{k_1} \cdots \mu_{k_m} \det^2\left[\,w_{k_1}\,\vdots\,\cdots\,\vdots\,w_{k_m}\,\right]. \tag{3.134}$$

Returning to the original indices, we obtain

$$\det(M(\xi))$$

$$= \frac{1}{m!} \sum_{i_1=1}^{\ell} \sum_{j_1=1}^{m} \cdots \sum_{i_m=1}^{\ell} \sum_{j_m=1}^{m} p_{i_1} \lambda_{i_1 j_1} \cdots p_{i_m} \lambda_{i_m j_m} \det^2\left[\,v_{i_1 j_1}\,\vdots\,\cdots\,\vdots\,v_{i_m j_m}\,\right]$$

$$= \sum_{i_1=1}^{\ell} \cdots \sum_{i_m=1}^{\ell} p_{i_1} \cdots p_{i_m}$$

$$\cdot \left\{ \frac{1}{m!} \sum_{j_1=1}^{m} \cdots \sum_{j_m=1}^{m} \lambda_{i_1 j_1} \cdots \lambda_{i_m j_m} \det^2\left[\,v_{i_1 j_1}\,\vdots\,\cdots\,\vdots\,v_{i_m j_m}\,\right] \right\}, \tag{3.135}$$

which is precisely the assertion of the lemma with the function h corresponding to the expression in the braces. ∎

We now prove a general convergence theorem due to Pázman [208, Prop. V.6, p. 139].

THEOREM 3.4
Assume that $\{\xi^{(k)}\}$ is a sequence of iterates constructed by Algorithm 3.2. Then the sequence $\{\det(M(\xi^{(k)}))\}$ is nondecreasing and

$$\lim_{k \to \infty} \det(M(\xi^{(k)})) = \sup_{\xi \in \Xi(X)} \det(M(\xi)). \tag{3.136}$$

PROOF Recall that any feasible design (3.92) can be interpreted as a discrete probability distribution with the outcomes being its support points $x^1, \ldots, x^\ell$ and the respective probabilities being the corresponding weights $p_1, \ldots, p_\ell$. We can therefore define independent discrete random variables $V_1, \ldots, V_m$ with the same probability mass function as ξ. From Lemma 3.7 we deduce that

$$\det(M(\xi)) = E_\xi[h], \tag{3.137}$$

where $E_\xi[h]$ denotes the expectation of h calculated for a given ξ and the probability distribution on the Cartesian product X_d^m, i.e., we have

$$P(V_1 = x^{i_1}, \dots, V_m = x^{i_m}) = p_{i_1} \cdots p_{i_m} \qquad (3.138)$$

for $1 \le i_1, \dots, i_m \le \ell$.

For any weight p_r we see that

$$\frac{\partial \ln \det(M(\xi))}{\partial p_r} = \text{tr}[M^{-1}(\xi) \Upsilon(x^r)] = \phi(x^r, \xi). \qquad (3.139)$$

On the other hand, from (3.137) it follows that

$$
\begin{aligned}
&\frac{\partial \ln \det(M(\xi))}{\partial p_r} \\
&= \frac{\partial \ln E_\xi[h]}{\partial p_r} = \frac{1}{E_\xi[h]} \frac{\partial E_\xi[h]}{\partial p_r} \\
&= \frac{1}{E_\xi[h]} \Bigg\{ \sum_{i_2=1}^\ell \cdots \sum_{i_m=1}^\ell h(x^r, x^{i_2}, \dots, x^{i_m}) p_{i_2} \cdots p_{i_m} \\
&\quad + \cdots + \sum_{i_1=1}^\ell \cdots \sum_{i_{m-1}=1}^\ell h(x^{i_1}, \dots, x^{i_{m-1}}, x^r) p_{i_1} \cdots p_{i_{m-1}} \Bigg\} \\
&= \frac{1}{E_\xi[h]} \{ E_\xi[h \mid V_1 = x^r] + \cdots + E_\xi[h \mid V_m = x^r] \},
\end{aligned}
\qquad (3.140)
$$

cf. Appendix B.10 for the definition and properties of conditional expectations. But

$$E_\xi[h \mid V_1 = x^r] = \cdots = E_\xi[h \mid V_m = x^r], \qquad (3.141)$$

since the function h is symmetric, so we conclude that

$$\frac{\partial \ln \det(M(\xi))}{\partial p_r} = \frac{m \, E_\xi[h \mid V_i = x^r]}{E_\xi[h]} \qquad (3.142)$$

for any $i \in \{1, \dots, m\}$. Consequently, (3.139) taken in conjunction with (3.142) gives

$$\frac{\phi(x^r, \xi)}{m} = \frac{E_\xi[h \mid V_i = x^r]}{E_\xi[h]}, \qquad 1 \le i \le m. \qquad (3.143)$$

Let ξ^+ be the design obtained from ξ by modifying the weights of ξ in accordance with (3.125). It follows that

$$
\begin{aligned}
\det(M(\xi^+)) \\
&= \sum_{i_1=1}^{\ell} \cdots \sum_{i_m=1}^{\ell} \frac{\phi(x^{i_1}, \xi)}{m} \cdots \frac{\phi(x^{i_m}, \xi)}{m} h(x^{i_1}, \ldots, x^{i_m}) p_{i_1} \cdots p_{i_m} \\
&= \sum_{i_1=1}^{\ell} \cdots \sum_{i_m=1}^{\ell} \frac{\mathrm{E}_\xi[h \mid V_1 = x^{i_1}]}{\mathrm{E}_\xi[h]} \cdots \frac{\mathrm{E}_\xi[h \mid V_m = x^{i_m}]}{\mathrm{E}_\xi[h]} \\
&\quad \cdot h(x^{i_1}, \ldots, x^{i_m}) p_{i_1} \cdots p_{i_m} \\
&= \mathrm{E}_\xi\left[h \frac{\mathrm{E}_\xi[h \mid V_1]}{\mathrm{E}_\xi[h]} \cdots \frac{\mathrm{E}_\xi[h \mid V_m]}{\mathrm{E}_\xi[h]} \right] \geq \mathrm{E}_\xi[h] = \det(M(\xi)),
\end{aligned}
\tag{3.144}
$$

the inequality resulting from Lemma B.1.

Therefore the sequence $\{\det(M(\xi^{(k)}))\}$ is nondecreasing. Note that it is also bounded by $\det(M(\xi^\star))$, where $\xi^\star$ is a D-optimum design. Hence $\{\det(M(\xi^{(k)}))\}$ converges to a finite limit. Below we shall show that this limit is just $\det(M(\xi^\star))$.

At each iteration, the weights are updated so that

$$
p_i^{(k+1)} = p_i^{(k)} \frac{\phi(x^i, \xi^{(k)})}{m}, \quad i = 1, \ldots, \ell.
\tag{3.145}
$$

Since the weights always belong to the canonical simplex $\mathbb{S}$ which is a compact set in $\mathbb{R}^\ell$, there exists a subsequence $\{p^{(k_j)}\}$ of $\{p^{(k)}\}$ such that

$$
p^{(k_j+1)} \xrightarrow[j \to \infty]{} \bar{p}, \quad p^{(k_j)} \xrightarrow[j \to \infty]{} \widetilde{p},
\tag{3.146}
$$

where $\bar{p}$ and $\widetilde{p}$ are some elements of $\mathbb{S}$.

Let $\bar{\xi}$ and $\widetilde{\xi}$ be the designs defined by $\bar{p}$ and $\widetilde{p}$, respectively. Substituting k_j for k in (3.145) and letting $j \to \infty$, we obtain

$$
\bar{p}_i = \widetilde{p}_i \frac{\phi(x^i, \widetilde{\xi})}{m}.
\tag{3.147}
$$

The continuity of the determinant implies that

$$
\det(M(\bar{\xi})) = \det(M(\widetilde{\xi})).
\tag{3.148}
$$

But then (3.144) yields

$$
\mathrm{E}_{\widetilde{\xi}}\left[h\, \mathrm{E}_{\widetilde{\xi}}[h \mid V_1] \ldots \mathrm{E}_{\widetilde{\xi}}[h \mid V_m] \right] = \left\{ \mathrm{E}_{\widetilde{\xi}}[h] \right\}^{m+1},
\tag{3.149}
$$

which can take place only if

$$
\mathrm{E}_{\widetilde{\xi}}[h \mid V_1] = \cdots = \mathrm{E}_{\widetilde{\xi}}[h \mid V_m] = \mathrm{E}_{\widetilde{\xi}}[h]
\tag{3.150}
$$

almost surely, cf. Lemma B.1. Clearly, from (3.143) we deduce that

$$\phi(x^i, \widetilde{\xi}) = m, \quad i = 1, \ldots, \ell \tag{3.151}$$

almost everywhere, i.e., at all support points with positive weights in $\widetilde{\xi}$. Theorem 3.3 now shows that $\widetilde{\xi}$ must be a D-optimum design.

As the sequences $\{\det(M^{(k)})\}$ and $\{\det(M^{(k_j)})\}$ cannot converge to different limits, we see that

$$\lim_{k \to \infty} \det(M(\xi^{(k)})) = \det(M(\widetilde{\xi})) = \sup\{\det(M(\xi)) : \xi \in \Xi(X)\}, \tag{3.152}$$

which is the desired conclusion.   ∎

3.2 Construction of minimax designs

So far, our attention has been focused on the construction and characterization of designs for smooth optimality criteria. In practice, however, it is sometimes desirable to consider the so-called worst-case design criteria, among which the E-optimality design criterion $\Psi_E[M] = \lambda_{\max}[M^{-1}]$ is the most representative in the context discussed in this chapter. This seemingly simple criterion becomes nondifferentiable as soon as the maximum eigenvalue in its definition comes to be repeated, cf. a detailed discussion of this topic in Appendix D. However, the situation is not that hopeless, as in what follows we shall show how to derive the corresponding equivalence theorem. We shall closely follow the approach originally proposed by Fedorov [85] and further extended by Wong [352] who embedded the E-optimality criterion into a broad class of nondifferentiable design criteria, where an optimal design

$$\xi^\star = \arg \min_{\xi \in \Xi(X)} \max_{a \in A} \Psi[M(\xi), a] \tag{3.153}$$

is to be determined, A being a given compact set of parameters which cannot be controlled by the experimenter.

Inevitably, in order to tackle the problem (3.153), Assumptions (A1)–(A6) have to be slightly modified. Namely, we orient the problem by imposing the following conditions:

(B1) X and A are compact sets.

(B2) $f \in C(X \times T; \mathbb{R}^m)$.

(B3) For each $a \in A$, the mapping $\Psi[\,\cdot\,, a]$ is convex.

(B4) If $M_1 \preceq M_2$, then $\Psi[M_1, a] \geq \Psi[M_2, a]$ for every $a \in A$.

(B5) There exists a finite real q such that

$$\{\xi : \Psi[M(\xi), a] \leq q < \infty, \ \forall a \in A\} = \widetilde{\Xi}(X) \neq \emptyset.$$

(B6) $\Psi[\cdot, \cdot]$ and $\partial\Psi[\cdot, \cdot]/\partial M$ are continuous on the set $\widetilde{\mathfrak{M}}(X) \times A$, where

$$\widetilde{\mathfrak{M}}(X) = \{M(\xi) : \xi \in \widetilde{\Xi}(X)\}.$$

Under the above assumptions, the following version of the equivalence theorem can be proven:

THEOREM 3.5
Let Assumptions (B1)–(B6) hold. Then a design $\xi^\star \in \widetilde{\Xi}(X)$ is optimal if and only if there exists a probability measure $\mu^\star$ defined on the set

$$\widehat{A}(\xi^\star) = \left\{ \hat{a} : \Psi[M(\xi^\star), \hat{a}] = \max_{a \in A} \Psi[M(\xi^\star), a] \right\} \tag{3.154}$$

such that

$$\max_{x \in X} \int_{\widehat{A}(\xi^\star)} \phi(x, \xi^\star, a)\, \mu^\star(da) \leq \int_{\widehat{A}(\xi^\star)} c(\xi^\star, a)\, \mu^\star(da), \tag{3.155}$$

where

$$c(\xi, a) = -\operatorname{tr}\left[\mathring{\Psi}(\xi, a) M(\xi)\right], \tag{3.156}$$

$$\phi(x, \xi, a) = -\operatorname{tr}\left[\mathring{\Psi}(\xi, a)\Upsilon(x)\right] = -\frac{1}{t_f}\int_0^{t_f} f^{\mathsf{T}}(x, t)\mathring{\Psi}(\xi, a)f(x, t)\, dt, \tag{3.157}$$

$$\mathring{\Psi}(\xi, a) = \left.\frac{\partial\Psi[M, a]}{\partial M}\right|_{M=M(\xi)}. \tag{3.158}$$

PROOF To check whether $\xi^\star$ is optimal over $\Xi(X)$ amounts to verification whether $M^\star = M(\xi^\star)$ is a global minimum of

$$J(M) = \max_{a \in A} \Psi(M, a) \tag{3.159}$$

over $\mathfrak{M}(X) = \{M(\xi) : \xi \in \Xi(X)\}$.

From Assumption (B3) and Theorem B.21 we see that J is convex. Moreover, by Theorem E.1, its one-sided directional differential $\delta_+ J(M^\star; M - M^\star)$ in the direction $M - M^\star$ exists for every $M \in \mathfrak{M}(X)$ and we have

$$\delta_+ J(M^\star; M - M^\star) = \max_{a \in \widehat{A}(\xi^\star)} \operatorname{tr}\left[\mathring{\Psi}(\xi^\star, a)(M - M^\star)\right]. \tag{3.160}$$

Since A is compact, from the continuity of Ψ it follows that so is $\widehat{A}(\xi^\star)$. Hence a necessary and sufficient condition for $M^\star$ to minimize J over $\mathfrak{M}(X)$ is

$$\delta_+ J(M^\star; M - M^\star) \geq 0, \quad \forall M \in \mathfrak{M}(X), \tag{3.161}$$

cf. Theorem B.25, or equivalently,

$$\min_{M \in \mathfrak{M}(X)} \delta_+ J(M^\star; M - M^\star) \geq 0. \tag{3.162}$$

The left-hand side of the latter inequality can be rewritten as

$$\min_{M \in \mathfrak{M}(X)} \max_{a \in \widehat{A}(\xi^\star)} \operatorname{tr}\left[\mathring{\Psi}(\xi^\star, a)(M - M^\star)\right]$$

$$= \min_{M \in \mathfrak{M}(X)} \max_{\mu \in \Sigma(\xi^\star)} \int_{\widehat{A}(\xi^\star)} \operatorname{tr}\left[\mathring{\Psi}(\xi^\star, a)(M - M^\star)\right] \mu(da), \tag{3.163}$$

where $\Sigma(\xi^\star)$ stands for the set of all probability measures on $\widehat{A}(\xi^\star)$. Indeed, choosing any $a_0 \in \widehat{A}(\xi^\star)$ at which $\operatorname{tr}\left[\mathring{\Psi}(\xi^\star, a)(M - M^\star)\right]$ is maximized, we have

$$\max_{a \in \widehat{A}(\xi^\star)} \operatorname{tr}\left[\mathring{\Psi}(\xi^\star, a)(M - M^\star)\right]$$

$$= \operatorname{tr}\left[\mathring{\Psi}(\xi^\star, a_0)(M - M^\star)\right]$$

$$= \max_{\mu \in \Sigma_0(\xi^\star)} \int_{\widehat{A}(\xi^\star)} \operatorname{tr}\left[\mathring{\Psi}(\xi^\star, a)(M - M^\star)\right] \mu(da)$$

$$\leq \max_{\mu \in \Sigma(\xi^\star)} \int_{\widehat{A}(\xi^\star)} \operatorname{tr}\left[\mathring{\Psi}(\xi^\star, a)(M - M^\star)\right] \mu(da) \tag{3.164}$$

$$\leq \max_{\mu \in \Sigma(\xi^\star)} \int_{\widehat{A}(\xi^\star)} \max_{a \in \widehat{A}(\xi^\star)} \operatorname{tr}\left[\mathring{\Psi}(\xi^\star, a)(M - M^\star)\right] \mu(da)$$

$$= \max_{a \in \widehat{A}(\xi^\star)} \operatorname{tr}\left[\mathring{\Psi}(\xi^\star, a)(M - M^\star)\right],$$

where $\Sigma_0(\xi^\star)$ is the set of all measures which put unit weight at only one point of $\widehat{A}(\xi^\star)$.

Furthermore,

$$\min_{M \in \mathfrak{M}(X)} \max_{\mu \in \Sigma(\xi^\star)} \int_{\widehat{A}(\xi^\star)} \operatorname{tr}\left[\mathring{\Psi}(\xi^\star, a)(M - M^\star)\right] \mu(da)$$

$$= \min_{\xi \in \Xi(X)} \max_{\mu \in \Sigma(\xi^\star)} \int_{\widehat{A}(\xi^\star)} \operatorname{tr}\left[\mathring{\Psi}(\xi^\star, a)(M(\xi) - M(\xi^\star))\right] \mu(da)$$

$$= \min_{\xi \in \Xi(X)} \max_{\mu \in \Sigma(\xi^\star)} \int_{\widehat{A}(\xi^\star)} \left[c(\xi^\star, a) - \int_X \phi(x, \xi^\star, a)\, \xi(dx)\right] \mu(da) \tag{3.165}$$

$$= \min_{\xi \in \Xi(X)} \max_{\mu \in \Sigma(\xi^\star)} \int_X \int_{\widehat{A}(\xi^\star)} \left[c(\xi^\star, a) - \phi(x, \xi^\star, a)\right] \xi(dx)\, \mu(da).$$

Note that the function $(x, a) \mapsto c(\xi^*, a) - \phi(x, \xi^*, a)$ is continuous on all of $X \times \widehat{A}(\xi^*)$ and recall that X and $\widehat{A}(\xi^*)$ are compact. Then a fundamental minimax result of game theory permits us to interchange the min and max operators in (3.165), cf. [71, p. 108] or [271]:

$$\min_{\xi \in \Xi(X)} \max_{\mu \in \Sigma(\xi^*)} \int_X \int_{\widehat{A}(\xi^*)} \left[c(\xi^*, a) - \phi(x, \xi^*, a) \right] \xi(dx) \, \mu(da)$$

$$= \max_{\mu \in \Sigma(\xi^*)} \min_{\xi \in \Xi(X)} \int_X \int_{\widehat{A}(\xi^*)} \left[c(\xi^*, a) - \phi(x, \xi^*, a) \right] \xi(dx) \, \mu(da) \qquad (3.166)$$

$$= \max_{\mu \in \Sigma(\xi^*)} \min_{x \in X} \int_{\widehat{A}(\xi^*)} \left[c(\xi^*, a) - \phi(x, \xi^*, a) \right] \mu(da),$$

the last inequality being established in much the same way as (3.163).

Consequently, ξ^* is optimal if and only if

$$\max_{\mu \in \Sigma(\xi^*)} \min_{x \in X} \int_{\widehat{A}(\xi^*)} \left[c(\xi^*, a) - \phi(x, \xi^*, a) \right] \mu(da) \geq 0, \qquad (3.167)$$

which is equivalent to the existence of a measure $\mu^* \in \Sigma(\xi^*)$ satisfying

$$\min_{x \in X} \int_{\widehat{A}(\xi^*)} \left[c(\xi^*, a) - \phi(x, \xi^*, a) \right] \mu^*(da) \geq 0. \qquad (3.168)$$

Expanding the left-hand side, we obtain

$$\int_{\widehat{A}(\xi^*)} c(\xi^*, a) \, \mu^*(da) - \max_{x \in X} \int_{\widehat{A}(\xi^*)} \phi(x, \xi^*, a) \, \mu^*(da) \geq 0, \qquad (3.169)$$

which is the desired conclusion. ∎

This theorem is indeed of great importance, as in the context of the linear regression model (3.1) it can be immediately used to determine necessary and sufficient E-optimality conditions for the design of the form

$$\Psi_E[M(\xi)] = \lambda_{\max}[M^{-1}(\xi)]. \qquad (3.170)$$

To see this, recall that from the symmetry of $M^{-1}(\xi)$ and Theorem B.4 we have

$$\lambda_{\max}[M^{-1}(\xi)] = \max_{a \in A} a^\top M^{-1}(\xi) a, \qquad (3.171)$$

where $A = \{a : \|a\| = 1\}$, which means that, as a matter of fact, we deal with a problem of the form (3.153) for which $\Psi[M(\xi), a] = a^\top M^{-1}(\xi) a$. The following theorem constitutes a generalization of the corresponding result for the case of static systems, cf. [263, Thm. 2.12].

THEOREM 3.6 Equivalence Theorem for E-Optimum Designs

Let $\xi^\star$ be a design measure for which $M(\xi^\star)$ is nonsingular and let $A^\star$ be the set of all normalized elements of the eigenspace associated with $\lambda_{\min}[M(\xi^\star)]$, i.e., $\|a\| = 1$ for every $a \in A^\star$. Then a necessary and sufficient condition for $\xi^\star$ to be E-optimal is the existence of a probability measure $\mu^\star$ on $A^\star$ such that

$$\max_{x \in X} \int_{A^\star} a^\mathsf{T} \Upsilon(x) a\, \mu^\star(\mathrm{d}a) \le \lambda_{\min}[M(\xi^\star)]. \qquad (3.172)$$

PROOF   By Theorem B.19, we have

$$\overset{\circ}{\Psi}(\xi^\star, a) = \left. \frac{\partial \left(a^\mathsf{T} M^{-1} a\right)}{\partial M} \right|_{M=M(\xi^\star)} = -M^{-1}(\xi^\star) a a^\mathsf{T} M^{-1}(\xi^\star) \qquad (3.173)$$

and hence

$$c(\xi^\star, a) = a^\mathsf{T} M^{-1}(\xi^\star) a, \qquad (3.174)$$

$$\phi(x, \xi^\star, a) = a^\mathsf{T} M^{-1}(\xi^\star) \Upsilon(x) M^{-1}(\xi^\star) a. \qquad (3.175)$$

Theorem B.4 then shows that $\widehat{A}(\xi^\star)$ of Theorem 3.5 reduces to $\{a \in \mathfrak{E} : \|a\| = 1\}$, where $\mathfrak{E}$ constitutes the eigenspace associated with $\lambda_{\max}[M^{-1}(\xi^\star)]$. What is more, for any point of this set we have

$$a^\mathsf{T} M^{-1}(\xi^\star) a = \lambda_{\max}[M^{-1}(\xi^\star)], \qquad (3.176)$$

$$M^{-1}(\xi^\star) a = \lambda_{\max}[M^{-1}(\xi^\star)] a. \qquad (3.177)$$

Given a probability measure μ on $\widehat{A}(\xi^\star)$, we see that

$$\int_{\widehat{A}(\xi^\star)} c(\xi^\star, a)\, \mu(\mathrm{d}a) = \int_{\widehat{A}(\xi^\star)} a^\mathsf{T} M^{-1}(\xi^\star) a\, \mu(\mathrm{d}a) = \lambda_{\max}[M^{-1}(\xi^\star)] \qquad (3.178)$$

and

$$\int_{\widehat{A}(\xi^\star)} \phi(x, \xi^\star, a)\, \mu(\mathrm{d}a) = \int_{\widehat{A}(\xi^\star)} a^\mathsf{T} M^{-1}(\xi^\star) \Upsilon(x) M^{-1}(\xi^\star) a\, \mu(\mathrm{d}a)$$

$$= \lambda_{\max}^2[M^{-1}(\xi^\star)] \int_{\widehat{A}(\xi^\star)} a^\mathsf{T} \Upsilon(x) a\, \mu(\mathrm{d}a). \qquad (3.179)$$

Hence from Theorem 3.5 it follows that a necessary and sufficient condition for $\xi^\star$ to be E-optimal is that there exists a probability measure $\mu^\star$ on $\widehat{A}(\xi^\star)$ such that

$$\int_{\widehat{A}(\xi^\star)} a^\mathsf{T} \Upsilon(x) a\, \mu^\star(\mathrm{d}a) \le \lambda_{\max}^{-1}[M^{-1}(\xi^\star)], \quad \forall x \in X. \qquad (3.180)$$

To complete the proof, it suffices to observe that

$$\lambda_{\max}^{-1}[M^{-1}(\xi^\star)] = \lambda_{\min}[M(\xi^\star)] \qquad (3.181)$$

and the eigenvectors associated with $\lambda_{\max}[M^{-1}(\xi^\star)]$ and $\lambda_{\min}[M(\xi^\star)]$ coincide. ∎

If the eigenvalue $\lambda^\star = \lambda_{\min}[M(\xi^\star)]$ is unique, then $A^\star$ reduces to a unique eigenvector $v_\star$ associated with $\lambda^\star$ and the equivalence condition (3.172) simplifies to

$$v_\star^{\mathsf{T}}\Upsilon(x)v_\star \leq \lambda^\star, \quad \forall x \in X. \tag{3.182}$$

If the smallest eigenvalue $\lambda^\star$ is repeated, (3.172) can be used to verify the optimality of $\xi^\star$ by integrating the integrand on the left-hand side of (3.172) with respect to any probability measure on $A^\star$ and observing whether (3.172) holds.

It goes without saying that, apart from the above characterizations, the following crucial question has also to be addressed: How to generate approximations to E-optimum designs? This is because Theorem 3.6 can only be used to check candidate solutions for optimality. Unfortunately, there appears to be no known specialized algorithm for constructing E-optimum designs such as those discussed in Section 3.1.3.1 for differentiable criteria. Some attempts at adapting a bundle trust method were reported in [226] and efficient algorithms of semidefinite programming can be applied to find optimal designs for a fixed set of support points, cf. Section 3.1.3.2, but nevertheless the problem is far from being satisfactorily solved.

From a practical point of view, a way out of this predicament can be the reduction of the original minimax problem to the minimization of its smooth convex approximation. This topic is examined in detail in Appendix D. Based on those results, after replacing the minimization of $\lambda_{\max}[M^{-1}(\xi)]$ by the equivalent problem of minimizing $J(\xi) = \lambda_{\max}[-M(\xi)]$ (this is to avoid matrix inversion), we can consider the smoothed convex approximated design criterion

$$\Psi_\epsilon[M(\xi)] = \epsilon \ln\left(\sum_{i=1}^{m} \exp\left(-\lambda_i[M(\xi)]/\epsilon\right)\right), \tag{3.183}$$

where $\lambda_i[\cdot]$ is the i-th eigenvalue of its matrix argument and $0 < \epsilon \ll 1$ stands for a fixed parameter steering the accuracy of such an approximation.

Then Algorithm 3.1 can be directly employed to find approximated E-optimum designs. While implementing this idea, it is necessary to use the information provided by the gradient which has the following form:

$$\frac{\partial\Psi_\epsilon(M)}{\partial M} = -\frac{\sum_{i=1}^{m} \exp\left(-\lambda_i(M)/\epsilon\right)v_i(M)v_i^{\mathsf{T}}(M)}{\sum_{i=1}^{m} \exp\left(-\lambda_i(M)/\epsilon\right)}, \tag{3.184}$$

where $v_i(M)$ is the normalized eigenvector corresponding to the eigenvalue $\lambda_i(M)$.

3.3   Continuous designs in measurement optimization

The introduction of continuous designs makes it possible, on one hand, to advance a very elegant theory based on convex optimization and, on the other hand, to develop very effective numerical algorithms which are implementable even on a low-cost PC. A natural question is therefore how to exploit all those benefits while designing a measurement system to estimate the unknown parameters of a given DPS as accurately as possible. Unfortunately, the answer is not as simple as that, since even if the model equation under consideration is linear in its parameters, the state depends linearly on those parameters only in exceptional cases and the rule is that this dependence is highly nonlinear, which causes severe difficulties and practically excludes direct analytical solutions in most interesting situations.

To settle this problem, it is customary to linearize the system response in the vicinity of a prior estimate θ^0 to the unknown parameter vector θ [282]. As a result, observations may be represented approximately as

$$z(t) \approx y_{\mathrm{m}}(t; \theta^0) + \left.\frac{\partial y_{\mathrm{m}}(t; \theta)}{\partial \theta}\right|_{\theta=\theta^0} (\theta - \theta^0) + \varepsilon_{\mathrm{m}}(t), \qquad (3.185)$$

where

$$y_{\mathrm{m}}(t; \theta^0) = \mathrm{col}[y(x^1, t; \theta^0), \ldots, y(x^N, t; \theta^0)], \qquad (3.186)$$

$$\frac{\partial y_{\mathrm{m}}(t; \theta)}{\partial \theta} = \begin{bmatrix} \dfrac{\partial y(x^1, t; \theta)}{\partial \theta_1} & \cdots & \dfrac{\partial y(x^1, t; \theta)}{\partial \theta_m} \\ \vdots & & \vdots \\ \dfrac{\partial y(x^N, t; \theta)}{\partial \theta_1} & \cdots & \dfrac{\partial y(x^N, t; \theta)}{\partial \theta_m} \end{bmatrix}, \qquad (3.187)$$

the last quantity being the Jacobian of the observation vector y_{m} with respect to parameter vector θ. After an obvious rearrangement, (3.185) is expressed in the form

$$z(t) - y_{\mathrm{m}}(t; \theta^0) + \left.\frac{\partial y_{\mathrm{m}}(t; \theta)}{\partial \theta}\right|_{\theta=\theta^0} \theta^0 = \left.\frac{\partial y_{\mathrm{m}}(t; \theta)}{\partial \theta}\right|_{\theta=\theta^0} \theta + \varepsilon_{\mathrm{m}}(t), \qquad (3.188)$$

so we loosely get the setting of (3.1) with $z(t)$ replaced by the left-hand side of (3.188) and

$$f(x^j, \cdot) = \left(\frac{\partial y(x^j, \cdot; \theta)}{\partial \theta}\right)^{\mathsf{T}}_{\theta=\theta^0}. \qquad (3.189)$$

Accordingly, from (3.16) it follows that the respective average 'FIM' which approximates (up to a constant multiplier) the dispersion matrix $\mathrm{cov}\{\hat{\theta}\}$ may be expressed as

$$M(\xi_N) = \sum_{i=1}^{\ell} p_i \frac{1}{t_f} \int_0^{t_f} \left(\frac{\partial y(x^i, t; \theta)}{\partial \theta} \right)^{\mathsf{T}} \left(\frac{\partial y(x^i, t; \theta)}{\partial \theta} \right) \Bigg|_{\theta = \theta^0} dt. \qquad (3.190)$$

It goes without saying that (3.190) is valid as long as the approximation (3.185) is warranted. One way or another, it is now evident that the FIM (3.190) will depend on the preliminary estimate θ^0 around which the model is linearized, so logically, the optimal sensor location can never be found at the design stage unless θ^0 is very close to the true parameters or the Jacobian (3.187) is insensitive to the values of the model parameters (in practice, the latter is unlikely in the considered applications).

It is worth pointing out that the delineated procedure is not only based in brute force and ignorance, as there have appeared many works regarding its statistical rationale. For more precise results on the consistency of least-squares estimates in DPSs and their asymptotic distributions, the interested reader should consult some painstaking works of Fitzpatrick [15, 88–90, 353]. At this juncture note that characterizing parameter uncertainty in a nonlinear-in-parameters model by the inverse of the FIM involves approximations to which few alternatives exist [349].

If we assume that both $y(\cdot, \cdot; \theta^0)$ and $\partial y(\cdot, \cdot; \theta^0)/\partial \theta_i$, $i = 1, \ldots, m$ are continuous in $\bar{\Omega} \times T$ (for some relaxation of this requirement, see Remark 3.1), then all the results of Section 3.1, starting from the notion of a continuous design, through all the lemmas and theorems, and finally inclusive of the outlined algorithms of first-order, can be directly employed without any changes, bearing in mind (3.189). For example, the counterpart to the FIM of (3.21) is given in terms of the sensitivity coefficients as

$$M(\xi) = \int_X \Upsilon(x) \, \xi(\mathrm{d}x), \qquad (3.191)$$

X being interpreted as an admissible region where we are allowed to place the sensors (a compact subset of $\bar{\Omega}$), and

$$\Upsilon(x) = \frac{1}{t_f} \int_0^{t_f} \left(\frac{\partial y(x, t; \theta)}{\partial \theta} \right)^{\mathsf{T}} \left(\frac{\partial y(x, t; \theta)}{\partial \theta} \right) \Bigg|_{\theta = \theta^0} dt.$$

At this very moment, some interpretation of the resulting optimal design of the form (3.30) would be relevant. Since we manipulate continuous designs, the products $N p_i$, $i = 1, \ldots, \ell$ are not necessarily integers. In the spatial setting, however, the number of sensors may be quite large and the set of candidate points is continuous so that we can expect that some rounding procedures [225] of the considered approximate designs calculated by the aforementioned algorithms will yield sufficiently good designs. Alternatively, some *exchange algorithms* can be adopted from the classical theory of optimal experiments if N is relatively small, but such a procedure does not change the underlying idea and therefore it will not be pursued.

An interesting interpretation of continuous designs in terms of the randomized choice is given in [238]. Namely, for ξ_N given by (3.15), if N sensors are randomly allocated to the points x^i, $i = 1, \ldots, \ell \leq N$ according to the distribution p_i, $i = 1, \ldots, \ell$ and such that the measurement process is repeated many times, then (3.190) is the expected value of the FIM. This justifies our results as theoretically exact from a slightly different point of view.

Some numerical examples have been solved to indicate the general nature of the results.

Example 3.2

Consider the one-dimensional parabolic equation

$$\frac{\partial y(x,t)}{\partial t} = \alpha \frac{\partial^2 y(x,t)}{\partial x^2} + \beta y(x,t), \quad x \in \Omega = (0,1), \quad t \in T = (0,1), \quad (3.192)$$

subject to the following initial and boundary conditions:

$$y(x,0) = \sin(\pi x) + \sin(2\pi x) \quad \text{in } \Omega, \tag{3.193}$$
$$y(0,t) = y(1,t) = 0 \quad \text{on } T, \tag{3.194}$$

respectively. The purpose here is to determine optimal measurement points so as to estimate the parameter vector $\theta = (\alpha, \beta)$ as accurately as possible based on continuous-time measurements of y.

Let us reformulate this problem as that of looking for the respective D-optimum design. In order to solve it, observe first that the solution to the problem (3.192)–(3.194) is of the form

$$y(x,t) = e^{(\beta - \alpha \pi^2)t} \sin(\pi x) + e^{(\beta - 4\alpha \pi^2)t} \sin(2\pi x). \tag{3.195}$$

Consequently, the function f in (3.189) takes the form

$$f(x,t) = \begin{bmatrix} -\pi^2 t e^{(\beta - \alpha \pi^2)t} \sin(\pi x) - 4\pi^2 t e^{(\beta - 4\alpha \pi^2)t} \sin(2\pi x) \\ t e^{(\beta - \alpha \pi^2)t} \sin(\pi x) + t e^{(\beta - 4\alpha \pi^2)t} \sin(2\pi x) \end{bmatrix}. \tag{3.196}$$

Clearly, it depends nonlinearly on the parameters α and β, so we have to assume some nominal values for them. In what follows, we adopt the values

$$\alpha^0 = 1/\pi^2, \quad \beta^0 = 1. \tag{3.197}$$

In practice, they may result, e.g., from a preliminary experiment. Thus we get

$$f(x,t) = \begin{bmatrix} -\pi^2 t \sin(\pi x) - 4\pi^2 t e^{-3t} \sin(2\pi x) \\ t \sin(\pi x) + t e^{-3t} \sin(2\pi x) \end{bmatrix}. \tag{3.198}$$

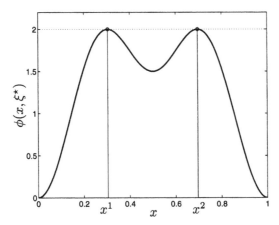

FIGURE 3.2
Sensitivity function for the D-optimal design of Example 3.2 (solid line).

Calculation of the outer product of f by itself and integration of the result over T leads to the following form of the information matrix:

$$M(\xi) = \sum_{i=1}^{\ell} p_i \widetilde{M}(x^i), \qquad (3.199)$$

where the matrix $\widetilde{M}(x)$ is symmetric with

$$\widetilde{M}_{11}(x) = \frac{1}{27} \pi^4 \sin^2(\pi x)\Big[9 - 272\, e^{-3} \cos(\pi x)$$
$$- 400\, e^{-6} \cos^2(\pi x) + 32 \cos(\pi x) + 16 \cos^2(\pi x)\Big], \qquad (3.200)$$

$$\widetilde{M}_{12}(x) = \frac{1}{27} \pi^2 \sin^2(\pi x)\Big[-9 + 170\, e^{-3} \cos(\pi x)$$
$$+ 100\, e^{-6} \cos^2(\pi x) - 20 \cos(\pi x) - 4 \cos^2(\pi x)\Big], \qquad (3.201)$$

$$\widetilde{M}_{22}(x) = \frac{1}{27} \sin^2(\pi x)\Big[9 - 68\, e^{-3} \cos(\pi x)$$
$$- 25\, e^{-6} \cos^2(\pi x) + 8 \cos(\pi x) + \cos^2(\pi x)\Big]. \qquad (3.202)$$

In what follows, we argue that the following design is optimal:

$$\xi^* = \left\{ \begin{matrix} \frac{1}{\pi} \arctan(\sqrt{2}), & 1 - \frac{1}{\pi} \arctan(\sqrt{2}) \\[2mm] \frac{1}{2}, & \frac{1}{2} \end{matrix} \right\}, \qquad (3.203)$$

i.e., the best solution is to use only two measurement points $x^1 = \arctan(\sqrt{2})/\pi = .3041$ and $x^2 = 1 - x^1 = .6959$, and to place one half of the total number of sensors at each of them. Indeed, the sensitivity function $\phi(x, \xi^*) = (1/t_f) \int_0^{t_f} f^\mathsf{T}(x, t) M^{-1}(\xi^*) f(x, t) \, dt$ simplifies to the form

$$\phi(x, \xi^*) = \frac{3}{2} \sin^2(\pi x) \left[1 + 3 \cos^2(\pi x) \right]. \tag{3.204}$$

It is very easy to see that on the interval $(0, 1)$ this function attains its maximum equal to two (note that so is the numer of parameters to be estimated) exactly at points x^1 and x^2, cf. Fig. 3.2. Accordingly, Theorem 3.3 guarantees that ξ^* is a D-optimum design in the case considered.   ☐

Example 3.3

The approach to the sensor placement developed in the previous section was applied to the optimal estimation of the spatially varying parameter $\kappa = \kappa(x)$ in the heat-conduction process through a thin flat isotropic plate whose flat surfaces were insulated and which occupied the region $\Omega = [0, 1]^2$ with boundary Γ along which heat was lost to the surroundings. The unsteady-state temperature $y = y(x, t)$ over the time horizon $T = (0, 1)$ was described by a linear parabolic equation of the form

$$\frac{\partial y(x, t)}{\partial t} = \frac{\partial}{\partial x_1} \left(\kappa(x) \frac{\partial y(x, t)}{\partial x_1} \right) + \frac{\partial}{\partial x_2} \left(\kappa(x) \frac{\partial y(x, t)}{\partial x_2} \right) \quad \text{in } \Omega \times T. \tag{3.205}$$

The initial and boundary conditions of (3.205) were

$$y(x, 0) = 5 \qquad \text{in } \Omega, \tag{3.206}$$
$$y(x, t) = 5(1 - t) \quad \text{on } \Gamma \times T. \tag{3.207}$$

In our simulation study, the following true parameter was considered:

$$\kappa(x) = \theta_1 + \theta_2 x_1 + \theta_3 x_2, \tag{3.208}$$

where $\theta_1 = 0.1$, $\theta_2 = \theta_3 = 0.3$. On the basis of simulated data generated with the specified κ, we tried to determine a continuous design over $X = \bar{\Omega}$ such that the D-optimality criterion for $\theta = (\theta_1, \theta_2, \theta_3)$ would be minimized.

In order to numerically solve the measurement location problem, a computer programme was written in Lahey/Fujitsu Fortran 95 Ver. 5.60a [159, 176] using a PC (Pentium IV, 2.40 GHz, 512 MB RAM) running Windows 2000. The state and sensitivity equations were first solved using the finite-element method on an even grid (with 15 divisions along each space axis and 30 divisions of the time interval). The sensitivity coefficients were then interpolated via tricubic spline interpolation (see Appendix G.2) and the corresponding spline parameters stored in computer memory. Finally, Fedorov's version of Algorithm 3.1 was applied with the ARS algorithm of p. 54 to maximize the determinant of the FIM. (The maximum number of evaluations for the performance index was 2000.)

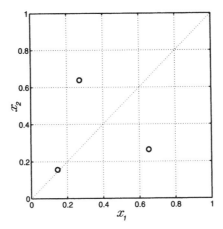

FIGURE 3.3
D-optimum location of the support points in the problem of Example 3.3 (the axis of symmetry is represented by a sloping dotted line).

Starting from the design

$$\xi^{(0)} = \left\{ \begin{array}{ccc} (0.6, 0.2), & (0.2, 0.5), & (0.1, 0.2) \\ 1/3 & 1/3 & 1/3 \end{array} \right\},$$

after 13 iterations (which took about 40 seconds), the following approximation to the optimal design was obtained:

$$\xi^\star = \left\{ \begin{array}{ccc} (0.65224, 0.26353), & (0.27083, 0.63834), & (0.14647, 0.15668) \\ 0.33570, & 0.33410, & 0.33019 \end{array} \right\}$$

for the tolerance $\eta = 10^{-2}$.

The design is concentrated at three support points with approximately equal weights, which means that if we are to locate N sensors, then we should strive to distribute them as evenly as possible between the three calculated potential locations (as outlined before, sensor clusterization is inherent to the approach due to the assumption that the measurements are independent even though some of the sensors take measurements at the same point).

Let us observe that the diffusivity coefficient κ together with the system of boundary and initial conditions assume one axis of symmetry, i.e., the line $x_2 = x_1$. We feel by intuition that this symmetry should also be preserved in a way in the optimal design. In fact, this is confirmed in Fig. 3.3 where the optimal sensor positions are displayed. They are slightly shifted towards the lower-left part of the system, at which place the diffusivity coefficient is smaller and the system output is the most sensitive to changes in θ.

As a supplement to the above results, our computer experiments were also extended to cover algorithms determining optimal weights for support points

TABLE 3.2

Optimum designs in Example 3.3 given the finite
set of candidate support points (3.209)

Optimality criterion	Support point	Weight
A	$(0.15, 0.15)$	0.33316
	$(0.25, 0.70)$	0.23050
	$(0.30, 0.70)$	0.10291
	$(0.70, 0.25)$	0.23050
	$(0.70, 0.30)$	0.10291
E	$(0.15, 0.15)$	0.28917
	$(0.25, 0.70)$	0.34013
	$(0.30, 0.70)$	0.01527
	$(0.70, 0.25)$	0.34013
	$(0.70, 0.30)$	0.01527
D	$(0.15, 0.15)$	0.33355
	$(0.25, 0.65)$	0.13957
	$(0.30, 0.65)$	0.19365
	$(0.65, 0.25)$	0.13957
	$(0.65, 0.30)$	0.19365

given *a priori*. To this end, a uniform grid of nodes

$$X_d = \big\{(i/20, j/20) : i, j = 0, \ldots, 20\big\} \tag{3.209}$$

was adopted as the known finite set of candidate points for inclusion in the
support of an optimal design.

The SDP approach was then applied to find A- and E-optimum designs
using the MATLAB-compatible solver SeDuMi Ver. 1.05 [280] based on the
primal-dual interior-point algorithm. For an accuracy of 10^{-9}, ultimate approximations to the corresponding solutions were obtained in several seconds
after 25 and 23 iterations, respectively. The optimal designs are shown in
Table 3.2. Their support points thus turn to be the same. They are shown
in Fig. 3.4(a). In turn, an approximated D-optimum design was sought by
implementing Algorithm 3.2 in MATLAB Ver. 6.5, Rel. 13, which converged to
the design shown in Table 3.2 in 2154 iterations (it took about half a minute)
starting from the same values of all weights for an accuracy of $\eta = 10^{-5}$. The
corresponding support points are shown in Fig. 3.4(b).

The slow convergence rate in this case is inherent to Algorithm 3.2, cf.
remarks in [349, p. 306], as only the monotonicity of of the determinant values
for the successive FIMs is guaranteed and in the space of weights, the update
(3.125) does not correspond to the direction of the steepest descent. Note
that the convergence can be significantly accelerated by online elimination of
candidate points for which some criteria exist which exclude their presence in
the optimal design, cf. [220]

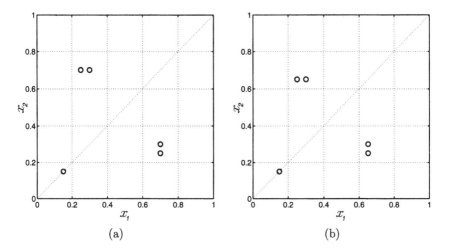

(a)
(b)

FIGURE 3.4
Optimum support points in the problem of Example 3.3 given the finite set of candidate nodes (3.209): (a) A- and E-optimum designs; (b) D-optimum design.

Note that all the results obtained using different algorithms and criteria are in good agreement and the differences are rather minor. Even though the number of support points is greater for designs with a finite set of candidate points, these points are situated in the close vicinity of the support of the design $\xi^\star$. In terms of the FIMs, the obtained results can hardly be considered as distinct. ☐

Example 3.4
In another simulation experiment, the spatiotemporal domain was the same as in Example 3.3 (similarly, the introduced discretization for numerical calculations was retained). This time, however, the diffusion equation contained a driving force, i.e.,

$$\frac{\partial y(x,t)}{\partial t} = \frac{\partial}{\partial x_1}\left(\kappa(x)\frac{\partial y(x,t)}{\partial x_1}\right) + \frac{\partial}{\partial x_2}\left(\kappa(x)\frac{\partial y(x,t)}{\partial x_2}\right)$$
$$+ 30\exp(-30\|x-a\|^2) \quad \text{in } \Omega \times T, \tag{3.210}$$

where $a = (0.75, 0.25)$, see Fig. 3.5. The boundary Γ was split into two subsets: $\Gamma_1 = \{(0, x_2) : 0 \le x_2 \le 1\}$ and $\Gamma_2 = \Gamma \setminus \Gamma_1$ so that the boundary conditions were

$$y(x,t) = \begin{cases} 6x_2(1-x_2) & \text{on } \Gamma_1 \times T, \\ 0 & \text{on } \Gamma_2 \times T. \end{cases} \tag{3.211}$$

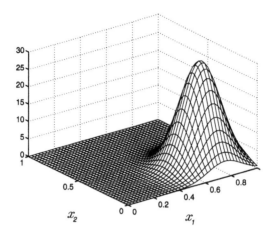

FIGURE 3.5
Driving force employed in Example 3.4.

The initial state was the same as before, i.e.,

$$y(x,0) = 5 \quad \text{in } \Omega. \tag{3.212}$$

Our task consisted in finding a best D-optimal design to identify a slightly changed diffusion coefficient

$$\kappa(x) = \theta_1 + \theta_2 x_1, \quad \theta_1 = 0.1, \quad \theta_2 = 0.3 \tag{3.213}$$

or, more precisely, the coefficients θ_1 and θ_2. All the other settings were the same as in Example 3.3.

Starting from the initial design

$$\xi_0 = \left\{ \begin{matrix} (0.6, 0.2), & (0.2, 0.5) \\ 1/2 & 1/2 \end{matrix} \right\},$$

after three iterations, we obtained the following approximation to the optimal design whose support is shown in Fig. 3.6:

$$\xi^\star = \left\{ \begin{matrix} (0.70166, 0.30929), & (0.16965, 0.53752) \\ 0.483938, & 0.516062 \end{matrix} \right\}.$$

On reflection, this result is not surprising. Indeed, in our system there exist two perturbations, i.e., the boundary excitation on Γ_1 and the impulse-like force concentrated around a, so logically the regions which provide most information about the system should lie in the vicinity of them. Since the corresponding weights are practically equal to each other, we should assign to each support point half of the available sensors. □

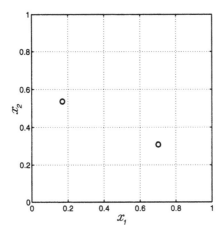

FIGURE 3.6
Optimal location of the support points in the problem of Example 3.4.

3.4 Clusterization-free designs

As pointed out in [84, 181], two special features distinguish the spatial data collection schemes from classical regression designs. First of all, spatial observations are often affected by local correlations which are unaccounted for by standard techniques of optimum experimental design. What is more, there is usually no possibility of replicated measurements, i.e., different sensors cannot take measurements at one point without influencing one another. Anyway, several sensors situated in the close vicinity of one another usually do not give more information than a single sensor. The assumption of independent observations is advantageous from a theoretical point of view, since it allows for direct use of sublime results of convex optimization, but it can hardly be justified when in the optimal solution some sensors are to take measurements near one another. This generates interest in the so-called *clusterization-free* designs where the distances between the sensors are long enough in order to guarantee the independence of their measurements. This is reminiscent of the idea of replication-free designs which have emerged relatively late in the context of spatial statistics (see the monograph by Müller [181], a survey by Fedorov [84], and a seminal work by Brimkulov *et al.* [31]).

In the literature, a typical motivation to work on replication-free designs is the situation when we observe the values of a random process or a random field at some times (at some points). When the mean of the process contains unknown parameters, we have a regression model, but typically without the possibility of replications, because just one realization of the process is allowed, and the experimental design consists of an adequate choice of times (points) of observation. This setup was considered, e.g., in [181–183, 209] where the concept of continuous designs was extended by the introduction of the so-called approximate information matrices. The preliminary results are quite promising, but the attendant derivations are rather awkward and lengthy, and the results themselves are obtained after a sequence of approximations.

An alternative approach to constructing replication-free designs was proposed by Fedorov [54, 83–85]. In spite of its somewhat abstract assumptions, the resulting algorithm of exchange type is very easy to implement. It turns out that Fedorov's approach can be adapted to the problems considered in our monograph with relative ease, which is going to be shown in what follows. Note, however, that this section constitutes only a brief introduction to the topic of clusterization-free designs. Their comprehensive treatment is included in the next chapter in the context of scanning observations.

The main idea here is to operate on the density of sensors (i.e., the number of sensors per unit area), rather than on the locations of the sensors, which is justified for a sufficiently large total number of sensors N. In contrast to the classical designs, however, we impose the crucial restriction that the density of sensor allocation must not exceed some prescribed level. For a design measure $\xi(\mathrm{d}x)$ this amounts to the condition

$$\xi(\mathrm{d}x) \leq \omega(\mathrm{d}x), \tag{3.214}$$

where $\omega(\mathrm{d}x)$ signifies the maximal possible 'number' of sensors per $\mathrm{d}x$ [85] such that

$$\int_X \omega(\mathrm{d}x) \geq 1. \tag{3.215}$$

Consequently, we are faced with the following optimization problem: Find

$$\xi^* = \arg \min_{\xi \in \Xi(X)} \Psi(\xi) \quad \text{subject to} \quad \xi(\mathrm{d}x) \leq \omega(\mathrm{d}x). \tag{3.216}$$

The design ξ^* above is then said to be a (Ψ, ω)-*optimal design*.

Apart from Assumptions (A1), (A2) of p. 38 and (A3)–(A6) of p. 41, a proper mathematical formulation calls for the following proviso:

(A7) $\omega(\mathrm{d}x)$ is atomless, i.e., for any $\Delta X \subset X$ there exists a $\Delta X' \subset \Delta X$ such that

$$\int_{\Delta X'} \omega(\mathrm{d}x) < \int_{\Delta X} \omega(\mathrm{d}x). \tag{3.217}$$

In what follows, we write $\bar{\Xi}(X)$ for the collection of all the design measures which satisfy the requirement*

$$\xi(\Delta X) = \begin{cases} \omega(\Delta X) & \text{for } \Delta X \subset \operatorname{supp}\xi, \\ 0 & \text{for } \Delta X \subset X \setminus \operatorname{supp}\xi. \end{cases} \tag{3.218}$$

---

*The support of a measure ξ is defined as the closed set $\operatorname{supp}\xi = X \setminus \bigcup\{G : \xi(G) = 0, \ G - \text{open}\}$, cf. [248, p.80].

The main feature of a measure ξ from $\bar{\Xi}(X)$ is that the design region X can be split up into two subsets for which ξ coincides either with zero or with the upper bound $\omega(\mathrm{d}x)$ [261] (on the analogy of the terminology established in optimal control, ξ could be called a bang-bang measure).

Given a design ξ, we will say that the function $\psi(\,\cdot\,,\xi)$ defined by (3.37) separates sets X_1 and X_2 with respect to $\omega(\mathrm{d}x)$ if for any two sets $\Delta X_1 \subset X_1$ and $\Delta X_2 \subset X_2$ with equal nonzero ω-measures we have

$$\int_{\Delta X_1} \psi(x,\xi)\,\omega(\mathrm{d}x) \leq \int_{\Delta X_2} \psi(x,\xi)\,\omega(\mathrm{d}x). \tag{3.219}$$

We can now formulate the main result which provides a characterization of (Ψ,ω)-optimal designs.

THEOREM 3.7
Let Assumptions (A1)–(A7) hold. Then:

(i) There exists an optimal design $\xi^\star \in \bar{\Xi}(X)$, and

(ii) A necessary and sufficient condition for $\xi^\star \in \bar{\Xi}(X)$ to be (Ψ,ω)-optimal is that $\psi(\,\cdot\,,\xi^\star)$ separates $X^\star = \mathrm{supp}\,\xi^\star$ and its complement $X \setminus X^\star$ with respect to the measure $\omega(\mathrm{d}x)$.

PROOF   This theorem constitutes a particular case of much more general results in Chapter 4, cf. Theorems 4.1 and 4.2, in which it suffices to set $R = 1$. Its proof is therefore omitted.   ∎

From a practical point of view, the above theorem means that at all the support points of an optimal design $\xi^\star$ the mapping $\psi(\,\cdot\,,\xi^\star)$ should be less than anywhere else, i.e., preferably $\mathrm{supp}\,\xi^\star$ should coincide with minimum points of $\psi(\,\cdot\,,\xi^\star)$ (let us note that for the D-optimality criterion this can be expressed as the situation when $\phi(\,\cdot\,,\xi^\star)$ is greater in $\mathrm{supp}\,\xi^\star$ than in the complement of $\mathrm{supp}\,\xi^\star$, which amounts to allocating observations to the points at which we know least of all about the system response, cf. the interpretation of $\phi(\,\cdot\,,\xi^\star)$ of p. 49).

If we were able to construct a design with this property, then it would be qualified as an optimal design. This conclusion forms a basis for numerical algorithms of constructing solutions to the problem under consideration.

As regards the interpretation of the resultant optimal designs (provided that we are in a position to calculate at least their approximations), one possibility is to partition X into subdomains ΔX_i of relatively small areas and then to allocate to each of them the number

$$N^\star(\Delta X_i) = \left\lceil N \int_{\Delta X_i} \xi^\star(\mathrm{d}x) \right\rceil \tag{3.220}$$

of sensors whose positions may coincide with nodes of some uniform grid [85] (here $\lceil \zeta \rceil$ is the smallest integer greater than or equal to ζ). Additionally, bear in mind that we must also have $\xi^\star(\mathrm{d}x) = \omega(\mathrm{d}x)$ in $X^\star$.

Clearly, unless the considered design problem is quite simple, we must employ a numerical algorithm to make the outlined conceptions useful. Since $\xi^\star(\mathrm{d}x)$ should be nonzero in the areas where $\psi(\,\cdot\,,\xi^\star)$ takes on a smaller value, the central idea is to move some measure from areas with higher values of $\psi(\,\cdot\,,\xi^{(k)})$ to those with smaller values, as we expect that such a procedure will improve $\xi^{(k)}$. This is embodied by an iterative algorithm presented below, being an adaptation of the algorithm presented in [83]:

ALGORITHM 3.3 Computation of clusterization-free designs

Step 1. *Guess a nondegenerate initial design $\xi^{(0)} \in \bar{\Xi}(X)$ (i.e., it is required that $\Psi[M(\xi^{(0)})] < \infty$). Set $k = 0$.*

Step 2. *Set $X_1^{(k)} = \operatorname{supp} \xi^{(k)}$ and $X_2^{(k)} = X \setminus X_1^{(k)}$. Determine*

$$x_1^{(k)} = \arg \max_{x \in X_1^{(k)}} \psi(x, \xi^{(k)}), \quad x_2^{(k)} = \arg \min_{x \in X_2^{(k)}} \psi(x, \xi^{(k)}).$$

If $\psi(x_1^{(k)}, \xi^{(k)}) < \psi(x_2^{(k)}, \xi^{(k)}) + \eta$, where $\eta \ll 1$, then STOP. Else, find two sets $S_1^{(k)} \subset X_1^{(k)}$ and $S_2^{(k)} \subset X_2^{(k)}$ such that $x_1^{(k)} \in S_1^{(k)}$, $x_2^{(k)} \in S_2^{(k)}$ and

$$\int_{S_1^{(k)}} \omega(\mathrm{d}x) = \int_{S_2^{(k)}} \omega(\mathrm{d}x) = \alpha_k$$

(i.e., the measures of $S_1^{(k)}$ and $S_2^{(k)}$ must be identical) for some $\alpha_k > 0$.

Step 3. *Construct $\xi^{(k+1)}$ such that*

$$\operatorname{supp} \xi^{k+1} = X_1^{(k+1)} = (X_1^{(k)} \setminus S_1^{(k)}) \cup S_2^{(k)}.$$

Increment k and go to Step 2.

Convergence is guaranteed if the sequence $\{\alpha\}_{k=0}^{\infty}$ satisfies [83]

$$\lim_{k \to \infty} \alpha_k = 0, \quad \sum_{k=0}^{\infty} \alpha_k = \infty. \tag{3.221}$$

Within the framework of sensor placement, we usually have $\omega(\mathrm{d}x) = \varrho(x)\,\mathrm{d}x$, where ϱ is a density function. But in this situation we may restrict our attention to constant ϱ's (indeed, in any case we can perform an appropriate change of coordinates). Moreover, while implementing the algorithm on a computer, all integrals are replaced by sums over some regular grid elements.

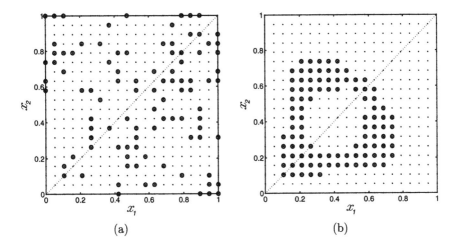

(a)   (b)

FIGURE 3.7
Calculation of the clusterization-free D-optimal design of Example 3.5: (a) Initial design; (b) Optimal solution.

Analogously, the sets X, $X_1^{(k)}$, $X_2^{(k)}$, $S_1^{(k)}$ and $S_2^{(k)}$ then simply consist of grid elements. Consequently, the above iterative procedure may be considered as an exchange-type algorithm with the additional constraint that every grid element must not contain more than one supporting point and the weights of all supporting points are equal to $1/N$. In practice, α_k is usually fixed and, what is more, one-point exchanges are most often adopted, i.e., $S_1^{(k)} = \{x_1^{(k)}\}$ and $S_2^{(k)} = \{x_2^{(k)}\}$, which substantially simplifies implementation. Let us note, however, that convergence to an optimal design is assured only for decreasing α_k's and hence some oscillations in $\Psi[M(\xi^{(k)})]$ may sometimes be observed. A denser spatial grid usually constitutes a remedy for this predicament [181].

Example 3.5
Having developed the algorithm for calculation of clusterization-free designs, we go straight to a demonstrative example. To this end, for the setting of Example 3.3, consider the problem of locating $N = 97$ sensors. A (20×20)-point uniform grid was introduced to approximate the design space and an initial design was generated by randomly selecting its support points. This situation is shown in Fig. 3.7(a), where dots represent the grid points (these were potential sites where the sensors could be placed, but at most one sensor at one point) and open circles indicate the actual sensor positions.

In order to calculate a D-optimal design, a simple one-point correction algorithm was employed ($\eta = 10^{-2}$) which produced after only 83 iterations (practically, in the blink of an eye) the solution displayed in Fig. 3.7(b). The

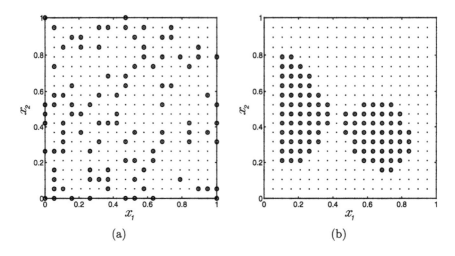

FIGURE 3.8
Calculation of the clusterization-free D-optimal design of Example 3.6: (a) Initial design; (b) Optimal solution.

interesting thing about this solution is that the sensors tend to assemble round the points calculated based on the continuous-design approach. Finally, note that, as expected, symmetry is perfectly retained (cf. the axis of symmetry expressed by the sloping dotted line).   ☐

Example 3.6
 The setting of Example 3.4 served as another test of the algorithm. The grid and parameters of Example 3.5 were left without changes and only a slightly altered number of sensors ($N = 100$) were used. Similarly, the initial design presented graphically in Fig. 3.8(a) was generated by randomly selecting its support points. The optimal design obtained after 68 iterations is shown in Fig. 3.8(b). As can be seen, the sensors split into two groups: the first reacts to the perturbation on the left boundary, while the other takes measurements in the zone where the driving force acts as another perturbation.   ☐

3.5   Nonlinear programming approach

When the total number of sensors to be located in a given domain is moderate, the very first idea, which suggests itself, is to exploit numerous well-known numerical techniques of constrained optimization. In principle, such an approach is not difficult to apply [304] and only computation of the gradient of the design criterion necessitates some comments if gradient methods are to be employed.

As in (2.12), write

$$s = (x^1, \dots, x^N). \tag{3.222}$$

Moreover, we denote $(\partial y / \partial \theta)^\mathsf{T}$ at $\theta = \theta^0$ briefly by g. Accordingly, the design criterion to be minimized may be rewritten as

$$J(s) = \Psi[M(s)], \tag{3.223}$$

where

$$M(s) = \frac{1}{Nt_f} \sum_{j=1}^{N} \int_0^{t_f} g(x^j, t) g^\mathsf{T}(x^j, t) \, \mathrm{d}t. \tag{3.224}$$

Using the chain rule, we get

$$\frac{\partial J(s)}{\partial s_r} = \operatorname{tr} \left\{ \mathring{\Psi}(s) \frac{\partial M(s)}{\partial s_r} \right\}, \tag{3.225}$$

where s_r stands for the r-th component of s, and

$$\mathring{\Psi}(s) = \left. \frac{\partial \Psi[M]}{\partial M} \right|_{M=M(s)}. \tag{3.226}$$

For most popular criteria we have, cf. Theorem B.19,

- If $\Psi(M) = -\ln \det(M)$, then

$$\mathring{\Psi}(s) = -M^{-1}(s).$$

- If $\Psi(M) = \operatorname{tr}(M^{-1})$, then

$$\mathring{\Psi}(s) = -M^{-2}(s).$$

- If $\Psi(M) = -\operatorname{tr}(M)$, then

$$\mathring{\Psi}(s) = -I,$$

where I is the identity matrix.

As for computation of $\partial M / \partial s_r$, let us observe first that s_r appears at only one term of the sum in (3.224), since s_r is just a spatial coordinate of one of the sensors. If we use the symbol j_r to denote the index of the corresponding sensor, then obviously we have

$$\frac{\partial M}{\partial s_r} = \frac{1}{Nt_f} \int_0^{t_f} \left\{ \frac{\partial g(x^{j_r}, t)}{\partial s_r} g^\mathsf{T}(x^{j_r}, t) + g(x^{j_r}, t) \frac{\partial g^\mathsf{T}(x^{j_r}, t)}{\partial s_r} \right\} \mathrm{d}t. \tag{3.227}$$

Hence, on account of the symmetry of $\overset{\circ}{\Psi}(s)$, it follows that

$$\frac{\partial J(s)}{\partial s_r} = \frac{2}{Nt_f} \operatorname{tr} \left\{ \overset{\circ}{\Psi}(s) \int_0^{t_f} \frac{\partial g(x^{jr}, t)}{\partial s_r} g^{\mathsf{T}}(x^{jr}, t) \, dt \right\}. \tag{3.228}$$

We see at once that calculation of $\nabla J(s)$ requires an efficient procedure to determine spatial derivatives of the sensitivity coefficients. Let us note, however, that this does not present a problem if we take advantage of spline interpolation (see Appendix G.2 for details).

Direct application of optimization techniques by no means excludes the phenomenon of clusterization. One way to attempt to avoid this undesirable effect is to include into the nonlinear programming formulation appropriate constraints on the admissible distances between the sensors. Since such a solution will be discussed in the next chapter within a more general framework of moving sensors, here we focus our attention on an alternative approach, which consists of taking account of mutual correlations between the measurements made by different sensors. In other words, this time we assume that the covariance matrix C in (3.3) may not be diagonal. For example, its elements could be of the following isotropic form [195]:

$$c_{ij} = \sigma^2 \exp(-\|x^i - x^j\|/\beta). \tag{3.229}$$

Occasionally, its extension

$$c_{ij} = \sigma(x^i)\sigma(x^j) \exp(-\|x^i - x^j\|/\beta) \tag{3.230}$$

is also used, which allows for different marginal variances.

Let us note that if any two sensors are placed at one point, then the corresponding columns (and rows) of C are identical, which means that C becomes singular.

It is easy to check, cf. Appendix C.2, that the average FIM is then given by

$$M(s) = \frac{1}{Nt_f} \sum_{i=1}^{N} \sum_{j=1}^{N} \int_0^{t_f} d_{ij}(s) g(x^i, t) g^{\mathsf{T}}(x^j, t) \, dt, \tag{3.231}$$

where d_{ij}'s are the elements if the inverse of C (i.e., $D = \begin{bmatrix} d_{ij} \end{bmatrix} = C^{-1}$). A first inconvenience is that the form of $M(s)$ is much more cumbersome than in the case of independent measurements. But a more severe difficulty is that the functional dependence of M on s is much less regular owing to the occurence that C may be singular or nearly singular, which necessitates the notion of pseudo-inverses and involves serious problems with differentiability and numerical stability. Consequently, in practice it is much easier to simply impose additional constraints on the distances between the sensors which will warrant the assumption of independent measurements.

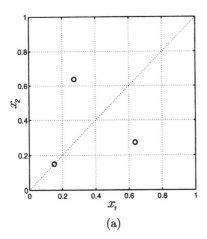 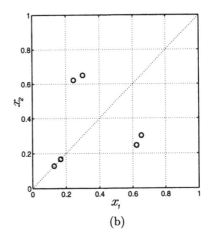

(a)   (b)

FIGURE 3.9
Optimal sensor location of Example 3.7 calculated via the direct approach:
(a) Independent measurements; (b) Correlated measurements.

Example 3.7
Consider anew the setting of Example 3.3 for which six sensors were to
be placed with the use of the direct nonlinear programming approach. At
first, the case of independent measurements was tested based on a sequential
constrained quadratic programming (SQP) method, cf. [22,178,262,275,276].
Starting from an initial solution generated via the ARS procedure of p. 54,
the SQP algorithm found the approximate optimal solution

$$s^\star = \begin{pmatrix} 0.1505197,\ 0.1505197,\ldots \\ 0.1505197,\ 0.1505197,\ldots \\ 0.2724469,\ 0.6376952,\ldots \\ 0.2724469,\ 0.6376952,\ldots \\ 0.6376952,\ 0.2724469,\ldots \\ 0.6376952,\ 0.2724469 \end{pmatrix}$$

shown in Fig. 3.9(a). This means that we have three pairs of sensors and
each of these pairs tends to measure the system state at the same point. In
principle, this result should not be surprising, since it was already predicted in
Example 3.3 where virtually the same support points were obtained. On the
other hand, it tallies with some results on replications of D-optimal designs
for nonlinear models [107].

The case of correlated observations was also tested for the model (3.229)
with $\beta = 10^{-2}$. Since the gradient algorithms are not appropriate for this
type of performance indices, the ARS technique was employed to assess the

optimal solution as

$$s^{\star}_{\text{corr}} = \begin{pmatrix} 0.128788, \, 0.128788, \dots \\ 0.167856, \, 0.167856, \dots \\ 0.246423, \, 0.621963, \dots \\ 0.301227, \, 0.650355, \dots \\ 0.621963, \, 0.246423, \dots \\ 0.650355, \, 0.301226 \end{pmatrix},$$

which is illustrated in Fig. 3.9(b). We see at once that the introduction of interrelations between the sensors results in removing the harmful clusterizaton. During experiments, however, some numerical instabilities were observed in addition to a considerably increased computational burden (three minutes versus half a minute for the correlation-free case). □

Example 3.8

Practically the same calculations as in the previous example were carried out for the setting of Example 3.4 and four sensors to be allocated. For independent measurements and the initial solution found with the use of the ARS algorithm, we obtained

$$s^{\star} = \begin{pmatrix} 0.176752, \, 0.476745, \dots \\ 0.176752, \, 0.476745, \dots \\ 0.701849, \, 0.309336, \dots \\ 0.701849, \, 0.309336 \end{pmatrix},$$

see Fig. 3.10(a), so we were faced again with the curse of clusterization. By allowing for correlations between the measurements of different sensors (according to the model (3.229) with $\beta = 10^{-2}$), this hurdle was avoided, since we got

$$s^{\star}_{\text{corr}} = \begin{pmatrix} 0.175956, \, 0.511760, \dots \\ 0.176134, \, 0.443545, \dots \\ 0.681270, \, 0.330796, \dots \\ 0.718677, \, 0.291468 \end{pmatrix},$$

see Fig. 3.10(b). □

3.6 A critical note on a deterministic approach

Independently of any statistical motivations, it happens that some authors are interested in choosing sensor positions which make an approximation H to

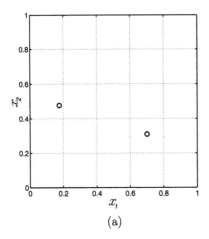

(a)

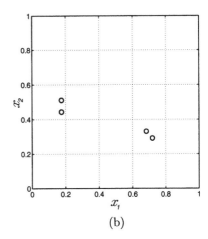

(b)

FIGURE 3.10
Optimal sensor location of Example 3.8 calculated via the direct approach:
(a) Independent measurements; (b) Correlated measurements.

the Hessian of the estimation cost well conditioned, as is pointed out in [349].
More precisely, minimization with respect to s of the Frobenius condition
number defined as

$$J(s) = \sqrt{\mathrm{tr}[H(s)]\,\mathrm{tr}[H^{-1}(s)]} \tag{3.232}$$

is considered. But as was already shown in Section 2.5, the second Gâteaux
derivative at a global minimum $\hat{\theta}$ of the least-squares criterion is approximately equal, up to a constant multiplier, to the corresponding FIM.

In consequence, an optimal sensor location can equivalently be determined
by choosing s which corresponds to a minimum of the criterion

$$\tilde{J}(s) = \frac{1}{m}\sqrt{\mathrm{tr}(M)\,\mathrm{tr}(M^{-1})}. \tag{3.233}$$

This form of the criterion to be used while looking for optimum experimental
designs was suggested in [348] by the name *Turing's measure of conditioning*.
It is used when it is desirable to obtain a confidence region for the parameters
as spherical as possible. It is easy to check that its minimum value is 1 and
it is obtained for spherical confidence regions.

It is a simple matter to show that for the criterion

$$\Psi(M) = \mathrm{tr}(M)\,\mathrm{tr}(M^{-1}) \tag{3.234}$$

whose minimization is equivalent to minimization of (3.233), we get

$$\overset{\circ}{\Psi} = -\mathrm{tr}(M)M^{-2} + \mathrm{tr}(M^{-1})I, \tag{3.235}$$

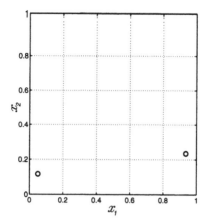

FIGURE 3.11
Optimal location of two sensors for Turing's measure of conditioning of Example 3.9.

where, as usual, I stands for the identity matrix.

Unfortunately, in spite of its clear rationale, the approach should be used with great care, as it only guarantees that the condition number is close to unity and no more than that [326]. This means that we might have a low value of J and at the same time little information about the parameters. This is confirmed by the following example.

Example 3.9

Let us consider again the situation of Example 3.4 with two sensors and criterion (3.233). Direct minimization by using the ARS procedure to produce the initial guess for the SQP algorithm yielded the ultimate solution (cf. Fig. 3.11)

$$s^\star = \begin{pmatrix} 0.050002, & 0.115412, \ldots \\ 0.936228, & 0.234803 \end{pmatrix}$$

for which Turing's measure of conditioning was 1.739576, which is not bad if we recall that the smallest possible value is one (which is not always attainable). Nevertheless, let us note that those sensor positions are rather poor from the point of view of parameter accuracy, as, e.g., the determinant of the FIM equals 1.7686, which is relatively small compared to the value 157.438 obtained in the identical conditions from the D-optimal solution for the same number of sensors. □

3.7 Modifications required by other settings

3.7.1 Discrete-time measurements

The theory that we have presented above for continuous-time measurements applies, with minor modifications, to discrete-time measurements of the form

$$z_m^j(t_r) = y(x^j, t_r) + \varepsilon(x^j, t_r), \quad r = 1, \ldots, R, \quad j = 1, \ldots, N, \quad (3.236)$$

where the t_r's are given time instants and the $\varepsilon(x^j, t_r)$'s constitute a sequence of zero-mean i.i.d. random variables. The only changes required are simply to replace integrations over the time horizon $[0, t_f]$ by summations over time instants $t_1, \ldots, t_R$ and to substitute the reciprocals of R for those of t_f. For example, the modifications imply the following form of the FIM:

$$M(\xi) = \int_X \Upsilon(x) \, \xi(\mathrm{d}x), \quad (3.237)$$

which constitutes the counterpart of (3.21) for

$$\Upsilon(x) = \frac{1}{R} \sum_{r=1}^{R} f(x, t_r) f^{\mathsf{T}}(x, t_r), \quad f(x, t) = \left(\frac{\partial y(x, t; \theta)}{\partial \theta} \right)^{\mathsf{T}}_{\theta = \theta^0}, \quad (3.238)$$

θ^0 being a preliminary estimate of θ.

3.7.2 Multiresponse systems and inaccessibility of state measurements

The conceptual framework within which this monograph has been placed is by no means limited to situations where we can measure directly a state having only one component. If a given DPS is described by a system of PDEs, then its state is a mapping $y : \Omega \times T \to \mathbb{R}^r$. What is more, it may happen that this state cannot be observed directly and we only have access to its indirect observations of the form

$$z_m^j(t) = h(y(x^j, t; \theta), t) + \varepsilon(x^j, t), \quad j = 1, \ldots, N, \quad (3.239)$$

where $h : \mathbb{R}^{r+1} \to \mathbb{R}^q$ is known, the notation $y(x, t; \theta)$ reflects the dependence of the state on a vector of constant parameters $\theta \in \mathbb{R}^m$, and the noise with values in $\mathbb{R}^q$ satisfies

$$E[\varepsilon(x^j, t)] = 0, \quad (3.240)$$

$$E[\varepsilon(x^i, \tau)\varepsilon^{\mathsf{T}}(x^j, t)] = \delta_{ij}\delta(t - \tau)C(x^i, t), \quad (3.241)$$

where $C(x^i, t) \in \mathrm{PD}(q)$, δ_{ij} and $\delta(\cdot)$ being the Kronecker and Dirac delta functions, respectively.

In this general situation, the weighted least-squares method reduces to minimization of the fit-to-data criterion

$$J(\theta) = \frac{1}{2} \sum_{j=1}^{N} \int_0^{t_f} \|z_m^j(t) - h(\hat{y}(x^j, t; \theta), t)\|_{C^{-1}(x^j, t)}^2 \, dt, \qquad (3.242)$$

where $\hat{y}(x^j, t; \theta)$ constitutes the solution to the respective state equation given a value of θ, and the notation $\|e\|_A^2$ means $e^{\mathsf{T}} A e$ for any $e \in \mathbb{R}^q$ and $A \in \mathrm{PD}(q)$.

It then follows that the relevant average FIM is

$$M(x^1, \dots, x^N) = \sum_{j=1}^{N} \Upsilon(x^j) \qquad (3.243)$$

or

$$M(\xi) = \int_X \Upsilon(x)\, \xi(\mathrm{d}x), \qquad (3.244)$$

for discrete and continuous designs, respectively, where

$$\Upsilon(x) = \frac{1}{t_f} \int_0^{t_f} \left(\frac{\partial y}{\partial \theta}\right)^{\mathsf{T}} \left(\frac{\partial h}{\partial y}\right)^{\mathsf{T}} C^{-1} \left(\frac{\partial h}{\partial y}\right) \left(\frac{\partial y}{\partial \theta}\right) \bigg|_{\theta=\theta^0} \, dt, \qquad (3.245)$$

θ^0 being a preliminary estimate of θ. All results of this chapter then remain unchanged provided that the above definition of $\Upsilon(x)$ is employed. A comprehensive treatment of this setting is contained in [206].

3.7.3 Simplifications for static DPSs

Theoretical results which have been presented so far include static systems as a particular case, i.e., they are more general than those presented in classical monographs and textbooks on the optimality theory of experimental designs, where continuous-time measurements are not considered. Specifically, the assumption that the state equation (2.1) does not contain time, i.e., that only x constitutes the independent variable, implies the description by a PDE of the form

$$\mathcal{F}\left(x, y, \frac{\partial y}{\partial x_1}, \frac{\partial y}{\partial x_2}, \frac{\partial^2 y}{\partial x_1^2}, \frac{\partial^2 y}{\partial x_2^2}, \theta\right) = 0, \quad x \in \Omega, \qquad (3.246)$$

supplemented with the boundary condition

$$\mathcal{E}\left(x, y, \frac{\partial y}{\partial x_1}, \frac{\partial y}{\partial x_2}, \theta\right) = 0, \quad x \in \partial\Omega, \qquad (3.247)$$

where $y = y(x; \theta)$ signifies the real-valued state variable corresponding to a given vector $\theta \in \mathbb{R}^m$ of constant parameter values, and $\mathcal{F}$ and $\mathcal{E}$ are some known functions.

Assuming noisy measurements made by N sensors located at spatial points $x^1, \ldots, x^N \in X \subset \Omega \cup \partial\Omega$, X being a given compact set, the observations are given by

$$z_m^j = y(x^j; \theta) + \varepsilon(x^j), \quad j = 1, \ldots, N, \quad (3.248)$$

where $\varepsilon(x)$ stands for the measurement noise such that $\{\varepsilon(x^j)\}$ forms a sequence of i.i.d. random variables. The respective FIM then has the form

$$M(x^1, \ldots, x^N) = \sum_{j=1}^{N} \Upsilon(x^j), \quad (3.249)$$

where

$$\Upsilon(x) = f(x)f^{\mathsf{T}}(x), \quad f(x) = \left(\frac{\partial y(x; \theta)}{\partial \theta} \right)^{\mathsf{T}}_{\theta=\theta^0}. \quad (3.250)$$

Just as in Section 3.1.1, we can then relax the notion of the admissible design to any probability measure ξ on the set X and consider the FIMs of the form (3.21) with $\Upsilon(\cdot)$ defined by (3.250). Formally, we may also adapt all the characterizations of the optimal solutions from the results derived in Section 3.1.2 by simply dropping the averaging with respect to time. Thus the counterparts of (3.38) and (3.39) are just

$$c(\xi) = -\mathrm{tr}\left[\overset{\circ}{\Psi}(\xi) M(\xi) \right] \quad (3.251)$$

and

$$\phi(x, \xi) = -\mathrm{tr}\left[\overset{\circ}{\Psi}(\xi) \Upsilon(x) \right] = -f^{\mathsf{T}}(x) \overset{\circ}{\Psi}(\xi) f(x), \quad (3.252)$$

respectively.

The formulations of Theorems 3.1 and 3.2 remain the same, while Theorem 3.3 can be rewritten as follows:

THEOREM 3.8
The following conditions are equivalent:

(i) the design $\xi^\star$ maximizes $\det(M(\xi))$,

(ii) the design $\xi^\star$ minimizes $\max_{x \in X} f^{\mathsf{T}}(x) M^{-1}(\xi) f(x)$, and

(iii) $\max_{x \in X} f^{\mathsf{T}}(x) M^{-1}(\xi^\star) f(x) = m$.

Clearly, the general schemes of Algorithm 3.1 and other algorithms for the sequential construction of Ψ-optimum designs discussed in this chapter remain the same, too. What is more, convenient formulae for updating matrix inverses and determinants can be applied here to considerably accelerate the computations. This is because, when looking for an improvement in the current design $\xi^{(k)}$, great efforts must be made to check trial designs of the form

$$\xi^+ = (1 - \alpha)\xi^{(k)} + \alpha\delta_{x^+}, \tag{3.253}$$

where $\alpha \in (0, 1)$ and $x^+ \in X$ and $\delta_{x^+} = \left\{ \begin{smallmatrix} x^+ \\ 1 \end{smallmatrix} \right\}$. Observing that

$$M(\xi^+) = (1 - \alpha)M(\xi^{(k)}) + \alpha f(x^+) f^{\mathsf{T}}(x^+), \tag{3.254}$$

i.e., that $M(\xi^+)$ results from modifying $M(\xi^{(k)})$ by a rank 1 correction (in fact, $\mathrm{rank}\big(f(x^+) f^{\mathsf{T}}(x^+)\big) = 1$), we can make use of the Sherman-Morrison formula, cf. (B.12), to obtain

$$M^{-1}(\xi^+)$$
$$= \frac{1}{1 - \alpha}\left[M^{-1}(\xi^{(k)}) - \frac{\alpha M^{-1}(\xi^{(k)}) f(x^+) f^{\mathsf{T}}(x^+) M^{-1}(\xi^{(k)})}{1 - \alpha + \alpha f^{\mathsf{T}}(x^+) M^{-1}(\xi^{(k)}) f(x^+)} \right], \tag{3.255}$$

which permits avoidance of the time-consuming matrix inversions. While determining D-optimum designs, the computation of the determinants can additionally be speeded up through the use of the formula

$$\det\big(M(\xi^+)\big) = (1 - \alpha)^m \left(1 + \frac{\alpha\phi(x^+, \xi^{(k)})}{1 - \alpha} \right) \det\big(M(\xi^{(k)})\big), \tag{3.256}$$

which results from (B.14). What is more, in this case it is a simple matter to check that the optimal value of the stepsize α during the line search (3.76) is given by

$$\alpha_k = \frac{\phi(x_0^{(k)}, \xi^{(k)}) - m}{\big[\phi(x_0^{(k)}, \xi^{(k)}) - 1\big]m}. \tag{3.257}$$

The D-optimum designs for static DPSs possess an interesting property, which we give as a theorem:

THEOREM 3.9
[208, Prop. V.7, p. 139] *If*

$$\xi^\star = \left\{ \begin{matrix} x^{1\star}, & \dots, & x^{\ell\star} \\ p_1^\star, & \dots, & p_\ell^\star \end{matrix} \right\} \tag{3.258}$$

is a D-optimum design, then

$$p_i^\star \le \frac{1}{m}, \quad i = 1, \dots, \ell, \tag{3.259}$$

where m is the number of the estimated parameters.

PROOF   Since (3.259) is satisfied for $p_i^\star = 0$, it suffices to investigate the situation when $p_i^\star > 0$. Choosing an $\epsilon \in (0, p_i^\star)$ and setting $\alpha = p_i^\star - \epsilon$, we can define the design

$$\tilde{\xi} = \left\{ \begin{array}{ccccccc} x^{1\star}, & \ldots, & x^{(i-1)\star}, & x^{i\star}, & x^{(i+1)\star}, & \ldots, & x^{\ell\star} \\ \dfrac{p_1^\star}{1-\alpha}, & \ldots, & \dfrac{p_{i-1}^\star}{1-\alpha}, & \dfrac{\epsilon}{1-\alpha}, & \dfrac{p_{i+1}^\star}{1-\alpha}, & \ldots, & \dfrac{p_\ell^\star}{1-\alpha} \end{array} \right\}. \qquad (3.260)$$

Obviously, we have $M(\xi^\star) \succ 0$, since $\det\big(M(\xi^\star)\big) \neq 0$. But the positive definiteness of $M(\xi^\star) = \sum_{j=1}^{\ell} p_j^\star \Upsilon(x^{i\star})$ will be preserved if we perturb the weights $p_1^\star, \ldots, p_\ell^\star$ such that they still remain positive. (Indeed, consider the sign of the quadratic form $\varphi(v) = v^\mathsf{T} M(\xi^\star) v = \sum_{j=1}^{\ell} p_j^\star \big[v^\mathsf{T} f(x^{j\star})\big]^2$ for any $0 \neq v \in \mathbb{R}^m$.) The weights of the design $\tilde{\xi}$ constitute such a perturbation, which thus implies $M(\tilde{\xi}) \succ 0$ and hence $M^{-1}(\tilde{\xi})$ exists.

Since $\xi^\star$ can be expressed as the convex combination

$$\xi^\star = (1 - \alpha)\tilde{\xi} + \alpha \delta_{x^{i\star}}, \qquad (3.261)$$

where $\delta_{x^{i\star}} = \left\{ \begin{array}{c} x_1^{i\star} \\ 1 \end{array} \right\}$, we see that

$$M(\xi^\star) = (1 - \alpha)M(\tilde{\xi}) + \alpha f(x^{i\star}) f^\mathsf{T}(x^{i\star}). \qquad (3.262)$$

From (3.255) we obtain

$$M^{-1}(\xi^\star) = \frac{1}{1-\alpha}\left[M^{-1}(\tilde{\xi}) - \frac{\alpha M^{-1}(\tilde{\xi}) f(x^{i\star}) f^\mathsf{T}(x^{i\star}) M^{-1}(\tilde{\xi})}{1 - \alpha + \alpha f^\mathsf{T}(x^{i\star}) M^{-1}(\tilde{\xi}) f(x^{i\star})} \right]. \qquad (3.263)$$

Multiplying the above equation by $f^\mathsf{T}(x^{i\star})$ from the left and by $f(x^{i\star})$ from the right, we conclude that

$$\phi(x^{i\star}, \xi^\star) = \frac{\phi(x^{i\star}, \tilde{\xi})}{1 - \alpha + \alpha \phi(x^{i\star}, \tilde{\xi})}. \qquad (3.264)$$

Hence

$$\underbrace{(p_i^\star - \epsilon)}_{\alpha} \phi(x^{i\star}, \xi^\star) = \frac{\alpha \phi(x^{i\star}, \tilde{\xi})}{\underbrace{1 - \alpha}_{>0} + \alpha \phi(x^{i\star}, \tilde{\xi})} < 1. \qquad (3.265)$$

But Theorem 3.8 implies that $\phi(x^{i\star}, \xi^\star) = m$, so we get

$$(p_i^\star - \epsilon)m < 1. \qquad (3.266)$$

Letting $\epsilon \downarrow 0$, we obtain

$$p_i^\star m \leq 1, \qquad (3.267)$$

which establishes the desired formula. ∎

The form of the FIM (3.249) combined with the fact that

$$\text{rank}\{f(x^j)f^{\mathsf{T}}(x^j)\} = 1 \tag{3.268}$$

for any $f(x^j) \neq 0$ implies that

$$\text{rank}(M(\xi)) = \text{rank}\Big(\sum_{j=1}^{N} f(x^j)f^{\mathsf{T}}(x^j)\Big) \leq \sum_{j=1}^{N} \text{rank}\big(f(x^j)f^{\mathsf{T}}(x^j)\big) = N,$$
$$\tag{3.269}$$

cf. Appendix B.1. This means that $M(\xi)$ will be singular, and thus not all the parameters will be identifiable, whenever $N < m$. Consequently, for static DPSs, the number of support points must not be less than the number of the parameters to be identified.

3.8 Summary

The results contained in this chapter show that some well-known methods of optimum experimental design for linear regression models can be easily extended to the setting of our sensor-location problem. The advantage of introducing continuous designs lies in the fact that the problem dimensionality is dramatically reduced. Moreover, with some minor changes, sequential numerical design algorithms, which have been continually refined since the early 1960s, can be employed here. Unfortunately, this approach does not prevent sensors from clustering, which is a rather undesirable phenomenon in potential applications. Note, however, that this does not necessarily mean that continuous designs are of little practical importance. First of all, they may serve to indicate spatial subdomains (i.e., the ones around the support points) which furnish much useful information on the estimated parameters. What is more, they simultaneously quantify the importance of these regions through the respective design weights. This can constitute a good starting point for further investigations. Apart from that, the weights can be interpreted as precision measures of measurements made at individual points. Indeed, it is a simple matter to check that the FIM corresponding to an optimal discrete design (3.30) does not change if, instead of several sensors placed at a support point $x^{i\star}$ such that their precision is characterized by the same variance σ^2, we use only one sensor for which the variance of measurements errors is $\tilde{\sigma}^2(x^{i\star}) = \sigma^2/\ell p_i^\star$, where ℓ signifies the number of support points. This means that a larger weight at a support point indicates the necessity of locating a more precise measurement device at that point.

Alternatively, we may seek to find an optimal design not within the class of all designs, but rather in a restricted subset of competing clusterization-free designs. To implement this idea, some recent advances in spatial statistics were employed, and in particular, Fedorov's idea of directly constrained design measures was adapted to our framework. As a consequence, this led to a very efficient and particularly simple exchange-type algorithm. Bear in mind, however, that this approach should in principle be used if the number of sensors is relatively high. If this is not the case, we can resort to standard optimization routines which ensure that the constraints on the design measure and region are satisfied (as indicated, computation of the required gradients does not present a problem). Although the numerical examples presented here are clearly not real-world problems and their purpose is primarily to illustrate our considerations in an easily interpretable manner, they are complex enough to provide evidence for the effectiveness of the proposed approaches.

4

Locally optimal strategies for scanning and moving observations

As was already emphasized, most of the contributions to the measurement optimization problem deal with the choice of stationary sensor positions. An alternative to such a strategy of taking measurements is to use movable sensors which offer additional degrees of freedom regarding optimality. Since systems with mobile observations are no doubt more flexible than those with nonmobile ones and their capabilities are wider, we can expect the minimal value of an adopted design criterion to be lower than the one for the stationary case. This is due to the fact that a nonmobile observation is a special case of a mobile one when all mobile observations are fixed (this implies, in turn, that the results of the theory for stationary observations must be contained in the more general mobile observation theory). Consequently, sensors are not assigned to positions which are optimal only on the average, but are capable of tracking points which provide at a given time moment the best information about the parameters to be identified. As indicated in [38], mobile observations may significantly alter fundamental system properties, e.g., systems which are not observable under nonmobile observations become observable after transition to mobile measurements.

A possibility of using moving observations does arise in a variety of applications, e.g., air pollutants in the environment are often measured using data gathered by monitoring cars moving in an urban area and atmospheric variables are measured using instruments carried in a satellite [187, 188]. Other examples include scanning measurement of a surface temperature by optical pyrometers and measurement of vibrations and strains in materials using optical registration [238]. The remainder of this chapter provides an exposition of basic systematic approaches to the design of scanning and moving sensor configurations.

4.1 Optimal activation policies for scanning sensors

Owing to recent advances in sensor technology, in an increasing number of industrial and environmental processes, application of switching measurement devices has become possible, thereby making a viable alternative to measurement strategies with fixed positions of the sensors which monitor system outputs [65, 66, 161]. The observation system then often comprises multiple sensors whose location is already specified and it is desired to activate only a subset of them during a given time interval while the other sensors remain dormant. A reason for not using all the available sensors could be the reduction of the observation system complexity and the costs of operation and maintenance [65]. For example, air pollutants in an environmental system are often estimated using only part of the data obtained from monitoring stations because of the high cost of processing huge amounts of data. As pointed out in [330], devices may be present that cannot be used at the same time, as different plant configurations, e.g., at startup and at full load, usually require different instrumentation.

The discussed scanning strategy of taking measurements can be also interpreted in terms of several sensors which are mobile in the sense that after performing measurements at fixed spatial positions (i.e., exactly in the same way as stationary sensors) on a time interval of nonzero length, they can change their locations to continue the measurements at more informative points, and the time necessary for this change may be neglected.

Beyond any doubt, one of the major difficulties in the scanning-sensor scheduling problem is its combinatorial nature. Even if only a finite number of given time instants are allowed for passing to new sensor configurations, a complete enumeration of all scheduling policies quickly becomes intractable, since the number of competing solutions grows exponentially with the numbers of sensors and switching instants. Complicated general search algorithms of discrete optimization can readily consume appreciable computer time and space, too.

Instead of the wasteful and tedious exhaustive search of the solution space, in this section we consider two efficient methods which construct solutions to the scanning sensor guidance problem. Specifically, a conversion to the problem of finding optimum sensor densities is considered in the following for the case of a relatively high number of measurement sensors and a predefined set of switching times. The solution is accomplished by generalizing the approach based on clusterization-free designs outlined in Section 3.4. This idea was first exposed in [329]. In the next subsection, we solve the scanning measurement problem with free switching times by treating it as an optimal discrete-valued control problem which is then transformed into an equivalent continuous-valued optimal-control formulation. In principle, the latter method is only applicable to situations when the number of scanning sensors to be scheduled is rather moderate, as it does not prevent the drawback of the "curse of dimensionality" inherent to the original combinatorial problem (enormous amounts of memory storage and time are still involved when the number of sensors is high). It offers, however, a possibility of selecting switching moments in continuous time based on the application of commonplace nonlinear programming algorithms.

4.1.1 Exchange scheme based on clusterization-free designs

4.1.1.1 Conversion to the problem of finding optimal sensor densities

Let us form an arbitrary partition of the time interval $T = [t_0, t_f]$ by choosing points $t_0 < t_1 < \cdots < t_R = t_f$ defining subintervals $T_r = [t_{r-1}, t_r)$, $r = 1, \ldots, R$. We then consider N scanning sensors which will possibly be changing their locations at the beginning of every time subinterval, but will be remaining stationary for the duration of each of the subintervals. Thus the instantaneous sensor configuration s can be viewed as follows:

$$s(t) = (x_r^1, \ldots, x_r^N) \quad \text{for } t \in T_r, \quad r = 1, \ldots, R, \tag{4.1}$$

where $x_r^j \in X \subset \mathbb{R}^d$ stands for the location of the j-th sensor on the subinterval T_r, X being the part of the spatial domain where the measurements can be taken.

Assume that the output equation has the form

$$z_{\mathrm{m}}^j(t) = y(x_r^j, t; \theta) + \varepsilon(x_r^j, t), \quad j = 1, \ldots, N \tag{4.2}$$

for $t \in T_r$, $r = 1, \ldots, R$. Then we can decompose the corresponding normalized FIM as follows:

$$M(s) = \frac{1}{N} \sum_{r=1}^{R} \sum_{j=1}^{N} \Upsilon_r(x_r^j), \tag{4.3}$$

where

$$\Upsilon_r(x) = \frac{1}{t_f} \int_{t_{r-1}}^{t_r} g(x, t) g^{\mathsf{T}}(x, t) \, \mathrm{d}t, \tag{4.4}$$

$$g(x, t) = \left(\frac{\partial y(x, t; \theta)}{\partial \theta} \right)^{\mathsf{T}}_{\theta = \theta^0} \tag{4.5}$$

for some preliminary estimate θ^0 of θ.

When the number of sensors N is large, which is rather a common situation in applications such as air-pollution monitoring networks or control architectures for smart material systems, the optimal sensor-location problem becomes extremely difficult from a computational point of view. It goes without saying that the situation becomes much more critical when scanning measurement devices are employed.

In order to overcome this predicament, in what follows we propose to operate on the spatial density of sensors (i.e., the number of sensors per unit area), rather than on the sensor locations, as was already the case in Section 3.4. This approach is proved reasonable for a sufficiently large N and potential solutions would be satisfactory for many technical processes. Performing such a conversion, we do not get rid of the discrete nature of the original formulation, and therefore the resultant computational problem is not still amenable to solution by standard optimization techniques. Thus we relax the definition of the set of admissible solutions by making use of the observation that the density of sensors over the subinterval T_r can be approximately described by a probability measure $\xi_r(\mathrm{d}x)$ on the space $(X, \mathcal{B})$, where $\mathcal{B}$ is the σ-algebra of all Borel subsets of X. The resulting extension of the set of feasible solutions makes it possible to apply convenient and efficient mathematical tools of convex programming theory. As regards the practical interpretation of the so produced solutions (provided that we are in a position to calculate at least their approximations), one possibility is to partition X into nonoverlapping subdomains ΔX_i of relatively small areas and then, on the subinterval T_r, to allocate to each of them the number

$$N_r(\Delta X_i) = \left\lceil N \int_{\Delta X_i} \xi_r(\mathrm{d}x) \right\rceil \tag{4.6}$$

of sensors (here $\lceil \rho \rceil$ is the smallest integer greater than or equal to ρ).

Thus our aim is to find probability measures ξ_r, $r = 1, \ldots, R$ over X which are absolutely continuous with respect to the Lebesgue measure and satisfy by definition the condition

$$\int_X \xi_r(\mathrm{d}x) = 1, \quad r = 1, \ldots, R. \tag{4.7}$$

For notational convenience, in what follows we shall briefly write

$$\xi = (\xi_1, \ldots, \xi_R) \tag{4.8}$$

and call ξ a *design*.

Such an extension of the concept of the sensor configuration allows us to replace (4.3) by

$$M(\xi) = \sum_{r=1}^{R} \int_X \Upsilon_r(x)\, \xi_r(\mathrm{d}x). \tag{4.9}$$

A rather natural additional assumption is that the density of sensors represented by $N_r(\Delta X_i)/N$ in a given part ΔX_i must not exceed some prescribed level. In terms of the probability measures, this amounts to imposing the conditions

$$\xi_r(\mathrm{d}x) \leq \omega(\mathrm{d}x), \quad r = 1, \ldots, R, \tag{4.10}$$

where $\omega(\mathrm{d}x)$ is a given measure satisfying $\int_X \omega(\mathrm{d}x) \geq 1$.

Defining $J(\xi) = \Psi[M(\xi)]$, we may phrase the scanning sensor-location problem as the selection of

$$\xi^\star = \arg \min_{\xi \in \Xi(X)} J(\xi), \tag{4.11}$$

where $\Xi(X)$ denotes the set of competing designs, i.e., the set of all R-tuples of the form (4.8) whose components satisfy (4.10) (note that $\Xi(X)$ is nonempty and convex). We call $\xi^\star$ the (Ψ, ω)-optimal solution.

In the sequel, we will need the following assumptions:

(A1) X is compact.

(A2) $g \in C(X \times T; \mathbb{R}^m)$.

(A3) Ψ is convex.

(A4) If $M_1 \preceq M_2$, then $\Psi(M_1) \geq \Psi(M_2)$ (monotonicity).

(A5) $\omega(dx)$ is atomless, i.e., for any $\Delta X \subset X$ there exists a $\Delta X' \subset \Delta X$ such that

$$\int_{\Delta X'} \omega(dx) < \int_{\Delta X} \omega(dx). \qquad (4.12)$$

(A6) There exists a finite real q such that

$$\{\xi = (\xi_1, \ldots, \xi_R) : J(\xi) \leq q < \infty, \ \xi_r(dx) \leq \omega(dx), \ r = 1, \ldots, R\}$$
$$= \widetilde{\Xi}(X) \neq \emptyset.$$

(A7) For any $\xi \in \widetilde{\Xi}(X)$ and $\bar{\xi} \in \Xi(X)$, we have

$$\delta_+ J(\xi; \bar{\xi} - \xi) = \sum_{r=1}^{R} \int_X \psi_r(x, \xi) \bar{\xi}_r(dx), \qquad (4.13)$$

where the quantity on the left-hand side stands for the one-sided directional derivative of J at ξ in the direction $\bar{\xi} - \xi$

$$\delta_+ J(\xi; \bar{\xi} - \xi) = \lim_{\alpha \downarrow 0} \frac{J(\xi + \alpha(\bar{\xi} - \xi)) - J(\xi)}{\alpha}$$
$$= \lim_{\alpha \downarrow 0} \frac{\Psi[M(\xi) + \alpha(M(\bar{\xi}) - M(\xi))] - \Psi[M(\xi)]}{\alpha}. \qquad (4.14)$$

These assumptions constitute rather obvious alterations of their counterparts used throughout Chapter 3. At this juncture, only Assumption (A7) calls for an appropriate comment. In practice, it simply amounts to the existence of the directional derivative of J at ξ in the direction of $\bar{\xi} - \xi$ whose form must be on one hand specific (reasoning just as in the proof of Lemma 3.6, we conclude that $\psi_r(\cdot, \xi)$, $r = 1, \ldots, R$ are automatically $C(X)$ functions), but on the other hand, for most practical criteria such a condition is not particularly restrictive.

In fact, requiring Ψ to be continuously differentiable with respect to individual elements of its matrix argument, we obtain

$$
\begin{aligned}
\mathrm{d}J(\xi; \bar{\xi} - \xi) &= \frac{\mathrm{d}}{\mathrm{d}\alpha}\Psi[M(\xi) + \alpha(M(\bar{\xi}) - M(\xi))]\Big|_{\alpha=0+} \\
&= \mathrm{tr}\Big[\mathring{\Psi}(\xi)(M(\bar{\xi}) - M(\xi))\Big] \\
&= \left\{\sum_{r=1}^{R}\int_{X}\mathrm{tr}\Big[\mathring{\Psi}(\xi)\Upsilon_r(x)\Big]\,\bar{\xi}_r(\mathrm{d}x)\right\} - \mathrm{tr}\Big[\mathring{\Psi}(\xi)M(\xi)\Big] \\
&= \sum_{r=1}^{R}\int_{X}\left\{\mathrm{tr}\Big[\mathring{\Psi}(\xi)\Upsilon_r(x)\Big] - \frac{1}{R}\mathrm{tr}\Big[\mathring{\Psi}(\xi)M(\xi)\Big]\right\}\,\bar{\xi}(\mathrm{d}x),
\end{aligned}
\tag{4.15}
$$

where

$$
\mathring{\Psi}(\xi) = \frac{\partial\Psi[M]}{\partial M}\Big|_{M=M(\xi)}.
$$

Therefore

$$
\psi_r(x, \xi) = c(\xi) - \phi_r(x, \xi),
\tag{4.16}
$$

with the functions c and ϕ_r, respectively, defined as

$$
c(\xi) = -\frac{1}{R}\mathrm{tr}\Big[\mathring{\Psi}(\xi)M(\xi)\Big]
\tag{4.17}
$$

and

$$
\phi_r(x, \xi) = -\mathrm{tr}\Big[\mathring{\Psi}(\xi)\Upsilon_r(x)\Big].
\tag{4.18}
$$

REMARK 4.1   Note that the quantity $\psi_r(x, \xi)$ constitutes the directional derivative of J at ξ in the direction of $\bar{\xi} - \xi$, where

$$
\bar{\xi} = (\underbrace{0, \ldots, 0}_{r-1\text{ zero}\atop\text{measures}}, \delta_x, 0, \ldots, 0)
\tag{4.19}
$$

and δ_x means the Dirac measure concentrated at point x.   ∎

The scanning sensor-location problem has been posed as an optimization one, that of minimizing a real-valued function Ψ defined on the set of FIMs, each parameterized by a design ξ. The indicated minimization is accomplished by the appropriate choice of an 'admissible' ξ. The purpose of the next subsection is to mathematically indicate such a choice. The task of actually constructing a solution $\xi^\star$ from the characterization given by the optimality conditions is the domain of Section 4.1.1.3.

In what follows, we write $\bar{\Xi}(X)$ for the collection of all the designs ξ whose components satisfy the requirement

$$\xi_r(\Delta X) = \begin{cases} \omega(\Delta X) & \text{for } \Delta X \subset \operatorname{supp} \xi_r, \\ 0 & \text{for } \Delta X \subset X \setminus \operatorname{supp} \xi_r. \end{cases} \tag{4.20}$$

The point of this definition is that the designs from $\bar{\Xi}(X)$ turn out to be vital while formulating optimality conditions. The main feature of a design $\xi \in \bar{\Xi}(X)$ is that for each of its components ξ_r the design domain X can be split into two subsets for which ξ_r coincides either with the zero-measure or with the upper bound ω.

4.1.1.2 Characterization of optimal designs

We begin with a fundamental result regarding the form of (Ψ, ω)-optimal designs.

THEOREM 4.1
Under Assumptions (A1)–(A7), a (Ψ, ω)-optimal design exists in $\bar{\Xi}(X)$.

PROOF This theorem is a fairly straightforward generalization of Theorem 1 in [83]. Since the proof is very technical, its details are omitted. Let us only note that Assumptions (A1) and (A2) guarantee that the set $\mathfrak{M}(X) = \{M(\xi) : \xi \in \Xi(X)\}$ is compact and convex. Moreover, the optimization process (4.11) can equivalently be seen as looking for a matrix M in $\mathfrak{M}(X)$ which minimizes $\Psi[M]$. The convexity of Ψ and Assumption (A6) then imply the existence of a matrix M^* being a minimum of Ψ on $\mathfrak{M}(X)$. Clearly, it corresponds to a design $\xi^* \in \Xi(X)$.

The fact that at least one optimal design must belong to $\bar{\Xi}(X)$ follows from the Lyapunov Theorem on the range of a vector measure [127, Thm.12.1, p. 266]. In particular, we can show that for any measure ξ_r satisfying (4.10) there is a measure $\bar{\xi}_r$ satisfying (4.20) such that $M_r(\xi_r) = M_r(\bar{\xi}_r)$. ∎

Consequently, we can focus our attention on designs from the set $\bar{\Xi}(X)$. Our goal now is to develop a method for checking whether a given design $\xi \in \bar{\Xi}(X)$ is (Ψ, ω)-optimal. The test stated below in Theorem 4.2 is based on the following notion of the separability of two sets:.

DEFINITION 4.1 *Given a design ξ, we will say that the function $\psi_r(\cdot, \xi)$ defined by (4.16) separates sets X_1 and X_2 with respect to $\omega(\mathrm{d}x)$ if for any two sets $\Delta X_1 \subset X_1$ and $\Delta X_2 \subset X_2$ satisfying*

$$\int_{\Delta X_1} \omega(\mathrm{d}x) = \int_{\Delta X_2} \omega(\mathrm{d}x) \tag{4.21}$$

we have

$$\int_{\Delta X_1} \psi_r(x, \xi)\, \omega(dx) \leq \int_{\Delta X_2} \psi_r(x, \xi)\, \omega(dx). \tag{4.22}$$

THEOREM 4.2
A necessary and sufficient condition for $\xi^* = (\xi_1^*, \ldots, \xi_R^*) \in \bar{\Xi}(X) \cap \tilde{\Xi}(X)$ *to be* (Ψ, ω)*-optimal is that the functions* $\psi_r(\cdot, \xi^*)$ *separate* $X_r^* = \text{supp}\, \xi_r^*$ *and* $X \setminus X_r^*$ *for* $r = 1, \ldots, R$.

PROOF   (Necessity) Choose $\kappa \in \{1, \ldots, R\}$. Assume that $\xi^* \in \bar{\Xi}(X)$ is (Ψ, ω)-optimal and $\text{supp}\, \xi_r^* = X_r^* \subset X$. Let the components of $\xi \in \bar{\Xi}(X)$ satisfy

$$\text{supp}\, \xi_r = \begin{cases} (X_r^* \setminus D_r) \cup E_r & \text{if } r = \kappa, \\ X_r^* & \text{if } r \neq \kappa, \end{cases} \tag{4.23}$$

where the sets $D_\kappa \subset X_\kappa^*$ and $E_\kappa \subset X \setminus X_\kappa^*$ are selected so as to fulfil the condition

$$\int_{D_\kappa} \omega(dx) = \int_{E_\kappa} \omega(dx) > 0. \tag{4.24}$$

It follows that

$$\sum_{r=1}^{R} \int_X \psi_r(x, \xi^*)\, \xi_r(dx)$$

$$= \left\{ \sum_{r=1}^{R} \int_{X_r^*} \psi_r(x, \xi^*)\, \omega(dx) \right\} + \int_{E_\kappa} \psi_r(x, \xi^*)\, \omega(dx)$$

$$- \int_{D_\kappa} \psi_r(x, \xi^*)\, \omega(dx) \tag{4.25}$$

$$= \left\{ \sum_{r=1}^{R} \int_X \psi_r(x, \xi^*)\, \xi_r^*(dx) \right\} + \int_{E_\kappa} \psi_r(x, \xi^*)\, \omega(dx)$$

$$- \int_{D_\kappa} \psi_r(x, \xi^*)\, \omega(dx).$$

But on account of Assumption (A7) and the optimality of ξ^*, we have

$$\sum_{r=1}^{R} \int_X \psi_r(x, \xi^*)\, \xi_r(dx) = \delta_+ J(\xi^*; \xi - \xi^*) \geq 0 \tag{4.26}$$

and

$$\sum_{r=1}^{R} \int_X \psi_r(x, \xi^*)\, \xi_r^*(dx) = \delta_+ J(\xi^*; \xi^* - \xi^*) = \delta_+ J(\xi^*; 0) = 0. \tag{4.27}$$

Consequently, we get

$$\int_{D_\kappa} \psi_r(x, \xi^\star)\, \omega(dx) \le \int_{E_\kappa} \psi_r(x, \xi^\star)\, \omega(dx). \tag{4.28}$$

Since the choice of κ, E_κ and D_κ was arbitrary, this proves the necessity.

(Sufficiency) For $\xi^\star \in \bar{\Xi}(X) \cap \widetilde{\Xi}(X)$, let the functions $\psi_r(\cdot, \xi^\star)$ separate $X_r^\star = \operatorname{supp}\xi_r^\star$ and $X \setminus X_r^\star$, $r = 1, \ldots, R$. Moreover, assume that $\xi^\star$ is not optimal, while $\xi^\circ \in \bar{\Xi}(X) \cap \widetilde{\Xi}(X)$ is, which implies that

$$J(\xi^\star) \ge J(\xi^\circ) + \eta \tag{4.29}$$

for some $\eta > 0$.

Consider $\bar{\xi} = (1 - \alpha)\xi^\star + \alpha\xi^\circ$ for some $\alpha \in [0, 1]$. From the convexity of J and (4.29), we deduce that

$$\begin{aligned} J(\bar{\xi}) = J((1 - \alpha)\xi^\star + \alpha\xi^\circ) &\le (1 - \alpha)J(\xi^\star) + \alpha J(\xi^\circ) \\ &\le (1 - \alpha)J(\xi^\star) + \alpha\{J(\xi^\star) - \eta\} = J(\xi^\star) - \alpha\eta. \end{aligned} \tag{4.30}$$

On the other hand, setting

$$X_r^\circ = \operatorname{supp}\xi_r^\circ, \quad D_r = X_r^\star \setminus X_r^\circ, \quad E_r = X_r^\circ \setminus X_r^\star, \tag{4.31}$$

we have

$$X_r^\circ = (X_r^\star \setminus D_r) \cup E_r, \quad r = 1, \ldots, R. \tag{4.32}$$

Since

$$\int_X \xi_r^\star(dx) = \int_{X_r^\star} \omega(dx) \tag{4.33}$$

and

$$\int_X \xi_r^\circ(dx) = \int_{X_r^\circ} \omega(dx) = \int_{X_r^\star} \omega(dx) - \int_{D_r} \omega(dx) + \int_{E_r} \omega(dx), \tag{4.34}$$

from the normalization condition (4.7) we see that

$$\int_X \xi_r^\star(dx) = \int_X \xi_r^\circ(dx) \tag{4.35}$$

and thus

$$\int_{D_r} \omega(dx) = \int_{E_r} \omega(dx). \tag{4.36}$$

Since, additionally, $D_r \subset X_r^\star$ and $E_r \subset X \setminus X_r^\star$, the separability of $X_r^\star$ and $X \setminus X_r^\star$ by $\psi_r(\cdot, \xi^\star)$ with respect to $\omega(dx)$ leads to

$$\int_{D_r} \psi_r(x, \xi^\star)\, \omega(dx) \le \int_{E_r} \psi_r(x, \xi^\star)\, \omega(dx), \quad r = 1, \ldots, R. \tag{4.37}$$

By Assumption (A7), we get

$$J(\bar{\xi}) = J((1-\alpha)\xi^* + \alpha\xi^\circ) = J(\xi^* + \alpha(\xi^\circ - \xi^*))$$
$$= J(\xi^*) + \alpha \sum_{r=1}^{R} \int_X \psi_r(x, \xi^*)\, \xi_r^\circ(\mathrm{d}x) + o(\alpha), \tag{4.38}$$

where o is the usual Landau symbol, i.e.,

$$\lim_{\alpha \downarrow 0} \frac{o(\alpha)}{\alpha} = 0. \tag{4.39}$$

Application of (4.31) and (4.32) yields

$$J(\bar{\xi}) = J(\xi^*) + \alpha \sum_{r=1}^{R} \int_{X_r^\circ} \psi_r(x, \xi^*)\, \omega(\mathrm{d}x) + o(\alpha)$$

$$= J(\xi^*) + \alpha \sum_{r=1}^{R} \left\{ \int_{X_r^*} \psi_r(x, \xi^*)\, \omega(\mathrm{d}x) + \int_{E_r} \psi_r(x, \xi^*)\, \omega(\mathrm{d}x) \right.$$
$$\left. - \int_{D_r} \psi_r(x, \xi^*)\, \omega(\mathrm{d}x) \right\} + o(\alpha)$$

$$= J(\xi^*) + \alpha \sum_{r=1}^{R} \left\{ \int_X \psi_r(x, \xi^*)\, \xi_r^*(\mathrm{d}x) \right\}$$

$$+ \alpha \sum_{r=1}^{R} \left\{ \int_{E_r} \psi_r(x, \xi^*)\, \omega(\mathrm{d}x) - \int_{D_r} \psi_r(x, \xi^*)\, \omega(\mathrm{d}x) \right\} + o(\alpha)$$

$$= J(\xi^*) + \alpha \sum_{r=1}^{R} \left\{ \int_{E_r} \psi_r(x, \xi^*)\, \omega(\mathrm{d}x) - \int_{D_r} \psi_r(x, \xi^*)\, \omega(\mathrm{d}x) \right\} + o(\alpha), \tag{4.40}$$

the last equality becoming clear after observing that

$$\sum_{r=1}^{R} \int_X \psi_r(x, \xi^*)\, \xi_r^*(\mathrm{d}x) = \delta_+ J(\xi^*; \xi^* - \xi^*) = \delta_+ J(\xi^*; 0) = 0. \tag{4.41}$$

On account of (4.37), we deduce that

$$J(\bar{\xi}) \geq J(\xi^*) + o(\alpha). \tag{4.42}$$

From (4.30) and (4.42) it follows that

$$J(\xi^*) + o(\alpha) \leq J(\xi^*) - \alpha\eta. \tag{4.43}$$

But this implies that

$$\frac{o(\alpha)}{\alpha} \leq -\eta \tag{4.44}$$

which is impossible. The proof is thus completed. ∎

As a companion to the above result, we next consider the special case where $w(\mathrm{d}x)$ has a continuous density ϱ. It is of paramount importance in applications.

COROLLARY 4.1
Let $\xi^* \in \bar{\Xi}(X) \cap \tilde{\Xi}(X)$ and $X_r^* = \operatorname{supp} \xi_r^*$, $r = 1, \ldots, R$. If $w(\mathrm{d}x) = \varrho(x)\,\mathrm{d}x$, where $\varrho(x)$ is a positive continuous function, then ξ^* is (Ψ, w)-optimal iff

$$\sup_{x \in X_r^*} \psi_r(x, \xi^*) \le \inf_{x \in X \setminus X_r^*} \psi_r(x, \xi^*), \qquad r = 1, \ldots, R. \tag{4.45}$$

PROOF (Sufficiency) Assuming (4.45) to hold, we choose $D_r \subset X_r^*$ and $E_r \subset X \setminus X_r^*$ such that

$$\int_{D_r} \varrho(x)\,\mathrm{d}x = \int_{E_r} \varrho(x)\,\mathrm{d}x, \qquad r = 1, \ldots, R. \tag{4.46}$$

Then

$$\int_{D_r} \psi_r(x, \xi^*)\varrho(x)\,\mathrm{d}x$$

$$\le \left\{ \sup_{x \in X_r^*} \psi_r(x, \xi^*) \right\} \int_{D_r} \varrho(x)\,\mathrm{d}x \tag{4.47}$$

$$\le \left\{ \inf_{x \in X \setminus X_r^*} \psi_r(x, \xi^*) \right\} \int_{E_r} \varrho(x)\,\mathrm{d}x \le \int_{E_r} \psi_r(x, \xi^*)\varrho(x)\,\mathrm{d}x,$$

i.e., X_r^* and $X \setminus X_r^*$ are separated by $\psi_r(\,\cdot\,, \xi^*)$ with respect to $\varrho(x)\,\mathrm{d}x$, which entails the optimality of ξ^*.

(Necessity) Assume that ξ^* is optimal. Fix $x_1 \in X_r^*$, $x_2 \in X \setminus X_r^*$ and a sufficiently small $\alpha > 0$. We will use the symbol B_{1r}^α to denote the intersection of X_r^* and the ball centred at x_1 with the radius adjusted so as to have $\int_{B_{1r}^\alpha} \varrho(x)\,\mathrm{d}x = \alpha$. Similarly, B_{2r}^α means the intersection of $X \setminus X_r^*$ and the ball centred at x_2 with the radius adjusted so that $\int_{B_{2r}^\alpha} \varrho(x)\,\mathrm{d}x = \alpha$.

Note that we have

$$\left| \frac{1}{\alpha} \int_{B_{1r}^\alpha} \psi_r(x, \xi^*)\varrho(x)\,\mathrm{d}x - \psi_r(x_1, \xi^*) \right|$$

$$= \left| \frac{1}{\alpha} \int_{B_{1r}^\alpha} [\psi_r(x, \xi^*) - \psi_r(x_1, \xi^*)]\varrho(x)\,\mathrm{d}x \right|$$

$$\le \left\{ \sup_{x \in B_{1r}^\alpha} |\psi_r(x, \xi^*) - \psi_r(x_1, \xi^*)| \right\} \frac{1}{\alpha} \int_{B_{1r}^\alpha} \varrho(x)\,\mathrm{d}x \tag{4.48}$$

$$= \sup_{x \in B_{1r}^\alpha} |\psi_r(x, \xi^*) - \psi_r(x_1, \xi^*)|.$$

Therefore

$$\frac{1}{\alpha} \int_{B_{1r}^{\alpha}} \psi_r(x, \xi^{\star}) \varrho(x) \, dx \xrightarrow[\alpha \to 0]{} \psi_r(x_1, \xi^{\star}) \qquad (4.49)$$

by the continuity of $\psi_r(\,\cdot\,, \xi^{\star})$. Similarly, we conclude that

$$\frac{1}{\alpha} \int_{B_{2r}^{\alpha}} \psi_r(x, \xi^{\star}) \varrho(x) \, dx \xrightarrow[\alpha \to 0]{} \psi_r(x_2, \xi^{\star}). \qquad (4.50)$$

The definition of the sets B_{1r}^{α} and B_{2r}^{α} implies that they must be separated by $\psi_r(\,\cdot\,, \xi^{\star})$, i.e.,

$$\int_{B_{1r}^{\alpha}} \psi_r(x, \xi^{\star}) \varrho(x) \, dx \leq \int_{B_{2r}^{\alpha}} \psi_r(x, \xi^{\star}) \varrho(x) \, dx. \qquad (4.51)$$

On account of (4.49) and (4.50), dividing both the sides of the above inequality by α and letting $\alpha \to 0$, we get

$$\psi_r(x_1, \xi^{\star}) \leq \psi_r(x_2, \xi^{\star}). \qquad (4.52)$$

Since x_1 and x_2 were arbitrary, this completes the proof. ∎

If Ψ is differentiable with respect to individual elements of its matrix argument, then (4.16) holds and we can rephrase Corollary 4.2 as follows.

COROLLARY 4.2
Under the assumptions of Corollary 4.1, if moreover $\partial \Psi[M]/\partial M \big|_{M=M(\xi^{\star})}$ *exists and is bounded, then $\xi^{\star}$ is (Ψ, ω)-optimal iff*

$$\inf_{x \in X_r^{\star}} \phi_r(x, \xi^{\star}) \geq \sup_{x \in X \setminus X_r^{\star}} \phi_r(x, \xi^{\star}), \qquad r = 1, \dots, R. \qquad (4.53)$$

According to the above result, the functions ϕ_r play a leading role in indicating spatial points which provide the most valuable information in terms of the adopted optimality criterion Ψ. They constitute a good starting point for constructing numerical procedures of determining the best configurations of scanning sensors in practice.

4.1.1.3 Numerical procedure of the exchange type

General algorithm. In spite of its seemingly abstract characters, Theorem 4.2 and Corollaries 4.1 and 4.2 form a basis for an efficient numerical algorithm of determining (Ψ, ω)-optimal designs. The main idea is to move some measure from an area of lower values of $\phi_r(\,\cdot\,, \xi^{(k)})$ to those with higher values, as we expect that such a procedure will improve the current design $\xi^{(k)}$. Details regarding this scheme are summarized as follows.

ALGORITHM 4.1 General scanning strategy algorithm

Step 1. *Guess an initial design $\xi^{(0)} \in \bar{\Xi}(X)$ which satisfies $\Psi[M(\xi^{(0)})] < \infty$. Select a tolerance $0 < \eta \ll 1$. Set $k = 0$.*

Step 2. *For $r = 1, \ldots, R$ separately set $X_{1r}^{(k)} = \operatorname{supp} \xi_r^{(k)}$ and $X_{2r}^{(k)} = \overline{X \setminus X_{1r}^{(k)}}$ (the bar over the symbol denoting a set stands for its closure), and determine*

$$x_{1r}^{(k)} = \arg \min_{x \in X_{1r}^{(k)}} \phi_r(x, \xi^{(k)}), \quad x_{2r}^{(k)} = \arg \max_{x \in X_{2r}^{(k)}} \phi_r(x, \xi^{(k)}). \quad (4.54)$$

If $\phi_r(x_{1r}^{(k)}, \xi^{(k)}) > \phi_r(x_{2r}^{(k)}, \xi^{(k)}) - \eta$ for all $r = 1, \ldots, R$, then STOP.

Step 3. *For $r = 1, \ldots, R$ proceed as follows: If $\phi_r(x_{1r}^{(k)}, \xi^{(k)}) > \phi_r(x_{2r}^{(k)}, \xi^{(k)}) - \eta$, then fix $S_{1r}^{(k)}(\alpha) = S_{2r}^{(k)}(\alpha) = \emptyset$. Otherwise define $S_{1r}^{(k)}(\alpha)$ as the intersection of $X_{1r}^{(k)}$ and the ball centred at $x_{1r}^{(k)}$ with a radius adjusted so that $\int_{S_{1r}^{(k)}(\alpha)} \rho(x) \, dx = \alpha$, and similarly, let $S_{2r}^{(k)}(\alpha)$ be the intersection of $X_{2r}^{(k)}$ and the ball centred at $x_{2r}^{(k)}$ with a radius adjusted so as to satisfy $\int_{S_{2r}^{(k)}(\alpha)} \rho(x) \, dx = \alpha$. Then construct*

$$\xi^{(k+1)} = (\xi_1^{(k)}(\alpha_k), \ldots, \xi_R^{(k)}(\alpha_k)) \in \bar{\Xi}(X) \quad (4.55)$$

by choosing $\alpha^{(k)}$ so that

$$\Psi[M(\xi_1^{(k)}(\alpha_k), \ldots, \xi_R^{(k)}(\alpha_k))]$$
$$= \min_{\alpha \in (0, \bar{\alpha}]} \Psi[M(\xi_1^{(k)}(\alpha), \ldots, \xi_R^{(k)}(\alpha))], \quad (4.56)$$

where

$$\operatorname{supp} \xi_r^{(k)}(\alpha) = \overline{(X_{1r}^{(k)} \setminus S_{1r}^{(k)}(\alpha)) \cup S_{2r}^{(k)}(\alpha)}, \quad (4.57)$$

and $\bar{\alpha} = \min\{1, \int_X \varrho(x) \, dx - 1\}$. Increment k and go to Step 2.

The properties of this computational algorithm can be considered in some detail, but in practice the scheme outlined in what follows is preferred as it is much easier to implement. Suffice it to say here that $\left\{\Psi[M(\xi^{(k)})]\right\}$ is a monotone-decreasing sequence. As for the convergence, it can be proven proceeding analogously to the main line of reasoning in the convergence analysis of feasible direction methods [22]. Since the proof is rather lengthy and highly technical, it is omitted.

Implementation issues. Within the framework of the sensor placement, we usually deal with a constant allowable sensor density $\varrho(x) = \text{const}$ (even if this is not the case, an appropriate change of coordinates transforms the original setting to this situation). Moreover, while implementing Algorithm 4.1 on a computer, all integrals are most often replaced by sums over some finite grid elements (the grid produced by the finite-element method can be employed for that purpose). Analogously, the sets X, $X_{1r}^{(k)}$, $X_{2r}^{(k)}$, $S_{1r}^{(k)}$ and $S_{2r}^{(k)}$ then simply consist of grid nodes. Thus Algorithm 4.1 can be interpreted as an exchange-type algorithm (in each iteration some points are deleted from the current design and replaced by the same number of vacant points). In practice, α_k is usually fixed and, what is more, one-point exchanges are most often adopted, i.e., $S_{1r}^{(k)} = \{x_{1r}^{(k)}\}$ and $S_{2r}^{(k)} = \{x_{2r}^{(k)}\}$, which substantially simplifies the implementation.

Taking account of the above remarks, the following computational scheme can be developed.

ALGORITHM 4.2 Practical scanning strategy algorithm

Step 1. *Construct a priori a sufficiently dense set of possible sensor locations $\tilde{X} = \{x^j\}_{j=1}^{N'}$ covering the domain X, where $N' > N$. For each node of this grid, determine and store the matrices $\Upsilon_r(x^j)$, $r = 1, \dots, R$. Select N-element sets $\tilde{X}_{1r}^{(0)} \subset \tilde{X}$, $r = 1, \dots, R$ which constitute initial guesses regarding the best sites for locating sensors over the consecutive subintervals T_r. They will possibly be improved in what follows. Set $k = 0$.*

Step 2. *Assemble the FIM:*

$$M^{(k)} = \sum_{r=1}^{R} \sum_{x \in \tilde{X}_{1r}^{(k)}} \Upsilon_r(x) \tag{4.58}$$

and compute

$$G^{(k)} = -\frac{\partial \Psi[M]}{\partial M}\bigg|_{M=M^{(k)}}. \tag{4.59}$$

For $r = 1, \dots, R$ separately, determine

$$x_{1r}^{(k)} = \arg\min_{x \in \tilde{X}_{1r}^{(k)}} \operatorname{tr}\{G^{(k)}\Upsilon_r(x)\}, \tag{4.60}$$

$$x_{2r}^{(k)} = \arg\max_{x \in \tilde{X} \setminus \tilde{X}_{1r}^{(k)}} \operatorname{tr}\{G^{(k)}\Upsilon_r(x)\}. \tag{4.61}$$

If $\operatorname{tr}\{G^{(k)}\Upsilon_r(x_{1r}^{(k)})\} \le \operatorname{tr}\{G^{(k)}\Upsilon_r(x_{2r}^{(k)})\} - \eta$, where $0 < \eta \ll 1$, then set $S_{1r}^{(k)} = \{x_{1r}^{(k)}\}$ and $S_{2r}^{(k)} = \{x_{2r}^{(k)}\}$, otherwise fix $S_{1r}^{(k)} = S_{2r}^{(k)} = \emptyset$.

Step 3. *If*

$$\text{tr}\big\{G^{(k)}\Upsilon_r(x_{1r}^{(k)})\big\} > \text{tr}\big\{G^{(k)}\Upsilon_r(x_{2r}^{(k)})\big\} - \eta \qquad (4.62)$$

for all $r = 1,\ldots, R$, *then STOP. Otherwise, set*

$$\tilde{X}_{1r}^{(k+1)} = (\tilde{X}_{1r}^{(k)} \setminus S_{1r}^{(k)}) \cup S_{2r}^{(k)}. \qquad (4.63)$$

Increment k *and go to Step 2.*

The integration required for determining the matrices $\Upsilon_r(x^j)$ can be performed using common quadratures (classical formulae for equally spaced abscissas, such as the trapezoidal or Simpson's rules, are usually sufficient).

Algorithm 4.2 performs well and turns out to be extremely fast despite the high dimensionality of the original problem. Switching from the formulation in terms of seeking the best sensor locations to that in terms of determining the best sensor densities makes it possible to avoid the complications caused by the inherent combinatorial nature of the sensor-location problem.

Apart from the decided advantages of the approach, two issues should be addressed as potential shortcomings. First of all, note that one-point exchanges in Algorithm 4.2, a simplified version of Algorithm 4.1, correspond to the situation in which all α_k's are the same, while the convergence of the proposed scheme is guaranteed only for a sequence of properly selected α_k's, cf. (4.56). As a result, some minor oscillations of the quantity $\Psi[M(\xi^{(k)})]$ may be observed after the initial stage of a monotonic decrease in its values. In practice, however, if the grid $\tilde{X}$ is sufficiently dense, the reduction in the value of the performance index is so significant that we may hope that the obtained designs do not deviate too much from the optimal ones.

Example 4.1
The following numerical example serves as a vehicle to test the solution techniques proposed in this chapter. The point of departure is the two-dimensional diffusion equation

$$\frac{\partial y(x,t)}{\partial t} = \frac{\partial}{\partial x_1}\left(\kappa(x)\frac{\partial y(x,t)}{\partial x_1}\right) + \frac{\partial}{\partial x_2}\left(\kappa(x)\frac{\partial y(x,t)}{\partial x_2}\right)$$
$$+ 20\exp\big(-50(x_1-t)^2\big), \quad x \in \Omega = (0,1) \times (0,1), \quad t \in (0,1)$$

subject to the conditions

$$y(x,0) = 0, \quad x \in \Omega,$$
$$y(x,t) = 0, \quad (x,t) \in \partial\Omega \times T.$$

The diffusion coefficient to be identified has the form

$$\kappa(x) = \theta_1 + \theta_2 x_1 + \theta_3 x_2, \quad \theta_1 = 0.1, \quad \theta_2 = -0.05, \quad \theta_3 = 0.2,$$

where the values θ_1, θ_2 and θ_3 are also treated as nominal and known to the experimenter prior to the identification itself.

As regards the forcing term in our model, it approximates the action of a line source whose support is constantly oriented along the x_2-axis and moves with constant speed from the left to the right boundary of Ω. Our purpose is to estimate κ (i.e., the parameters θ_1, θ_2 and θ_3) as accurately as possible based on the measurements made by 100 scanning sensors. In the case considered, the D-optimum design criterion was chosen as the measure of the estimation accuracy. A uniform mesh of 20×20 points was thus assumed as the set of places where sensors could be potentially placed (they are marked with points in Fig. 4.1).

As for technicalities, in order to numerically solve the measurement location problem, a computer programme was written in Lahey/Fujitsu Fortran 95 Ver. 5.60a using a PC (Pentium IV, 2.40 GHz, 512 MB RAM) running Windows 2000. The state and sensitivity equations were solved with the finite-element method. The sampling interval and coordinate divisions were $\Delta t = 0.0125$ and $\Delta x_1 = \Delta x_2 = 0.025$, respectively. About twenty seconds of CPU time were used to determine the optimal configurations shown in Fig. 4.1 (open circles denote the best points for locating measurement sensors).

Let us note that the diffusion coefficient values in the upper left of Ω are greater than those in the lower right. This means that the state changes during the system evolution are quicker when we move up and to the left (on the other hand, the system would have reached the steady state there earlier). This fact explains the form of the configurations obtained — the sensors switch to new positions so as to follow the moving source while measuring the state in the regions where the DPS is the most sensitive with respect to the unknown parameter κ, finally terminating the movement in the lower right part of the domain Ω.   ⬚

4.1.2 Scanning sensor scheduling as a constrained optimal control problem

Although the approach based on clusterization-free designs proposed in the preceding section turns out to be extremely efficient in practice, its main limitation is that it can be used only when the number of sensors is relatively high [329]. If the number of sensors is too low to justify this method, we have to return to operating on sensor positions. But then the combinatorial nature of such a formulation creates a confounded nuisance. It is compounded further if the sensor switchings are allowed to take place in continous time, i.e., the switching moments are not fixed arbitrarily. Additionally, it appears that specialized software must be employed for numerical solution of practical sensor placement problems.

In [153] a similar problem was considered in the context of state estimation. The authors proposed an approach making use of some recent results on discrete-valued optimal-control problems. By applying the transformation set

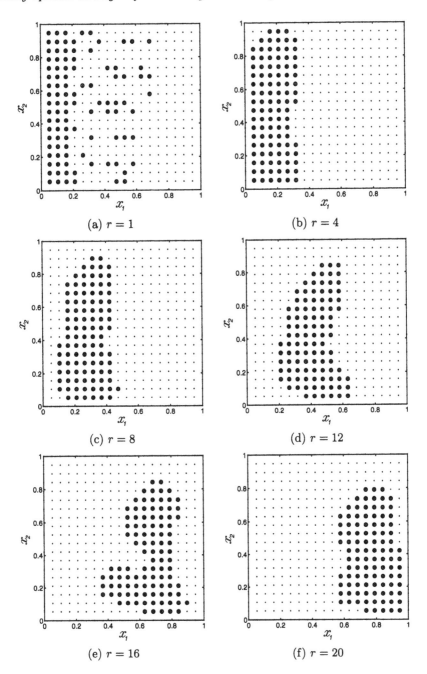

FIGURE 4.1
D-optimal clusterization-free configurations of scanning sensors at selected time subintervals, cf. Example 4.1.

forth in [154] they showed that the original discrete-valued optimal-control problem with variable switching times can be converted into an equivalent continuous-valued optimal-control problem which can then be solved using readily available nonlinear programming solvers. It what follows, it is shown how to adapt that approach to determination of optimal switching schedules for parameter estimation. This does not constitute a trivial task, as the specific features of the sensor-location problems for state and parameter estimation are rather different.

4.1.2.1 Optimal sensor scheduling problem

Suppose that there are N stationary sensors located at given spatial points $x^1, \ldots, x^N \in \Omega \cup \partial\Omega$ and, for simplicity, assume that at a given time moment only one sensor may be active while the others remain dormant. Then the sensor activation schedule can be represented by a function [153]

$$u : T = [0, t_f] \rightarrow \Lambda = \{1, \ldots, N\}. \tag{4.64}$$

In particular, $u(t) = j$ means that only the j-th sensor is active at time t. Therefore the set of admissible sensor schedules $\mathcal{U}$ consists of all measurable functions of the form (4.64). The currently active sensor takes continuous-time noise-corrupted measurements of the state, which can formally be represented as

$$z_{\mathrm{m}}(t) = \sum_{j=1}^{N} \chi_{\{u(t)=j\}}(t) \big[y(x^j, t; \theta) + \varepsilon^j(t) \big] \tag{4.65}$$

for $t \in T$, where

$$\chi_{\{u(t)=j\}}(t) = \begin{cases} 1 & \text{if } u(t) = j, \\ 0 & \text{otherwise,} \end{cases} \tag{4.66}$$

and $\varepsilon^j(\cdot)$ denotes the measurement noise assumed to be zero-mean, Gaussian and uncorrelated in both time and space.

The accuracy of the least-squares estimates of θ is then quantified via the average FIM

$$M(u) = \frac{1}{t_f} \sum_{j=1}^{N} \int_0^{t_f} \chi_{\{u(t)=j\}}(t) g(x^j, t) g^{\mathsf{T}}(x^j, t) \, \mathrm{d}t. \tag{4.67}$$

The optimum sensor scheduling problem is to find $u \in \mathcal{U}$ to minimize $\Psi[M(u)]$.

4.1.3 Equivalent Mayer formulation

Note that the foregoing sensor activation problem can be cast as a discrete-valued optimal-control problem in Mayer form through defining the quantity

$$\Pi(t) = \frac{1}{t_f} \sum_{j=1}^{N} \int_0^t \chi_{\{u(\tau)=j\}}(\tau) g(x^j, \tau) g^{\mathsf{T}}(x^j, \tau) \, \mathrm{d}\tau. \tag{4.68}$$

Then we get

$$M(u) = \Pi(t_f), \tag{4.69}$$

so that the sought $u^* \in \mathcal{U}$ minimizing $\Psi[M(u)]$ also solves the following problem: Minimize the performance index

$$J(u) = \Psi[\Pi(t_f)], \tag{4.70}$$

subject to the constraint in the form of the system of nonlinear ODEs:

$$\begin{cases} \dfrac{d}{dt}\Pi(t) = \dfrac{1}{t_f} \sum_{j=1}^{N} \chi_{\{u(t)=j\}}(t) g(x^j, t) g^{\mathsf{T}}(x^j, t), \\ \Pi(0) = 0. \end{cases} \tag{4.71}$$

This is an optimal-control problem in which the main difficulty stems from the fact that the set of admissible control values is discrete and hence not convex. Furthermore, choosing the appropriate elements from the control set in an appropriate order is, in fact, a nonlinear combinatorial optimization problem.

4.1.4 Computational procedure based on the control parameterization-enhancing technique

One solution to settle the above discrete optimal-control problem is to reduce it to that of determining the switching points of the discrete-valued control profile, but this may lead to severe numerical problems since the gradient of the performance index with respect to the switching times is discontinuous, the number of decision variables changes whenever adjacent switching times coalesce, and the involved piecewise integration over intervals with varying end-points is rather difficult to perform. A problem transformation called the Control Parameterization-Enhancing Technique (CPET) was proposed in [153, 154] to address these difficulties. Under the CPET, the switching points are mapped onto the integers, and the transformed problem is just an ordinary optimal-control problem with known and fixed switching points. It can then be readily solved numerically by many existing numerical procedures.

As in [153], where the CPET was employed to construct an optimal schedule for finding the most accurate estimates of the system state, we introduce a new time scale τ which varies from 0 to $Q = \kappa N^2$, where κ is an assumed upper bound imposed on the number of times each sensor can be selected. Let $\mathcal{V}$ denote the class of nonnegative piecewise constant real-valued functions defined on $[0, Q]$ with fixed knots located at $\{1, \ldots, Q - 1\}$. The CPET transformation from $t \in [0, t_f]$ to $\tau \in [0, Q]$ is defined by the differential equation

$$\frac{dt}{d\tau} = v(\tau), \quad t(0) = 0, \tag{4.72}$$

where the scalar function $v \in \mathcal{V}$ is called the enhancing control. Observe that we have

$$\int_0^Q v(\tau)\,\mathrm{d}\tau = t_f. \tag{4.73}$$

Furthermore, we introduce a fixed function $\mu : [0, Q] \to \Lambda$,

$$\mu(\tau) = (i \bmod N) + 1, \quad \tau \in [i, i+1) \tag{4.74}$$

for $i = 0, 1, \dots, Q - 1$. The idea of this transformation is to let any $u \in \mathcal{U}$ be naturally represented by a $v \in \mathcal{V}$. Substituting (4.72) into (4.70) and (4.71) while setting $P(\tau) = \Pi(t(\tau))$, we obtain the following problem: Choose $v \in \mathcal{V}$ to minimize

$$\mathcal{J}(v) = \Psi[P(Q)], \tag{4.75}$$

subject to

$$\begin{cases} \dfrac{\mathrm{d}}{\mathrm{d}\tau} P(\tau) = v(\tau) \dfrac{1}{t_f} \displaystyle\sum_{j=1}^{N} \chi_{\{\mu(\tau)=j\}}(\tau) g(x^j, t(\tau)) g^{\mathsf{T}}(x^j, t(\tau)), \\[2mm] P(0) = 0. \end{cases} \tag{4.76}$$

and (4.73). It can be easily proven that the original and reformulated problems are equivalent provided that in the optimal solution of the original problem the maximal number of selections (i.e., activations) of any sensor $j \in \Lambda$ does not exceed the assumed number κ.

The reformulated problem (4.75), (4.76) and (4.73) can be solved with relative ease, as the switching points of the original problem are mapped into the set of integers in chronological order. Piecewise integration can now be easily performed since discontinuity points in the τ-domain are known and fixed.

As for a practical solution of this problem, note that controls v are piecewise constant, i.e., $v(\tau) = v_i$ if $\tau \in [i, i+1)$ for $i = 0, 1, \dots, Q-1$. Consequently, the only unknowns to be determined are reals $v_0, \dots, v_{Q-1}$. Their values should be selected so that they are nonnegative and sum up to t_f while simultaneously minimizing $\mathcal{J}(v)$ subject to (4.76). But this problem can be solved using any nonlinear programming routine for minimization of functions subject to linear constraints.

The delineated technique can be easily extended to the case of several scanning sensors. The only change we have to make is the definition of Λ. Specifically, if n sensors are to be used, there are altogether $N' = \binom{N}{n}$ ways of activating n sensors from among a total of N sensors. Then we can define $\Lambda = \{1, \dots, N'\}$ and the meaning of each $j \in \Lambda$ is to be understood as one of the resulting sensor combinations.

Another question is the determination of an optimal number of switchings, as in the above method a fixed maximal number of switchings κ is assumed. A heuristic method proposed in [154] to address this problem is as follows: First, the sensor scheduling problem is solved for an arbitrarily guessed κ. Then

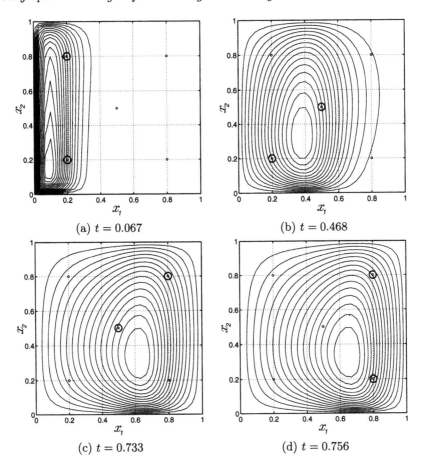

(a) $t = 0.067$ (b) $t = 0.468$

(c) $t = 0.733$ (d) $t = 0.756$

FIGURE 4.2
Optimal sensor activations versus contour plots of the state in Example 4.2.

the maximal number of switchings is incremented by one and the scheduling problem is solved again. If there is no improvement in the optimal cost, the former value of κ is assumed as the optimal number of switchings, otherwise the procedure of incrementing κ and computing the corresponding switching schedule is repeated over and over.

Example 4.2

The settings of Example 4.1 served as a test for the delineated algorithm. Five sensors were assumed to be placed at points listed in Table 4.1 and the purpose was to find their D-optimal activations such that exactly two

TABLE 4.1

Scanning sensor locations in Example 4.2

Sensor	S_1	S_2	S_3	S_4	S_5
Coordinates	$(0.2, 0.2)$	$(0.2, 0.8)$	$(0.5, 0.5)$	$(0.8, 0.2)$	$(0.8, 0.8)$

TABLE 4.2

Two-element subsets of the set $\{S_1, \ldots, S_5\}$ and the
corresponding control values

Combination of active sensors	Control signal value
$\{S_1, S_2\}$	1
$\{S_1, S_3\}$	2
$\{S_1, S_4\}$	3
$\{S_1, S_5\}$	4
$\{S_2, S_3\}$	5
$\{S_2, S_4\}$	6
$\{S_2, S_5\}$	7
$\{S_3, S_4\}$	8
$\{S_3, S_5\}$	9
$\{S_4, S_5\}$	10

sensors would be activated at each time instant. In order to apply the CPET approach, the individual two-element combinations of the five sensors were coded as admissible control values shown in Table 4.2.

The sensitivity coefficients were determined using MATLAB's PDE Toolbox (30 divisions along the time interval, a 30 × 30 rectangular, structured spatial mesh) on a PC equipped with Pentium IV 1.7 GHz, 768 MB RAM running Windows 2000. The CPET algorithm was implemented entirely in MATLAB. The system of ODEs (4.76) was solved by the **ode45** routine from the ODE suite, whereas the optimization process was performed using the **fmincon** procedure from the Optimization Toolbox.

The initial switching schedule was generated at random, and the maximal allowable number of sensor switchings was set as $\kappa = 2$. After six minutes, the following optimal switching schedule was obtained:

$$u^\star(t) = \begin{cases} 1 & \text{if } 0.000 \leq t < 0.468, \\ 2 & \text{if } 0.468 \leq t < 0.733, \\ 9 & \text{if } 0.733 \leq t < 0.756, \\ 10 & \text{if } 0.756 \leq t \leq 1.000. \end{cases}$$

The results conform to those obtained in Example 4.1, which is clearly evidenced in Fig. 4.2, where optimal pairs of activated sensors are indicated at successive time moments using open circles on the contour plots of the state variable y. ▯

4.2 Adapting the idea of continuous designs for moving sensors

The very first idea, which suggests itself while attempting to address the problem of how to construct optimal sensor trajectories, is to establish a connection between this problem and the Kiefer-Wolfowitz theory of experimental design for regression problems. It goes without saying that its implementation is not straightforward as the main difficulty lies in the necessity of operating on mappings with values being Radon probability measures on the Borel sets of a given admissible compact set instead of the trajectories themselves, but nevertheless this can still be achieved and it is shown in what follows.

4.2.1 Optimal time-dependent measures

The approach outlined below was originally developed in [327] and is based on the idea reported in [238] which has already become a classical reference work on moving sensors. As before, we will denote by $y = y(x, t; \theta)$ the scalar state of a DPS at a spatial point $x \in \bar{\Omega}$ (the closure of Ω) and time $t \in T = [0, t_f]$, which depends on a vector $\theta \in \mathbb{R}^m$ of unknown constant parameters. Furthermore, we will use the letter X to denote a compact set in which the observations of y can be made. Our main aim here is to study the optimal measurement scheduling problem for estimating θ in the case when the available observations are provided by N moving pointwise sensors, namely

$$z_m^j(t) = y(x^j(t), t; \theta) + \varepsilon(x^j(t), t), \quad t \in T, \quad j = 1, \ldots, N, \qquad (4.77)$$

where z_m^j is a scalar output, x^j stands for an observation curve (measurable in general) for the j-th sensor, so that

$$x^j(t) \in X \quad \text{a.e. on } T. \qquad (4.78)$$

Here ε signifies an additive disturbance being a realization of a white Gaussian random field whose statistics are given by

$$\mathrm{E}\{\varepsilon(x, t)\} = 0, \quad \mathrm{E}\{\varepsilon(x, t)\varepsilon(x', t')\} = \sigma^2 \delta(x - x')\delta(t - t'). \qquad (4.79)$$

Our basic assumption is that the function $y(x, t; \cdot)$ is continuously differentiable in a neighbourhood of some known preliminary estimate θ^0 of θ. We then define

$$g(x, t) = \left(\frac{\partial y(x, t; \theta)}{\partial \theta} \right)^{\mathsf{T}}_{\theta = \theta^0} \qquad (4.80)$$

and require g to be continuous in $\bar{Q} = \bar{\Omega} \times T$.

On the above assumptions, the average FIM is of the form

$$M = \frac{1}{Nt_f} \sum_{j=1}^{N} \int_0^{t_f} g(x^j(t), t) g^\mathsf{T}(x^j(t), t) \, dt. \tag{4.81}$$

The independence of the measurements made by different sensors implies, however, that at some time moments we admit clusterization, i.e., several sensors may measure the system state at the same points. We take account of this phenomenon through relabelling the sensors so that $x^i(t) \neq x^j(t)$ if $i \neq j$, where $1 \leq i, j \leq \ell(t)$, $\ell(t)$ being the number of sensors which are at different locations at time $t \in T$. Consequently, at a given time moment we may introduce the so-called exact design

$$\xi_N(t) = \begin{Bmatrix} x^1(t), \, x^2(t), \, \ldots, \, x^{\ell(t)}(t) \\ p_1(t), \, p_2(t), \, \ldots, \, p_{\ell(t)}(t) \end{Bmatrix}, \tag{4.82}$$

where $p_i(t) = r_i(t)/N$ and $r_i(t)$ denotes the number of sensors which occupy the position $x^i(t)$. In terms of this modified notation, the FIM is

$$M(\xi_N) = \frac{1}{t_f} \int_0^{t_f} \left\{ \sum_{j=1}^{\ell(t)} p_i(t) g(x^i(t), t) g^\mathsf{T}(x^i(t), t) \right\} dt. \tag{4.83}$$

In much the same way as in Section 3.1, we may then extend the notion of the exact design to a more general concept of a randomized design which is to be understood as a mapping

$$\xi : T \ni t \mapsto \xi_t \in \Xi_t(X), \tag{4.84}$$

where $\Xi_t(X)$ is the set of all probability measures for all Borel sets of X including single points. Clearly, some measurability conditions about the mapping ξ must be imposed, but this topic is beyond the scope of this monograph and the interested reader is referred, e.g., to [350], where a similar reasoning is conducted in the context of relaxed controls. In the sequel, $\Xi(X)$ denotes the set of all such mappings ξ.

It follows that the corresponding FIM is of the form

$$M(\xi) = \frac{1}{t_f} \int_0^{t_f} \int_X g(x, t) g^\mathsf{T}(x, t) \, \xi_t(dx) \, dt. \tag{4.85}$$

A Ψ-optimal design will be a design $\xi^\star$ such that $\Psi[M(\xi)]$ is minimized at $\xi^\star$. We assume that Ψ satisfies (A3)–(A5) of p. 41 and, in place of (A6), the following qualification:

(A6') For any $\xi \in \tilde{\Xi}(X) = \{\xi : \Psi[M(\xi)] \leq q < \infty\}$ and $\bar{\xi} \in \Xi(X)$, we have

$$\Psi[M(\xi) + \alpha(M(\bar{\xi}) - M(\xi))]$$

$$= \Psi[M(\xi)] + \alpha \frac{1}{t_f} \int_0^{t_f} \int_X \psi(x, t, \xi) \, \bar{\xi}_t(dx) \, dt + o(\alpha; \xi, \bar{\xi}), \tag{4.86}$$

where the scalar q is so chosen that $\widetilde{\widetilde{\Xi}}(X) \neq \emptyset$.

It turns out that (A6') is by no means restrictive, as for the differentiable criteria Ψ we have

$$
\begin{aligned}
\delta_+ & \Psi(M(\xi), M(\bar{\xi}) - M(\xi)) \\
&= \mathrm{tr}\left[\overset{\circ}{\Psi}(\xi)(M(\bar{\xi}) - M(\xi))\right] \\
&= \frac{1}{t_f} \int_0^{t_f} \int_X \left\{ g^\mathsf{T}(x,t)\overset{\circ}{\Psi}(\xi)g(x,t) - \mathrm{tr}\left[\overset{\circ}{\Psi}(\xi)M(\xi)\right] \right\} \bar{\xi}_t(\mathrm{d}x)\,\mathrm{d}t,
\end{aligned}
\tag{4.87}
$$

where

$$
\overset{\circ}{\Psi}(\xi) = \left.\frac{\partial\Psi(M)}{\partial M}\right|_{M=M(\xi)}.
$$

Hence

$$
\psi(x,t,\xi) = g^\mathsf{T}(x,t)\overset{\circ}{\Psi}(\xi)g(x,t) - \mathrm{tr}\left[\overset{\circ}{\Psi}(\xi)M(\xi)\right],
\tag{4.88}
$$

or alternatively

$$
\psi(x,t,\xi) = c(\xi) - \phi(x,t,\xi),
\tag{4.89}
$$

where

$$
\phi(x,t,\xi) = -g^\mathsf{T}(x,t)\overset{\circ}{\Psi}(\xi)g(x,t)
\tag{4.90}
$$

and

$$
c(\xi) = -\mathrm{tr}\left[\overset{\circ}{\Psi}(\xi)M(\xi)\right].
\tag{4.91}
$$

The next result provides a characterization of the optimal designs, cf. [327].

THEOREM 4.3
A design $\xi^\star$ is optimal iff

$$
\int_0^{t_f} \min_{x \in X} \psi(x,t,\xi^\star)\,\mathrm{d}t = 0.
\tag{4.92}
$$

PROOF From the convexity of Ψ it follows that a necessary and sufficient condition for optimality of $\xi^\star$ is

$$
\inf_{\bar{\xi}\in\Xi(X)} \delta_+\Psi(M(\xi^\star), M(\bar{\xi}) - M(\xi^\star)) \geq 0,
\tag{4.93}
$$

or equivalently

$$
\inf_{\bar{\xi}\in\Xi(X)} \delta_+\Psi(M(\xi^\star), M(\bar{\xi}) - M(\xi^\star)) = 0,
\tag{4.94}
$$

which is easy to check if we take $\bar{\xi} = \xi^\star$ on the left-hand side of (4.93).

If we set $x_0(t) = \arg \min\limits_{x \in X} \psi(x, t, \xi^\star)$, then

$$
\begin{aligned}
\int_0^{t_f} \psi(x_0(t), t, \xi^\star) \, dt &= \int_0^{t_f} \int_X \psi(x, t, \xi^\star) \delta(x - x_0(t)) \, dx \, dt \\
&\geq \inf_{\bar\xi} \int_0^{t_f} \int_X \psi(x, t, \xi^\star) \, \bar\xi_t(dx) \, dt \\
&\geq \inf_{\bar\xi} \int_0^{t_f} \int_X \min_{x \in X} \psi(x, t, \xi^\star) \, \bar\xi_t(dx) \, dt \\
&= \int_0^{t_f} \psi(x_0(t), t, \xi^\star) \, dt
\end{aligned}
\tag{4.95}
$$

and therefore

$$
\inf_{\bar\xi} \int_0^{t_f} \int_X \psi(x, t, \xi^\star) \, \bar\xi_t(dx) \, dt = \int_0^{t_f} \psi(x_0(t), t, \xi^\star) \, dt, \tag{4.96}
$$

which gives (4.92) when combined with (4.87) and (4.94). ∎

It is now a simple matter to deduce the respective form of the equivalence theorem:

COROLLARY 4.3
The following are equivalent:

(i) $\xi^\star$ minimizes $\Psi[M(\xi)]$,

(ii) $\xi^\star$ minimizes $\dfrac{1}{t_f} \displaystyle\int_0^{t_f} \max_{x \in X} \phi(x, t, \xi) \, dt - c(\xi)$, and

(iii) $\dfrac{1}{t_f} \displaystyle\int_0^{t_f} \max_{x \in X} \phi(x, t, \xi^\star) \, dt = c(\xi^\star)$.

This constitutes a generalization of Theorem 1 of [238] where only D-optimal designs were considered. In that case Corollary 4.3 takes a particularly simple form.

COROLLARY 4.4
Let $\xi^\star$ be a D-optimal design. Then the following are equivalent:

(i) $\xi^\star$ maximizes $\det M(\xi)$,

(ii) $\xi^\star$ minimizes $\dfrac{1}{t_f} \displaystyle\int_0^{t_f} \max_{x \in X} g^\mathsf{T}(x, t) M^{-1}(\xi) g(x, t) \, dt$, and

(iii) $\dfrac{1}{t_f} \displaystyle\int_0^{t_f} \max_{x \in X} g^{\mathsf{T}}(x,t) M^{-1}(\xi^{\star}) g(x,t)\, \mathrm{d}t = m.$

In [238] some sufficient conditions for optimality are further given (e.g., a *quasi-maximum* principle). Owing to their use, the computational task reduces to solving at each time moment a separate optimization problem reminiscent of the classical D-optimum experimental design problem for which many numerical algorithms exist. The idea is very elegant, but from a practical point of view it can be applied only in relatively simple situations, as the attendant calculations are very time-consuming. Furthermore, only measurability of the resulting trajectories can be guaranteed, which may cause some difficulties when trying to apply the solutions in the setting of a real process. These complications can be sometimes avoided by suitably parameterizing the trajectories. The dimension of the optimization problem is thus reduced and we may impose any regularity conditions on the solutions. This constitutes the subject of the next section.

4.2.2 Parameterization of sensor trajectories

From now on we make the assumption that the trajectories of the available sensors can be represented as parametric curves of the form

$$x^j(t) = \eta(t, \beta^j), \quad t \in T, \tag{4.97}$$

where η denotes a given function such that $\eta(\,\cdot\,, \beta^j)$ is continuous for each fixed β^j and $\eta(t, \,\cdot\,)$ is continuous for each fixed t, the constant parameter vector β^j ranging over a compact set $A \subset \mathbb{R}^p$. Since only the trajectories lying entirely in an admissible compact set X are interesting, we introduce the set

$$B = \{\beta \in A : \eta(t, \beta) \in X,\ \forall t \in T\} \tag{4.98}$$

and assume that it is nonempty. A trivial verification shows that B is also compact.

If there are N sensors at our disposal and they move along the paths (4.97), then the resulting average FIM can be written down as

$$M = \frac{1}{Nt_f} \sum_{j=1}^{N} \int_0^{t_f} g(\eta(t, \beta^j), t) g^{\mathsf{T}}(\eta(t, \beta^j), t)\, \mathrm{d}t. \tag{4.99}$$

Owing to the assumption of independent measurements, some trajectories may coincide and therefore we relabel the sensors so as to distinguish only $\ell \leq N$ different paths. Proceeding in this manner, we must also rewrite the FIM, which gives

$$M(\xi_N) = \sum_{i=1}^{\ell} p_i \left\{ \frac{1}{t_f} \int_0^{t_f} g(\eta(t, \beta^i), t) g^{\mathsf{T}}(\eta(t, \beta^i), t)\, \mathrm{d}t \right\}, \tag{4.100}$$

where

$$\xi_N = \left\{ \begin{matrix} \beta^1, \ \beta^2, \ \ldots, \ \beta^\ell \\ p_1, \ p_2, \ \ldots, \ p_\ell \end{matrix} \right\}, \tag{4.101}$$

$p_i = r_i/N$, r_i is the number of sensors which follow the i-th path.

Accordingly, the experimental setting ξ_N can be interpreted as a discrete probability distribution. Removing the restriction that p_i's are multiples of $1/N$, we can extend the idea and think of a design as a probability measure ξ for all Borel sets of B including single points. With respect to such a modification, we can define the FIM analogous to (4.100) for a design ξ:

$$M(\xi) = \int_B \Upsilon(\beta) \, \xi(\mathrm{d}\beta), \tag{4.102}$$

where

$$\Upsilon(\beta) = \frac{1}{t_f} \int_0^{t_f} g(\eta(t,\beta),t) g^\mathsf{T}(\eta(t,\beta),t) \, \mathrm{d}t. \tag{4.103}$$

The optimal design $\xi^\star$ is such that it minimizes the design criterion $\Psi[M(\xi)]$. It is easily seen that the form of this reformulated problem is practically the same as that of the main problem of Section 3.1, p. 38 (it suffices to replace x, X and $f(x,t)$ with β, B and $g(\eta(t,\beta),t)$, respectively). Consequently, the corresponding results are also valid here. For instance, the probability measure $\xi(\mathrm{d}\beta)$ can be chosen to be purely discrete, i.e., nonzero for a finite number of β values (strictly speaking, this number is guaranteed to be less than or equal to $m(m+1)/2$). Similarly, the numerical schemes briefly delineated at the end of Section 3.1 can be employed to find approximations to the optimal solutions.

Let us note, however, that the simplicity of the presented approach is only limited to the applied idea and the resulting computational burden may be still quite heavy. Indeed, as was already emphasized, the most cumbersome part of the algorithm for obtaining Ψ-optimal designs consists in repeatedly solving a global nonlinear programming problem with constraints. This problem can be settled with relative ease if stationary sensors are to be placed, as the dimension of the corresponding decision vector equals the number of space coordinates (i.e., 1, 2 or 3 in practical situations). In order for $\eta(\cdot,\beta)$ to be adjustable enough to approximate satisfactorily any trajectory x^j (i.e., to form a sufficiently rich family of feasible paths so as to be of any practical use), the size of β usually has to be much larger, which essentially complicates computations. But such a cost is unavoidable while attempting to increase the degree of freedom in solving any optimization problem (cf. parameter optimization problems versus optimization problems for dynamic systems).

4.3 Optimization of sensor trajectories based on optimal-control techniques

If the number of moving sensors is imposed *a priori* (this is often encountered in practice and results from high costs of such measurement equipment), the dynamics of the vehicles carrying the sensors must be taken into account and various geometric constraints are put on sensor movements (induced, e.g., by the admissible measurement regions and allowable distances between the sensors), then the only systematic and computationally tractable approach is to convert the problem to an optimal-control formulation and then to attempt to solve it numerically. Such an idea has already been successfully applied in the context of state estimation [39, 130, 187, 188], but those results can hardly be exploited in the framework considered here as those authors make extensive use of some specific features of the addressed problem (e.g., the linear dependence of the current state on the initial state for linear systems). To the best of our knowledge, no approaches have been proposed so far as regards this line in the context of parameter estimation. Our purpose here is thus to describe some original results concerning numerical methods for the off-line determination of moving sensor positions which maximize parameter-identification accuracy subject to various constraints imposed on the motions of the sensors. The technique employed is to transform the problem to an optimal-control one in which both the control forces of the sensors and initial sensor positions are optimized.

4.3.1 Statement of the problem and notation

4.3.1.1 Equations of sensor motion

For simplicity of notation, let us write

$$s(t) = (x^1(t), x^2(t), \ldots, x^N(t)), \quad \forall t \in T = [0, t_f] \qquad (4.104)$$

and set $n = \dim s(t)$. We assume that the sensors are conveyed by vehicles which are described by equations of motion of the form

$$\dot{s}(t) = f(s(t), u(t)) \quad \text{a.e. on } T, \quad s(0) = s_0, \qquad (4.105)$$

where a given function $f : \mathbb{R}^n \times \mathbb{R}^r \to \mathbb{R}^n$ is required to be continuously differentiable, $s_0 \in \mathbb{R}^n$ defines an initial sensor configuration, and $u : T \to \mathbb{R}^r$ is a measurable control function which satisfies

$$u_l \le u(t) \le u_u \quad \text{a.e. on } T \qquad (4.106)$$

for some constant vectors u_l and u_u. The last condition is quite sensible as in general the controls are limited for technical and/or economic reasons (strictly speaking, it is also of paramount importance in proving the existence of a solution to the optimal control problem described subsequently).

Given any initial sensor configuration s_0 and any control function, there is a unique, absolutely continuous function $s : T \to \mathbb{R}^n$ which satisfies (4.105) a.e. on T. In what follows, we will call it the state trajectory corresponding to s_0 and u. Various particular choices are proposed for the 'state' equation (4.105), including the following:

- first-order linear equation [130]

$$\dot{s}(t) = C(t)s(t) + D(t)u(t), \quad s(0) = s_0,$$

  where C and D are (continuous) matrices,

- second-order linear equation [39]

$$E\ddot{s}(t) + F\dot{s}(t) = Gu(t), \quad \dot{s}(0) = 0, \quad s(0) = s_0,$$

  where E and F are diagonal matrices, $E \succ 0$, $F \succeq 0$ (this case reduces to (4.105) after extending the state vector and an obvious change of variables), and

- the case where we do not attach importance to the dynamics of the vehicles carrying the sensors and the only interest is in the trajectories themselves [187]:

$$\dot{s}(t) = u(t), \quad s(0) = s_0.$$

REMARK 4.2   Clearly, if the sensor dynamics is not of primary concern, in lieu of the last model above, we might simply use the description

$$s(t) = u(t),$$

i.e., directly optimize sensor positions. Note, however, that such an approach necessitates additional assumptions about the regularity of s. Moreover, some supplementary constraints should also be introduced to guarantee a proper mathematical formulation and, as a consequence, the existence of solutions (e.g., constraints on the maximal lengths of the trajectories or on maximal speeds of sensor movements). In such a case, computational methods of calculus of variations can be exploited in order to find optimal trajectories. The optimal-control approach outlined in what follows is beyond doubt more general and flexible, and therefore we shall restrict our discussion only to this topic.   ∎

4.3.1.2   Induced pathwise state inequality constraints

If we intend to design sensor movements for a real application, some restrictions on the motions have to be inevitably included in our optimal-control formulation. First of all, all sensors should stay within an admissible region X (a given compact set) where measurements can be made. In what follows, it is convenient to choose a general form

$$X = \left\{ x \in \bar{\Omega} : b_i(x) \leq 0,\ i = 1, \ldots, I \right\}, \qquad (4.107)$$

where b_i's are given continuously differentiable functions. Accordingly, the conditions

$$\alpha_{ij}(s(t)) = b_i(x^j(t)) \leq 0, \quad \forall t \in T \qquad (4.108)$$

must be fulfilled, where $1 \leq i \leq I$ and $1 \leq j \leq N$.

Furthermore, to alleviate problems with sensor clusterization, we introduce constraints to restrict the admissible distances between the sensors. In the present approach, they are of the form

$$\beta_{ij}(s(t)) = R^2 - \|x^i(t) - x^j(t)\|^2 \leq 0, \quad \forall t \in T, \qquad (4.109)$$

where $1 \leq i < j \leq N$ and R stands for a minimum allowable distance which guarantees that the measurements taken by the sensors can be considered as independent.

To shorten notation, after relabelling, we rewrite constraints (4.108) and (4.109) in the form

$$\gamma_\ell(s(t)) \leq 0, \quad \forall t \in T, \qquad (4.110)$$

where γ_ℓ, $\ell = 1, \ldots, IN$ tally with (4.108), whereas γ_ℓ, $\ell = IN + 1, \ldots, [I + (N-1)/2]N$ coincide with (4.109). In the sequel, $\bar{\nu}$ stands for the set of indices $\{1, \ldots, \nu\}$, $\nu = [I + (N-1)/2]N$.

4.3.1.3 Optimal measurement problem

The goal in the optimal measurement problem is to determine the forces (controls) applied to each vehicle conveying a sensor, which minimize a design criterion $\Psi[M(s)]$ defined on the set of all real-valued information matrices of the form

$$M(s) = \frac{1}{Nt_f} \sum_{j=1}^{N} \int_0^{t_f} g(x^j(t), t) g^{\mathsf{T}}(x^j(t), t)\, \mathrm{d}t, \qquad (4.111)$$

where g is defined in (4.80), under the constraints (4.106) on the magnitude of the controls and induced state constraints (4.110). In order to increase the degree of optimality, in our approach we will regard s_0 as a control parameter vector to be chosen in addition to the control function u.

Evidently, in order to guarantee the correctness of such a formulation and further derivations, it is necessary to put some restrictions on the sensitivity coefficients g. In the remainder of this chapter, we require g and $\partial g/\partial x$ to be continuous functions.

Since sensor trajectories s are unequivocally determined as solutions to the state equation (4.105), the above control problem can be interpreted as an optimization problem over the set of feasible pairs

$$\mathcal{P} = \big\{(s_0, u) : s_0 \in X^N, \, u : T \to \mathbb{R}^r \text{ is measurable,}$$
$$u_l \le u(t) \le u_u \text{ a.e. on } T\big\}, \quad (4.112)$$

Because of this, here and subsequently we will also make the following notational convention: if s appears without mention in a formula, it is always understood that a control u and initial condition s_0 have been specified and s is the trajectory corresponding to u and s_0 through (4.105).

Consequently, we wish to solve the following problem:

$$\min_{(s_0, u) \in \mathcal{P}} J(s_0, u) \qquad (4.113)$$

subject to the inequality constraint

$$h(s_0, u) \le 0, \qquad (4.114)$$

where

$$J(s_0, u) = \Psi[M(s)], \qquad (4.115)$$
$$h(s_0, u) = \max_{(\ell, t) \in \bar{\nu} \times T} \big\{ \gamma_\ell(s(t)) \big\}. \qquad (4.116)$$

Clearly, this highly nonlinear problem is very complicated and we are not capable of finding closed-form formulae for its solution. Accordingly, we must resort to numerical techniques. A number of possibilities exist in this respect [105, 215], but prior to the presentation of a pertinent method, let us notice that in spite of its nonclassical form the resulting optimal-control problem can be easily cast as a classical Mayer problem where the performance index is defined only via terminal values of state variables [91].

4.3.2 Equivalent Mayer problem and existence results

To set forth our basic idea, define first the quantities

$$\chi_{ij}(s(t), t) = \frac{1}{Nt_f} \sum_{\ell=1}^{N} g_i(x^\ell(t), t) g_j(x^\ell(t), t) \qquad (4.117)$$

and then the matrix $\Pi(t) = \{\varpi_{ij}(t)\} \in \mathbb{R}^{m \times m}$ with components

$$\varpi_{ij}(t) = \int_0^t \chi_{ij}(s(\tau), \tau) \, d\tau \qquad (4.118)$$

for $1 \le i, j \le m$. This clearly forces

$$M(s) = \Pi(t_f) \qquad (4.119)$$

and

$$\dot{\varpi}_{ij}(t) = \chi_{ij}(s(t), t), \quad \forall t \in T. \tag{4.120}$$

Hence introducing

$$v(t) = \begin{pmatrix} s(t) \\ \varpi_{11}(t) \\ \vdots \\ \varpi_{m1}(t) \\ \varpi_{12}(t) \\ \vdots \\ \varpi_{mm}(t) \end{pmatrix}, \quad v_0 = \begin{pmatrix} s_0 \\ 0 \\ \vdots \\ 0 \\ 0 \\ \vdots \\ 0 \end{pmatrix}, \quad F(v(t), u(t), t) = \begin{pmatrix} f(s(t), u(t)) \\ \chi_{11}(s(t), t) \\ \vdots \\ \chi_{m1}(s(t), t) \\ \chi_{12}(s(t), t) \\ \vdots \\ \chi_{mm}(s(t), t) \end{pmatrix} \tag{4.121}$$

yields

$$\dot{v}(t) = F(v(t), u(t), t) \quad \text{a.e. on } T, \quad v(0) = v_0, \tag{4.122}$$

which allows us to treat v and (4.122) as a new extended state vector and a new state equation, respectively. Accordingly, the optimal control problem (4.113),(4.114) can be rewritten as

$$\min_{(v_0, u) \in \bar{\mathcal{P}}} \bar{J}(v_0, u) \tag{4.123}$$

subject to

$$\bar{h}(v_0, u) \leq 0, \tag{4.124}$$

where

$$\bar{\mathcal{P}} = \left\{ (v_0, u) : v_0 = (v_{01}, \ldots, v_{0n}, \underbrace{0, \ldots, 0}_{m^2 \text{ times}}), \ (v_{01}, \ldots, v_{0n}, u) \in \mathcal{P} \right\},$$

$$\bar{J}(v_0, u) = G(v(t_f)) \overset{\text{def}}{=} \Psi(\Pi(t_f)), \quad \bar{h}(v_0, u) = h(v_{01}, \ldots, v_{0n}, u).$$

In this way, we are faced with a Mayer form of the performance index, which leads to a standard problem studied extensively in most works on optimal control in the presence of state inequality constraints. Moreover, a basic assertion is that the Mayer problem is equivalent to the Lagrange and Bolza ones in that each can be formulated as one of the other forms [91].

At this juncture, it should be underlined that optimal control problems with state-variable inequality constraints are not easy to solve and even the theory is not unambiguous, since there exist various forms of the necessary and sufficient optimality conditions [109]. On rather strong regularity assumptions, standard existence theorems only provide the existence of an optimal *measurable* control. In our case such assumptions are not satisfied and measurable optimal solutions may fail to exist. This is because the set of admissible controls is not sequentially compact with respect to the L^∞ norm. A usual remedy to this predicament is to embed the considered set of controls into some larger topological space in which its closure is sequentially compact. This closure is usually called the class of *relaxed* (or *generalized*) controls [290]. Relaxed optimal controls always exist in the setting of our Mayer problem (for details, see [109, 228, 229]).

4.3.3 Linearization of the optimal-control problem

Since at each iteration of the numerical algorithm delineated in the next section improvements of the current approximations to the optimal initial sensor configuration and optimal controls are calculated by solving an optimization problem in which the sensor dynamics, performance index and constraint functional are replaced by their first-order approximations around the current pair (s_0, u), in what follows we give some details about the technique of such a linearization.

Let us consider an initial state s_0 and a control u which is admissible in the sense of satisfying (4.106). The corresponding state vector is denoted by s. We assume that u is perturbated by a small function (variation) $\delta u \in L^\infty(T; \mathbb{R}^r)$ and s_0 is perturbated by a small vector $\delta s_0 \in \mathbb{R}^n$ such that $\|\delta u\|_{L^\infty(T;\mathbb{R}^r)}$ and $\|\delta s_0\|_{\ell_n^\infty}$ are sufficiently small, which warrants the correctness of the presented method. To shorten notation, here and subsequently, we write $\|\cdot\|$ instead of $\|\cdot\|_{L^\infty(T;\mathbb{R}^r)}$ and $\|\cdot\|_{\ell_n^\infty}$ when no confusion can arise.

From Equation (4.105) which relates s to s_0 and u, we obtain the variational time-varying linear tangent system

$$\delta \dot{s}(t) = f_s(t)\delta s(t) + f_u(t)\delta u(t), \quad \delta s(0) = \delta s_0, \tag{4.125}$$

where

$$f_s(t) = \left(\frac{\partial f}{\partial s}\right)_{\substack{s=s(t)\\u=u(t)}}, \quad f_u(t) = \left(\frac{\partial f}{\partial u}\right)_{\substack{s=s(t)\\u=u(t)}},$$

which relates variations in s to variations in s_0 and u.

Based on the derivations presented in Appendix H.1, it may be concluded that the Fréchet differential of J at s_0 and u with increments δs_0 and δu, respectively, (the first variation of J due to variations in s_0 and u) is of the form

$$\delta J(s_0, u; \delta s_0, \delta u) = \langle \zeta(0), \delta s_0 \rangle + \int_0^{t_f} \langle f_u^\mathsf{T}(t)\zeta(t), \delta u(t) \rangle \, dt, \tag{4.126}$$

where the adjoint mapping ζ solves the Cauchy problem

$$\dot{\zeta}(t) + f_s^\mathsf{T}(t)\zeta(t) = -\sum_{i=1}^m \sum_{j=1}^m c_{ij}\left(\frac{\partial \chi_{ij}}{\partial s}\right)_{s=s(t)}^\mathsf{T}, \quad \zeta(t_f) = 0. \tag{4.127}$$

Here $\langle\,\cdot\,,\,\cdot\,\rangle$ stands for the inner product in the appropriate Euclidean space, χ_{ij}'s are defined in (4.117), and c_{ij}'s are the components of the matrix

$$\overset{\circ}{\Psi}(s) = \{c_{ij}\}_{m\times m} = \left.\frac{\partial\Psi[M]}{\partial M}\right|_{M=M(s)}. \qquad (4.128)$$

As for the state inequality constraints (4.114), owing to the fact that the functional h is not Fréchet differentiable (this is because the max function is nondifferentiable), in order to approximate its increments, we resort to the notion of the Gâteaux differential which is of the form (see Appendix H.2)

$$\delta h(s_0^0, u^0; \delta s_0, \delta u)$$
$$= \max_{(\ell,t)\in S}\left\{\langle\zeta_h^\ell(0;t),\delta s_0\rangle + \int_0^{t_f}\langle f_u^\mathsf{T}(\tau)\zeta_h^\ell(\tau;t),\delta u(\tau)\rangle\,d\tau\right\}, \qquad (4.129)$$

where $S = \{(\ell,t)\in\bar{\nu}\times T : \gamma_\ell(s(t)) = h(s_0,u)\}$ and $\zeta_h^\ell(\,\cdot\,;t)$ is the solution to the Cauchy problem

$$\frac{d\zeta_h^\ell(\tau;t)}{d\tau} + f_x^\mathsf{T}(\tau)\zeta_h^\ell(\tau;t) = -\left(\frac{\partial\gamma_\ell}{\partial s}\right)_{s=s(\tau)}^\mathsf{T}\delta(\tau-t), \quad \zeta_h^\ell(t_f;t) = 0. \quad (4.130)$$

4.3.4 A numerical technique for solving the optimal measurement problem

Owing to the complexity of the problem of minimizing the performance index (4.113) subject to pathwise inequality constraints (4.114), we have to resort to numerical techniques. Luckily, there exist numerous methods which can be exploited here, as the problem is frequently encountered in applications, e.g., in mechanics, aerospace engineering, econometrics or robotics [33,34,109]. In this context, we distinguish direct and indirect methods [164, 264]. With *direct* methods the optimal-control problem is treated directly as a minimization problem, i.e., the method is started with an initial approximation to the solution, which is improved iteratively by minimizing the performance index along a search direction, which is obtained via linearization of the problem. State constraints are often treated via a penalty function approach, i.e., a term which is a measure for the violation of the state constraints is added to the performance index. With *indirect* methods the optimality conditions, which must hold for a solution to the optimal-control problem, are used to derive a multipoint boundary-value problem. Solutions to the optimal-control problem will also be solutions to this multipoint boundary-value problem and hence the numerical solution to the multipoint boundary-value problem yields a candidate for the solution to the optimal-control problem [164]. The main drawback to indirect methods is their extreme lack of robustness: the iterations of an indirect method must start close, sometimes very close, to a local solution in order to solve the pertinent boundary-value problem. Additionally, since first-order optimality conditions are satisfied by maximizers and saddle points as well as minimizers, there is no reason, in general, to expect solutions obtained by indirect methods to be minimizers.

A survey and comparison of numerical methods for state-constrained optimal-control problems can be found, e.g., in [33, 34, 116, 164, 215, 227]. As regards our measurement problem, in what follows we adopt a relatively unfamiliar method delineated long ago [68, 81], but implemented only recently [93–95] for planning optimal motions of redundant manipulators. It is based on the so-called *negative formulation* of the Pontryagin Maximum Principle and leads to an iterative algorithm for improving estimates of the control histories u and initial states s_0 so as to decrease the value of the performance index J and to satisfy the imposed control and state constraints. Each iteration amounts, in turn, to linearization of the problem in the vicinity of the control approximation from the previous step and then solving the resulting linear-programming problem to modify the solution until a desired accuracy is achieved. It appears that this procedure is extremely suited for numerically solving the sensor-location problem formulated in Section 4.3.1, as the Fréchet derivative of the performance index can be determined with reasonable computational burden and the state inequality constraints are taken into account by the method with relative ease, especially those induced by the conditions of preserving safe distances between the sensors (in the jargon of robotics, these are collision-avoidance conditions with moving obstacles).

The method is very similar to the feasible-direction algorithm proposed and thoroughly studied by Pytlak [227–229], as both the methods originate from ideas given by Fedorenko. They also share the same characteristics which promote their efficient implementation, e.g., that the improvements generated by the algorithms drive state trajectories into the interior of the state constraint region [228]. Moreover, as opposed to the penalty-function method, the method presented here does not require knowledge of an initial solution satisfying the constraint (4.114), which makes the implementation much easier.

For properly selected variations δs_0 and δu, the differentials of J and h can approximate the exact increments of these functionals with any desired accuracy. Given an initial state s_0^0 and a control u^0 satisfying (4.106), consider now the linearized problem: Find δs_0^1 and δu^1 to minimize the truncated functional

$$J(s_0^0, u^0) + \delta J(s_0^0, u^0; \delta s_0^1, \delta u^1) \quad (\approx J(s_0^0 + \delta s_0^1, u^0 + \delta u^1)) \qquad (4.131)$$

subject to the constraints

$$\begin{cases} h(s_0^0, u^0) + \delta h(s_0^0, u^0; \delta s_0^1, \delta u^1) \leq 0, \\ u_l \leq u^0 + \delta u^1 \leq u_u, \\ \|\delta u^1\| \leq \rho, \quad \|\delta s_0^1\| \leq \eta, \end{cases} \qquad (4.132)$$

where ρ and η are sufficiently small positive numbers.

According to the *negative formulation of the Pontryagin Maximum Principle* [81,93], the assumption of nonoptimality of s_0^0 and u^0 implies the existence of an initial state $s_0^0 + \delta s_0^1$ and a control $u^0 + \delta u^1$ for (4.131), (4.132) such that $J(s_0^0 + \delta s_0^0, u^0 + \delta u^1) < J(s_0^0, u^0)$. A new initial state $s_0^1 = s_0^0 + \delta s_0^0$ and a new control $u^1 = u^0 + \delta u^1$ result in this way. The process of minimization is then rerun for s_0^1 and u^1 instead of s_0^0 and u^0, respectively. This procedure of linearization and minimization is thus repeated over and over. Sequences of pairs $\{(s_0^k, u^k)\}$ and the corresponding state trajectories $\{s^k\}$ are thus obtained. It is known that $\{s^k\}$ is convergent (strictly speaking, it possesses a uniformly convergent subsequence). The corresponding proof proceeds on the same lines as in [93].

As regards computational aspects of the problem (4.131), (4.132), its finite-dimensional approximation leads to a very effective and simple numerical procedure. The process of approximation can be accomplished by forming a partition on T by choosing points $t_k = kt_f/K$, $k = 0, 1, \ldots, K$ and then considering u^0 and δu^1 in the class of piecewise linear polynomials, i.e., we take

$$u^0(t) = \sum_{k=0}^{K} u_k^0 \varphi_k(t), \quad \delta u^1(t) = \sum_{k=0}^{K} \delta u_k^1 \varphi_k(t), \tag{4.133}$$

where the φ_k's can be, e.g., piecewise linear spline basis functions. Accordingly, the problem of determining the variations δs_0^1 and δu^1 reduces to solving the finite-dimensional linear-programming problem:

$$\mathfrak{J}(\delta s_0^1, \delta u_0^1, \ldots, \delta u_K^1) = \langle \zeta(0), \delta s_0^1 \rangle$$
$$+ \sum_{k=0}^{K} \delta u_k^1 \int_0^{t_f} \langle f_u^{\mathsf{T}}(t)\zeta(t), \varphi_k(t) \rangle \, dt \longrightarrow \min \tag{4.134}$$

subject to

$$h(s_0^0, u^0) + \langle \zeta_h^\ell(0; t_{k_i}), \delta s_0 \rangle + \sum_{k=0}^{K} \delta u_k^1 \int_0^{t_f} \langle f_u^T(t)\zeta_h^\ell(\tau; t_{k_i}), \varphi_k(\tau) \rangle \, d\tau \leq 0 \tag{4.135}$$

for $(\ell, t_{k_i}) \in S_d$,

$$u_{l,k} \leq u_k^0 + \delta u_k^1 \leq u_{u,k}, \quad k = 0, \ldots, K, \tag{4.136}$$

$$\|\delta u_k^1\| \leq \rho, \quad \|\delta s_0^1\| \leq \eta, \tag{4.137}$$

where $S_d = \{(\ell, t_k) : \gamma_\ell(s^0(t_k)) \geq h(s_0^0, u^0) - \epsilon |h(s_0^0, u^0)|\}$ and ϵ is a small positive number.

In spite of its simplicity, the method turns out to be extremely efficient, as was demonstrated while treating various aspects of the optimal measurement problem with moving sensors, see, e.g., [135, 299, 301, 303, 310, 311, 321, 322, 324, 328].

REMARK 4.3   Let us note that the same method (and, in particular, the same computer code) can be used to find optimal locations of stationary sensors. For that purpose, it suffices to set the initial control u^0 as zero and then to maintain the zero control variation (i.e., $\rho = 0$) during calculations. This forces the algorithm to improve the solution only by changing s_0.   ∎

REMARK 4.4   The proposed method can also be generalized to consider more sophisticated dynamic models of the vehicles carrying the sensors and/or assume various constraints imposed on the sensor motions, e.g., the existing obstacles (fixed or mobile), sensors' geometrical dimensions, etc. As regards computational aspects, a considerable speedup can be achieved when the so-called upper-bounding version of the simplex algorithm, which exploits the special form of the constraints (4.132), is used to solve the linear-programming subproblems [212]. Furthermore, some decomposition-coordination techniques can also be applied to parallelize the computations [95, 169].   ∎

Example 4.3
The DPS of Example 4.1 constitutes a good reference point for performance tests of the proposed solution technique. Our purpose now is to estimate κ (i.e., the parameters θ_1, θ_2 and θ_3) as accurately as possible based on the measurements made by three moving sensors. Accordingly, D- and A-optimum design criteria are primarily chosen as the measures of the estimation accuracy, but to make a comparison, the sensitivity criterion is also considered.

Assuming that the sensor dynamics is not of primary concern, we adopt the simple model

$$\dot{s}(t) = u(t), \quad s(0) = s_0$$

Moreover, we impose the following constraints on u:

$$|u_i(t)| \leq 0.7, \quad \forall t \in T, \quad i = 1, \ldots, 6$$

As for technicalities, in order to numerically solve the measurement location problem, a computer programme was written in Lahey/Fujitsu Fortran 95 Ver. 5.60a using a PC (Pentium IV, 2.40 GHz, 512 MB RAM) running Windows 2000. During simulations the velocities themselves were considered in the class of piecewise linear polynomials ($K = 40$). On the other hand, the state and sensitivity equations were solved with the finite-element method. The sampling interval and coordinate divisions were $\Delta t = 0.0125$ and $\Delta x_1 = \Delta x_2 = 0.025$, respectively. The parameters ρ and η were gradually decreased from 0.05 to 0.01. The Cauchy convergence criterion for the sequence $J(s_0^k, u^k)$ was set as 10^{-3}. For simplicity, the constraints on the minimum allowable distance between the sensors were not considered and only the state constraints forcing the sensors to remain in Ω were imposed. On aggregate, approximately seven minutes of CPU time were used to complete the simulations.

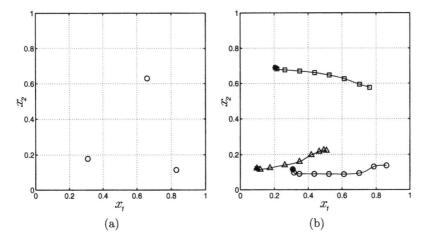

(a) (b)

FIGURE 4.3
D-optimum positions of stationary sensors (a) versus D-optimum trajectories of moving sensors (b).

Figures 4.3 and 4.4 show the optimal sensor trajectories obtained after a couple of trials with different initial guesses regarding s_0^0 and u^0 (to escape entrapment in a local minimum). Symbols like circles, squares and triangles denote consecutive sensor positions. Furthermore, the sensors' positions at $t = 0$ are marked by asterisks. As was already indicated, the diffusion coefficient values in the upper left of Ω are greater than those in the lower right. This means that the state changes during the system evolution are quicker when we move up and to the left (on the other hand, the system would have reached the steady state there earlier). This fact explains the form of the trajectories obtained — the sensors tend to measure the state in the regions where the distributed system is the most sensitive with respect to the unknown parameter κ, i.e., in the lower right. Figure 4.3(a) shows the D-optimum positions of stationary sensors ($\det(M) = 417.3$). A considerable gain in the accuracy of the parameter estimates is expected if moving observations are allowed, cf. Fig. 4.3(b), as in this case the value of the performance index is virtually four times as large as that for the stationary case, i.e., we have $\det(M) = 1563$. In turn, Fig. 4.4(a) shows the calculated trajectories for the A-optimality criterion whose ultimate value is $\text{tr}(M^{-1}) = 0.6101$. (In principle, the shapes of the trajectories are similar to those for the D-optimality counterpart.) On the other hand, Fig. 4.4(b) presents the results for the sensitivity criterion. In this case it turns out that the sensors strive to measure the system state very closely to one another and, in spite of a large value of the performance index ($\text{tr}(M) = 258.82$), a comparatively low value of the FIM determinant is obtained ($\det(M) = 13.65$), which suggests that the resulting measurement setting may cause some problems regarding identifiability. ⌷

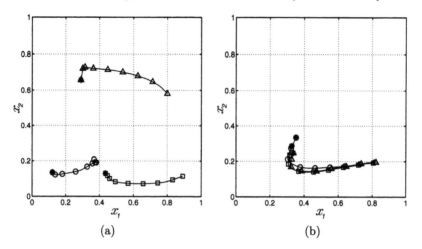

FIGURE 4.4
Optimum sensor trajectories: A-optimum criterion (a) versus sensitivity criterion (b).

4.3.5 Special cases

4.3.5.1 Optimal planning of sensor movements along given paths

Consider motion planning for sensors which can move only along prescribed paths being smooth curves parameterized, e.g., by their lengths. A motivation to study this kind of problem is the situation when the sensors used in a system of monitoring and prediction of air pollution in a city may move only along given streets or roads. Accordingly, the motions of N sensors are restricted to given paths which are smooth curves $[0, \lambda^j_{\max}] \ni \lambda^j \mapsto w_j(\lambda^j) \in \bar{\Omega}$ parameterized by their lengths. Let $\lambda^j : T \to [0, \lambda^j_{\max}]$ be the trajectory of the j-th sensor. Without loss of generality we assume that the state can be measured directly, i.e., the observations are of the form

$$z^j_m(t) = y(w_j(\lambda^j(t)), t; \theta) + \varepsilon(w_j(\lambda^j(t)), t), \quad t \in T, \tag{4.138}$$

for $j = 1, \ldots, N$.

Let the motion of the j-th sensor along the j-th trajectory be described by the equation

$$\dot{\lambda}^j(t) = u_j(t) \quad \text{a.e. on } T, \quad \lambda^j(0) = \lambda^j_0, \tag{4.139}$$

where u_j denotes the velocity of the j-th sensor, $j = 1, \ldots, N$. In the case considered here, the FIM can be written down as

$$M(\lambda_0, u) = \frac{1}{Nt_f} \sum_{j=1}^{N} \int_0^{t_f} g(w_j(\lambda^j(t)), t) g^{\mathsf{T}}(w_j(\lambda^j(t)), t) \, dt, \tag{4.140}$$

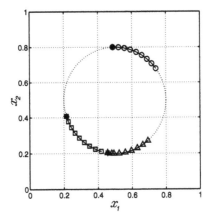

FIGURE 4.5
Optimal sensor trajectories along the circumference of a circle.

where $\lambda_0 = \mathrm{col}[\lambda_0^1, \dots, \lambda_0^N]$, $u(t) = \mathrm{col}[u_1(t), \dots, u_N(t)]$. Optimal sensor trajectories for system identification can be found by choosing λ_0 and u so as to minimize some scalar function Ψ of the informations matrix.

We see at once that the situation is by no means more difficult than in the setting of Section 4.3.1, as the only difference is that here the state is λ and g depends on λ only indirectly, via ω_j's. But while calculating the Fréchet derivative of J this complication is removed through the use of the Chain Rule (for details, see [302]).

Example 4.4

Consider the process of Example 4.3, but this time with the sensor motions restricted to the circumference of the circle with centre $(0.5, 0.5)$ and radius 0.3 (parameterized by its length). The sensor velocities (i.e., the control variables in the model (4.139)) were restricted so as to satisfy

$$|u_j(t)| \leq 0.3, \quad j = 1, 2, 3.$$

When planning, the D-optimum design procedure was adopted. During the calculation, all the other numerical parameters of Example 4.3 were retained.

Figure 4.5 shows the optimal trajectories of the sensors obtained for several trials with different initial guesses regarding s_0^0 and u^0. Squares, circles and triangles denote consecutive sensor positions. Furthermore, sensors' positions at $t = 0$ are marked by asterisks. The form of the trajectories is in agreement with our earlier findings, cf. Example 4.3. ☐

4.3.5.2 Measurement optimization with minimax criteria

In [323] the outlined method of sensor motion planning was extended to include two widely used minimax criteria, viz. those of MV- and E-optimality. As is well known, they are not Fréchet differentiable (in general, only their directional derivatives exist), which highly complicates their use and hinders direct application of the foregoing algorithms.

The MV-optimality criterion is as follows:

$$\Psi(M) = \max_{1 \leq i \leq m} d_{ii}(M), \qquad (4.141)$$

where $d_{ii}(M)$ stands for the i-th element on the diagonal of M^{-1}. In turn, the E-optimality criterion was extensively discussed in Section 3.2. Both the criteria have clear statistical interpretation. In the MV-optimum design, the maximal variance of the estimates $\hat{\theta}_1, \ldots, \hat{\theta}_m$ is minimized. On the other hand, while minimizing the E-optimality criterion, the length of the largest principal axis of the uncertainty ellipsoid of the estimates is suppressed.

The MV-optimality criterion is not Fréchet differentiable (this is because the max function is nondifferentiable), which essentially complicates its minimization. To overcome this difficulty, an additional control parameter c is introduced and the equivalent problem of minimizing

$$J(s_0, u, c) = c, \qquad (4.142)$$

is considered subject to the additional inequality state constraint

$$\tilde{h}(s_0, u, c) = \max_{1 \leq i \leq m} d_{ii}(M) - c \leq 0, \qquad (4.143)$$

The following dependence is helpful while derivations of the expressions for the differentials:

$$\frac{\partial d_{ii}(M)}{\partial M} = -d^{(i)} d^{(i)\mathrm{T}} \qquad (4.144)$$

where $d^{(i)}$ is the i-th column of M^{-1}.

In the case of minimizing the E-optimality criterion, we cannot use gradient methods, since in general the eigenvalues of M are not Fréchet differentiable, cf. Section 3.2. To overcome this difficulty, this time we propose to make use of the dependence

$$\left\{ \mathrm{tr}(M^{-\mu}) \right\}^{1/\mu} \xrightarrow[\mu \to \infty]{} \lambda_{\max}(M^{-1}), \qquad (4.145)$$

which is valid for any FIM, cf. (2.17) and to replace minimization of the largest eigenvalue by that of the 'smooth' functional

$$J_\mu(s_0, u) = \mathrm{tr}(M^{-\mu}) \qquad (4.146)$$

for a sufficiently large μ. For such a regularized criterion, we have

$$\frac{\partial \operatorname{tr}(M^{-\mu})}{\partial M} = -\mu M^{-\mu-1}. \tag{4.147}$$

This constitutes an alternative to the smooth convex approximation discussed in Section 3.2.

Example 4.5
As an example of the application of the proposed algorithm, the two-dimensional heat equation

$$\frac{\partial y(x,t)}{\partial t} = \frac{\partial}{\partial x_1}\left(\kappa(x)\frac{\partial y(x,t)}{\partial x_1}\right) + \frac{\partial}{\partial x_2}\left(\kappa(x)\frac{\partial y(x,t)}{\partial x_2}\right),$$

$$x \in \Omega = (0,1) \times (0,1), \quad t \in (0,1)$$

is considered, subject to the conditions

$$y(x,0) = 50, \qquad x \in \Omega,$$
$$y(x,t) = 50\,(1-t), \quad (x,t) \in \partial\Omega \times T.$$

The diffusion coefficient to be identified has the form

$$\kappa(x) = \theta_1 + \theta_2 x_1 + \theta_3 x_2 + \theta_4 x_1 x_2,$$
$$\theta_1 = 0.1, \quad \theta_2 = 0.3, \quad \theta_3 = 0.1, \quad \theta_4 = 0.3,$$

where the values $\theta_1, \ldots, \theta_4$ are also treated as nominal.

The problem is to estimate the thermal diffusivity coefficient κ (i.e., the parameters θ_i, $i = 1, \ldots, 4$) as accurately as possible, based on the measurements of the state made by four moving sensors. For that purpose, the MV- and E-optimum design procedures are adopted. As regards the sensor dynamics, we consider the simple model

$$\dot{s}(t) = u(t), \quad s(0) = s_0,$$

where the sensor velocities (controls) are limited according to

$$|u_i(t)| \le 0.3, \quad i = 1, \ldots, 8,$$

During simulations, the velocities were approximated by piecewise linear polynomials ($K = 40$). The parameters ρ, η and an additional parameter ς introduced to limit increments in c were changed to speed convergence in a manner similar to that used to adapt the learning rate in the training procedure for backpropagation neural networks [193]. The Cauchy convergence criterion for the sequence $J(s_0^k, u^k, c^k)$ equals 10^{-4}.

Figure 4.6(a) shows the MV-optimal trajectories of the sensors obtained after several trials with different initial guesses regarding s_0^0 and u^0 (to avoid

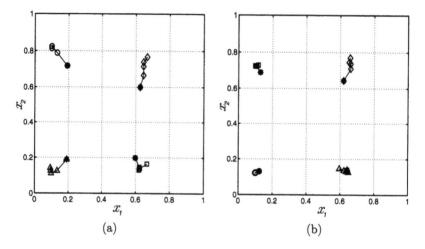

(a)   (b)

FIGURE 4.6
Optimal sensor trajectories: MV-optimality criterion (a) and approximated
E-optimality criterion (b).

getting stuck in a local minimum). Squares, circles, triangles and diamonds
denote consecutive sensor positions. Furthermore, the sensors' positions at
$t = 0$ are marked by asterisks.

Let us note that the diffusion coefficient values in the upper right of Ω
are greater than those in the lower left. This means that the state changes
during the system evolution are quicker when we move up and to the right
(on the other hand, the system reaches the steady state there earlier). This
fact explains the form of the obtained trajectories — the sensors follow the
regions where the distributed system is the most sensitive with respect to the
unknown parameter κ. This region shifts to the lower left as time elapses and
this is also reflected by the form of the trajectories. Let us note that one of
the sensors practically stays at one point for most of the time and thus its
behaviour is much like that of a stationary one.

For comparison, Fig. 4.6(b) shows the (sub-)optimal trajectories corresponding to the E-optimal criterion, which was in this case approximated
by (4.146) with $\mu = 7$. The ultimate value of $\lambda_{\max}(M^{-1})$ was 0.4287. As can
be seen, the results obtained for the MV- and E-optimality criteria are quite
similar. In optimum experimental design for static linear regression models,
some equivalence conditions for both the designs can even be formulated [78],
so it would be reasonable to expect some similarities also in the nonlinear
dynamic case. In this context, our results are not surprising. □

4.3.5.3 Minimal number of sensors

Since the number of sensors is generally governed by economic considerations, it is desirable to reduce their number to as few as possible provided that this new number of sensors still guarantees an acceptable level of precision for the estimates. The accuracy of parameter estimates for a fixed number of sensors is approximately described by the diagonal of the inverse of the FIM

$$M(s) = \frac{1}{\sigma^2} \sum_{j=1}^{N} \int_0^{t_f} \left(\frac{\partial y(x^j, t; \theta)}{\partial \theta} \right)^{\mathsf{T}} \left(\frac{\partial y(x^i, t; \theta)}{\partial \theta} \right) \bigg|_{\theta=\theta^0} dt, \qquad (4.148)$$

so we can define an error bound, e.g., of the form

$$\mathfrak{J}(N) = \frac{1}{m} \sum_{i=1}^{m} \frac{d_{ii}}{\theta_i^0} \leq \epsilon, \qquad (4.149)$$

where d_{ii} denotes the i-th element of the diagonal of $D = M^{-1}$, θ_i^0 is a prior estimate of the i-th parameter (obtained, e.g., from a preliminary experiment or by physical analysis), and ϵ stands for the desired estimation accuracy. This approach is similar to that used in the context of state estimation [11,138,197]. Clearly, there exist many alternative choices of the criterion (4.149). The particular form presented here takes into account the relative errors of the estimates and improves the accuracy of the small parameters, which is of interest, e.g., when parameters with very different magnitudes are to be found simultaneously.

The minimal number of sensors is equal to the minimum number N for which the condition (4.149) is satisfied. The fulfilment of (4.149) is checked after each optimization stage consisting in minimizing a selected design criterion for a given fixed number of sensors. If (4.149) is not satisfied, then a larger number of sensors should be used; otherwise, we should reduce the number of sensors to a value which has not been examined yet. Note that the right-hand side of (4.149) can be rewritten in the form $\mathrm{tr}(LM^{-1})$, where $L = \mathrm{Diag}[1/(m\theta_1^0), \ldots, 1/(m\theta_m^0)]$, which corresponds to an L-optimal criterion [224,311].

Example 4.6
In order to study the applicability and performance of the proposed approach, let us consider the situation of Example 4.5 with a slightly changed diffusivity coefficient, i.e.,

$$\kappa(x) = \theta_1 + \theta_2 x_1 + \theta_3 x_2, \quad \theta_1 = 0.1, \quad \theta_2 = 0.3, \quad \theta_3 = 0.1,$$

Our purpose is to estimate κ (i.e., the parameters θ_1, θ_2 and θ_3) as accurately as possible based on the measurements made by N moving sensors. To this end, for each fixed N, the L-optimum design procedure is adopted, where

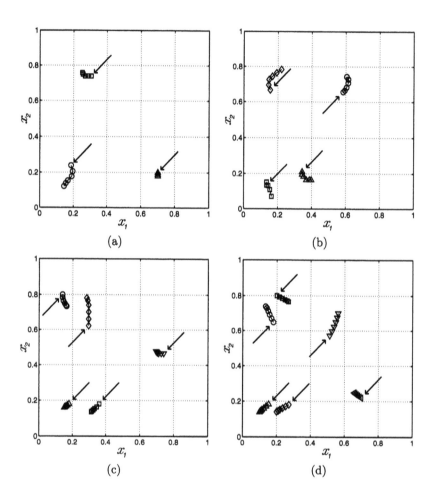

FIGURE 4.7
Optimum sensor trajectories of Example 4.6: (a) $N = 3$, (b) $N = 4$, (c) $N = 5$, (d) $N = 6$.

L is selected as described above, i.e., the criteria J and $\mathfrak{J}$ are the same. Let us note that we could also select other design criteria at this stage. The criterion (4.149) is only used to validate the results after optimization of sensor locations for a given N. If the results are not satisfactory, we test another number of sensors.

It is required that the sensor velocities be bounded in accordance with the conditions

$$|u_j(t)| \leq 0.2, \quad j = 1, \ldots, 2N,$$

As regards technicalities, during simulations the velocities were considered in the class of piecewise linear polynomials ($K = 50$). The parameters ρ and η were gradually decreased from 0.05 to 0.01. The Cauchy convergence criterion for the sequence $J(x_0^k, u^k)$ equals 10^{-4}. Furthermore, the minimum allowable distance between the sensors is $R = 0.1$.

In order to select a minimal number of sensors, we assume that $\sigma = 0.1$ and $\epsilon = 0.0025$ (this amounts to the average relative error for the parameters of about 5%). Figure 4.7 shows the optimal sensor trajectories obtained for $N = 3, 4, 5, 6$ after several trials with different initial guesses regarding s_0^0 and u^0. As usual, symbols like circles, squares and triangles denote consecutive sensor positions. Furthermore, the sensors' positions at $t = 0$ are marked by the arrows.

The following values of the criterion (4.149) were obtained:

$$\mathfrak{J}(3) = 0.0051, \quad \mathfrak{J}(4) = 0.004, \quad \mathfrak{J}(5) = 0.0029, \quad \mathfrak{J}(6) = 0.0024,$$

which indicates that the minimum number of sensors is equal to six.   ☐

4.4   Concluding remarks

In this chapter, the main principles of motion planning for scanning and mobile sensors have been presented. First, the problem of finding optimal activation policies for scanning devices was examined. Ordinarily, the task is reduced to seeking best subsets of locations from among all the possible ones. Numerical algorithms for the construction of optimum sensor configurations by searching over a list of candidate locations customarily involve an iterative improvement of the initial sensor configuration. The combinatorial nature of the problem so formulated implies that with a long list of candidate points, complicated search algorithms can readily consume appreciable computer time and space. In contrast to this approach, the key idea of the approach proposed here is to operate on the density of sensors per unit area instead of the positions of individual sensors. Such conditions allow us to relax the discrete optimization problem in context and to replace it by its continuous approximation. Mathematically, this procedure involves looking for a family of 'optimal' probability measures defined on subsets of the set of feasible measurement points. In spite of its somewhat abstract assumptions, the resulting algorithm of the exchange type is very easy to implement and extremely efficient. Note, however, that the underlying assumptions of the algorithm involve its main limitation which is that it can be used only when the number of sensors is relatively high. Consequently, an approach based on some recent results of discrete-valued optimal control was suggested for situations when the number of scanning sensors is small. By introducing the CPET transformation, it was shown that the original discrete-valued control problem with variable switching times can be transformed into an equivalent continuous-valued optimal control problem, which can then be solved using classical optimal-control techniques. Consequently, the combinatorial nature of the original problem is circumvented to some extent and the resulting transformed optimal control problem can be readily solved by common nonlinear-programming algorithms.

Afterwards, some fundamental results of modern optimum experimental design theory were extended to the framework of moving sensors following the ideas presented in the seminal paper [238]. The implication is that the problem reduces to solving at each time moment a separate optimization task to which classical optimum-experimental design algorithms can be applied. Apart from the fact that the computing power necessary to solve all the resulting subproblems is enormous, a major drawback to this method is that only measurability of the trajectories can be guaranteed. As was shown, these inconveniences can be somewhat alleviated by suitably parameterizing the trajectories, but the main disadvantage to the approaches based on continuous designs, i.e., sensor clusterization, still persists and restricts the spectrum of potential applications. On the other hand, the results obtained provide evidence for close relations between classical optimum experimental design for regression problems and motion planning for multiple sensors, and indicate some directions for future research (especially in connection with recent advances in spatial statistics).

Special attention has been paid to the problem of planning optimal motions for a given number of pointwise sensors which are to provide measurement data for parameter estimation of a general distributed system. Based on a scalar measure of performance defined on the Fisher information matrix related to the unknown parameters, the problem was formulated as an optimal-control one with state-variable inequality constraints representing geometric constraints induced by the admissible measurement regions and allowable distances between the sensors. Taking account of the dynamic models of the vehicles carrying the sensors, the problem was finally reduced to determination of both the control forces of the sensors and initial sensor positions. We showed that the resulting problem can be converted to an equivalent classical Mayer problem which is thoroughly treated in optimal control theory and for which numerous efficient algorithms exist. Accordingly, we applied one of them, i.e., a method of successive linearizations, to construct a quite efficient numerical scheme of determining optimal sensor trajectories. This scheme was verified through application to a two-dimensional parabolic equation. Simulation experiments validate the fact that making use of moving sensors may lead to dramatic gains in the values of the adopted performance indices, and hence to a much greater accuracy in the resulting parameter estimates.

The approach suggested here has the advantage that it is independent of a particular form of the partial-differential equation describing the considered distributed system. The only requirement is the existence of sufficiently regular solutions to the state and sensitivity equations, and consequently nonlinear systems can also be treated within the same framework practically without any changes. Furthermore, the optimal-control approach proposed here allows for a variety of possible sensor motion models and motion constraints to be directly considered. Apart from the constraints preventing clusterization and measurements outside the imposed admissible zones, we might also include those induced by the existing obstacles (stationary or mobile), sensors' geometrical dimensions, etc. Moreover, the approach can be easily generalized to three spatial dimensions and the only limitation is the amount of required computations.

Clearly, the method of successive linearizations, which has been used to numerically calculate approximations to the optimal sensor trajectories, constitutes only one of many possible choices and other algorithms for problems with state inequality constraints could have been employed for that purpose. Its decided advantages, however, are that the improvements generated by the algorithm drive the state trajectories into the interior of the state constraint region and that it does not require knowledge of an initial solution satisfying the state constraints. Moreover, the specific form of the resulting linear-programming subtasks makes the method particularly suited for parallel implementations.

5

Measurement strategies with alternative design objectives

The previous chapters of this book have been primarily concerned with the problem of extracting information from a given set of data provided by measurement sensors for accurate determination of parameter values which may have some physical significance. This gave rise to the use of various 'alphabetical' criteria related to the highest probability density region for the parameters. Nevertheless, other experiment design criteria are also possible and in order to demonstrate this issue, we devote the present chapter to a study of the design of the experimental conditions so that the experiment is maximally informative with respect to response prediction and the effectiveness of tests for model structure determination.

5.1 Optimal sensor location for prediction

In some applications, the reliability of model predictions is sometimes more important than the accuracy of model parameters, because the ultimate objective in modelling is the prediction or forecast of the system states [181]. The topic was discussed to some extent in [282, p. 201], but without connection to constructive solution methods. This failing constitutes the main motivation for the study undertaken in order to extend sensor motion planning techniques set forth in [312]. For that purpose, three output criteria were proposed as measures of the prediction accuracy. The main difficulty in solving the resulting optimal-control problem is that two criteria are not Fréchet differentiable. In what follows, some details on this approach are given.

5.1.1 Problem formulation

Consider a scalar DPS defined in a fixed, bounded, open spatial set $\Omega \subset \mathbb{R}^2$ with sufficiently smooth boundary $\partial\Omega$. It evolves over a time horizon $T = [0, t_f]$, and at each time moment t and each spatial point $x = (x_1, x_2)$ its state is denoted by $y(x, t; \theta)$. The notation emphasizes the fact that y depends on the vector $\theta \in \mathbb{R}^m$ of unknown parameters to be estimated from measurements made by N moving pointwise sensors.

Let $x^j : T \longrightarrow X \subset \bar{\Omega} = \Omega \cup \partial\Omega$ be the trajectory of the j-th sensor, where X signifies an admissible region (a compact set) in which measurements can be made. Our basic assumption is that the observations are of the form

$$z_m^j(t) = y(x^j(t), t; \theta) + \varepsilon(x^j(t), t), \quad t \in T, \quad j = 1, \dots, N, \quad (5.1)$$

where $\varepsilon = \varepsilon(x, t)$ is a Gaussian white noise process whose statistics are

$$\mathrm{E}\{\varepsilon(x, t)\} = 0, \quad \mathrm{E}\{\varepsilon(x, t)\varepsilon(x', t')\} = \sigma^2 \delta(x - x')\delta(t - t'), \quad (5.2)$$

$\sigma > 0$ being a constant and δ the Dirac delta function.

We assume that the parameter estimate $\hat{\theta}$, defined as the solution to the usual output least-squares formulation of the parameter estimation problem, is to provide a basis for prediction of certain variables depending on spatial location and/or time. Since in general the conditions applied for prediction may differ from the conditions of the experiment, the prediction equations need not be the same as the state equation, nor need the variables to be predicted coincide with the state y. Let the solution to the prediction problem in context be a scalar quantity $q = q(x, t; \theta)$. We are interested in selecting the sensors' trajectories in such a way as to maximize the accuracy of q in a given compact spatiotemporal domain $\mathcal{Q} = \mathcal{X} \times \mathcal{T}$. Clearly, in order to compare different trajectories, a quantitative measure of the 'goodness' of particular trajectories is required. A logical approach is to choose a measure related to the expected accuracy of prediction.

For a given $(x, t) \in \mathcal{Q}$, the variance of q obtained by a first-order expansion around a preliminary estimate θ^0 of θ [174] has the form

$$
\begin{aligned}
\mathrm{var}(q(x, t; \hat{\theta})) &= \mathrm{E}\left\{ \left[q(x, t; \theta) - q(x, t; \hat{\theta}) \right]^2 \right\} \\
&\approx \frac{\partial q(x, t; \theta^0)}{\partial \theta} \, \mathrm{cov}(\hat{\theta}) \left(\frac{\partial q(x, t; \theta^0)}{\partial \theta} \right)^{\mathsf{T}}.
\end{aligned}
\quad (5.3)
$$

As usual, it is customary to choose θ^0 as a nominal value of θ or a result of a preliminary experiment.

As for $\mathrm{cov}(\hat{\theta})$, under the appropriate assumptions it can be approximated by the inverse of the FIM whose normalized version can be written down as

$$M(s) = \frac{1}{Nt_f} \sum_{j=1}^{N} \int_0^{t_f} g(x^j(t), t) g^{\mathsf{T}}(x^j(t), t) \, \mathrm{d}t, \quad (5.4)$$

where

$$s(t) = (x^1(t), x^2(t), \dots, x^N(t)) \quad (5.5)$$

and

$$g(x, t) = \left(\frac{\partial y(x, t; \theta^0)}{\partial \theta} \right)^{\mathsf{T}} \quad (5.6)$$

(we require g and $\partial g/\partial x$ to be continuous).

Consequently, we get

$$\text{var}(q(x,t;\hat{\theta})) \sim \frac{\partial q(x,t;\theta^0)}{\partial \theta} M^{-1}(s) \left(\frac{\partial q(x,t;\theta^0)}{\partial \theta}\right)^{\mathsf{T}}. \tag{5.7}$$

Some criteria may now be set up such that the 'optimal' trajectories $x^1, \ldots,$ x^N minimize $\text{var}(q(x,t;\hat{\theta}))$ over $\mathcal{Q}$. Based on the suggestions of [85, p. 25], in the sequel the following choices are considered:

$$\Psi_1[M(s)] = \max_{(x,t)\in\mathcal{Q}} \text{var}(q(x,t;\hat{\theta})) \quad = \max_{(x,t)\in\mathcal{Q}} \text{tr}\left\{A(x,t)M^{-1}(s)\right\}, \tag{5.8}$$

$$\Psi_2[M(s)] = \max_{x\in\mathcal{X}} \int_T \text{var}(q(x,t;\hat{\theta}))\,dt = \max_{x\in\mathcal{X}} \text{tr}\left\{B(x)M^{-1}(s)\right\}, \tag{5.9}$$

$$\Psi_3[M(s)] = \iint_{\mathcal{Q}} \text{var}(q(x,t;\hat{\theta}))\,dx\,dt = \text{tr}\left\{CM^{-1}(s)\right\}, \tag{5.10}$$

where

$$A(x,t) = \left(\frac{\partial q(x,t;\theta^0)}{\partial \theta}\right)^{\mathsf{T}} \frac{\partial q(x,t;\theta^0)}{\partial \theta}, \tag{5.11}$$

$$B(x) = \int_T \left(\frac{\partial q(x,t;\theta^0)}{\partial \theta}\right)^{\mathsf{T}} \frac{\partial q(x,t;\theta^0)}{\partial \theta}\,dt, \tag{5.12}$$

$$C = \iint_{\mathcal{Q}} \left(\frac{\partial q(x,t;\theta^0)}{\partial \theta}\right)^{\mathsf{T}} \frac{\partial q(x,t;\theta^0)}{\partial \theta}\,dx\,dt. \tag{5.13}$$

5.1.2 Optimal-control formulation

As in Section 4.3, we assume that the dynamics of sensors motions is given by

$$\dot{s}(t) = f(s(t), u(t)) \quad \text{a.e. on } T, \quad s(0) = s_0, \tag{5.14}$$

where $f : \mathbb{R}^{n+r} \to \mathbb{R}^n$ ($n = \dim s(t) = 2N$) is required to be continuously differentiable, $s_0 \in \mathbb{R}^n$, $u : T \to \mathbb{R}^r$ is a measurable control which satisfies

$$u_l \leq u(t) \leq u_u \quad \text{a.e. on } T \tag{5.15}$$

for fixed $u_l, u_u \in \mathbb{R}^r$. Given any s_0 and u, there is a unique absolutely continuous trajectory $s : T \to \mathbb{R}^n$ which satisfies (5.14) a.e. on T. Additionally, several pathwise state inequality constraints have to be included, e.g., the requirement of keeping sensors away from one another (for independence of measurements) while staying within the region X where measurements are allowed. Hence, for the general case, introduce

$$\gamma_\ell(s(t)) \leq 0, \quad \forall t \in T \tag{5.16}$$

for $\ell = 1, \ldots, \nu$, where the γ_ℓ's are continuously differentiable, cf. (4.110). In the sequel, we set $\bar{\nu} = \{1, \ldots, \nu\}$.

Consequently, we wish to solve the following problem:

$$\min_{(s_0, u) \in \mathcal{P}} J(s_0, u), \tag{5.17}$$

subject to the inequality constraint

$$h(s_0, u) \le 0, \tag{5.18}$$

where

$$J(s_0, u) = \Psi_i[M(s)], \tag{5.19}$$

$$h(s_0, u) = \max_{(\ell, t) \in \bar{\nu} \times T} \{\gamma_\ell(s(t))\}, \tag{5.20}$$

$$\mathcal{P} = \{(s_0, u) \mid s_0 \in X^N, \ u : T \to \mathbb{R}^r \text{ is measurable}, \tag{5.21}$$

$$u_l \le u(t) \le u_u \text{ a.e. on } T\}. \tag{5.22}$$

A severe difficulty is that Ψ_1 and Ψ_2 are not Fréchet differentiable. In order to reduce this case to the framework handled by the algorithm outlined in what follows (which necessitates such differentiability of the performance index), an additional control parameter w is introduced for $i = 1$ or 2 and the equivalent problem of minimizing

$$\mathcal{J}(s_0, u, w) = w \tag{5.23}$$

is considered subject to (5.18) and the additional inequality state constraint

$$e(s_0, u, w) = \Psi_i[M(s)] - w \le 0. \tag{5.24}$$

(Due to the differentiability of Ψ_3, its minimization can be performed in much the same way as in Section 4.3.)

5.1.3 Minimization algorithm

Given $(s_0^0, u^0) \in \mathcal{P}$, which determines a trajectory s^0, consider the linearized problem: Find $\delta s_0^1 \in \mathbb{R}^n$, $\delta u^1 \in L^\infty(T; \mathbb{R}^r)$ and $\delta w^1 \in \mathbb{R}$ to minimize

$$\mathcal{J}(s_0^0 + \delta s_0^1, u^0 + \delta u^1, w^0 + \delta w^1) = w^0 + \delta w^1, \tag{5.25}$$

subject to the constraints

$$\begin{cases} h(s_0^0, u^0, w^0) + \delta h(s_0^0, u^0, w^0; \delta s_0^1, \delta u^1, \delta w^1) \le 0, \\ e(s_0^0, u^0, w^0) + \delta e(s_0^0, u^0, w^0; \delta s_0^1, \delta u^1, \delta w^1) \le 0, \\ u_l \le u^0 + \delta u^1 \le u_u, \\ \|\delta u^1\| \le \eta_u, \quad \|\delta s_0^1\| \le \eta_s, \quad |\delta w^1| \le \eta_w \end{cases} \tag{5.26}$$

where $w^0 = J(s_0^0, u^0)$, and η_u, η_s and η_w are sufficiently small positive numbers. Here δh and δe denote the Gâteaux differentials of h and e, respectively. Applying the Lagrangian approach to calculation of reduced gradients, cf. Appendix H, we conclude that

$$\delta h(s_0^0, u^0, w^0; \delta s_0, \delta u, \delta w)$$

$$= \max_{(\ell, t) \in S} \left\{ \langle \zeta_h^\ell(0; t), \delta s_0 \rangle + \int_0^{t_f} \langle f_u^\mathsf{T}(\tau) \zeta_h^\ell(\tau; t), \delta u(\tau) \rangle \, d\tau \right\}, \quad (5.27)$$

where $S = \{(\ell, t) \in \bar{\nu} \times T : \gamma_\ell(s(t)) = h(s_0, u)\}$, $\zeta_h^\ell(\cdot; t)$ is the solution to

$$\frac{d\zeta_h^\ell(\tau; t)}{d\tau} + f_s^\mathsf{T}(\tau) \zeta_h^\ell(\tau; t) = -\delta(\tau - t) \left(\frac{\partial \gamma_\ell}{\partial s} \right)^\mathsf{T}_{s = s^0(\tau)}, \quad \zeta_h^\ell(t_f; t) = 0, \quad (5.28)$$

with $f_s(t) = \partial f(s^0(t), u^0(t))/\partial s$, $f_u(t) = \partial f(s^0(t), u^0(t))/\partial u$.
In much the same way, for the criterion Ψ_1 it can be shown that

$$\delta e(s_0^0, u^0, w^0; \delta s_0, \delta u, \delta w)$$

$$= -\delta w + \max_{(x, t) \in S_1} \left\{ \langle \zeta_e(0; x, t), \delta s_0 \rangle + \int_0^{t_f} \langle f_u^\mathsf{T}(\tau) \zeta_e(\tau; x, t), \delta u(\tau) \rangle \, d\tau \right\},$$

$$(5.29)$$

where $S_1 = \{(x, t) \in Q : \operatorname{tr}\left[A(x, t) M^{-1}(s^0)\right] = e(s_0^0, u^0, w^0)\}$, $\zeta_e(\cdot; x, t)$ denotes the solution to the Cauchy problem:

$$\frac{d\zeta_e(\tau; x, t)}{d\tau} + f_s^\mathsf{T}(\tau) \zeta_e(\tau; x, t) = -\sum_{i=1}^m \sum_{j=1}^m d_{ij} \left(\frac{\partial \chi_{ij}(s, \tau)}{\partial s} \right)^\mathsf{T}_{s = s^0(\tau)}, \quad (5.30)$$

with $\zeta_e(t_f; x, t) = 0$, $D_1(s, x, t) = [d_{ij}] = -M^{-1}(s) A(x, t) M^{-1}(s)$,

$$\chi_{ij}(s(t), t) = \frac{1}{N t_f} \sum_{\ell=1}^N g_i(x^\ell(t), t) g_j(x^\ell(t), t). \quad (5.31)$$

In the case of Ψ_2, it suffices to replace D_1 by $D_2 = -M^{-1}(s)B(x)M^{-1}(s)$ and S_1 by $S_2 = \{x \in X : \operatorname{tr}\left[B(x)M^{-1}(s^0)\right] = e(s_0^0, u^0, w^0)\}$.

The nonoptimality of s_0^0 and u^0 implies the existence of $s^1 = s_0^0 + \delta s_0^1$ and $u^1 = u^0 + \delta u^1$ such that $J(s_0^1, u^1) < J(s_0^0, u^0)$. The process of minimization then resumes with s_0^1 and u^1 instead of s_0^0 and u^0, respectively. By repeating this procedure over and over, a sequence $\{(s_0^k, u^k)\}$ is thus obtained. Proceeding on the same lines as in [96], we can prove that the sequence of the corresponding trajectories is convergent.

Through a finite-dimensional approximation (e.g., with controls being piecewise linear polynomials), the algorithm reduces to solving a sequence of finite-dimensional linear-programming problems.

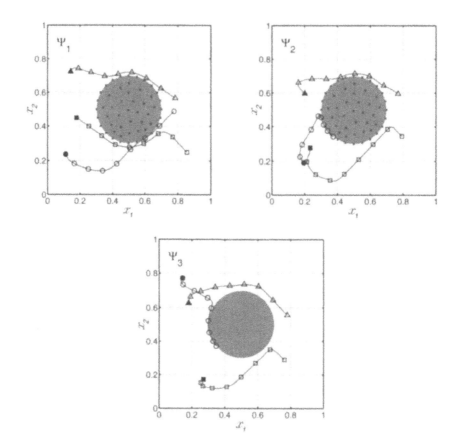

FIGURE 5.1
Optimal sensor trajectories for different criteria. The lightly shaded circle constitutes the forbidden region $\mathcal{X}$ where spatial prediction is required.

Example 5.1

Consider anew the two-dimensional diffusion equation of Example 4.1 subject to both the same initial and boundary conditions, and the same form of the diffusion coefficient κ. Our purpose is to estimate the state y inside the circle $\mathcal{X}$ with centre $(0.5, 0.5)$ and radius 0.2 during the time horizon $\mathcal{T} = T$, but in the course of the experiment, the sensors must not enter $\mathcal{X}$.

Assuming that the sensor dynamics is not of primary concern, we set

$$\dot{s}(t) = u(t), \quad s(0) = s_0. \tag{5.32}$$

Moreover, we impose the following constraints on u:

$$|u_i(t)| \leq 0.7, \quad \forall t \in T, \quad i = 1, \ldots, 6. \tag{5.33}$$

The controls were considered in the class of piecewise linear polynomials (with a uniform partition of T into 40 subintervals). The parameters η_u, η_s and η_w were gradually decreased from 0.05 to 0.01. For simplicity, the constraints on the minimum allowable distance between the sensors were not considered and only the constraints forcing the sensors to remain in $X = \bar{\Omega}$ were imposed.

Figure 5.1 shows the calculated approximations to the optimal motions of the sensors. Fifty crosses in $\mathcal{X}$ denote the discretization points introduced to numerically solve the minimax problems; additionally, ten points forming a uniform partition on $\mathcal{T}$ were considered for criterion Ψ_1 to discretize the time horizon. Open circles, squares and triangles indicate the consecutive sensor positions at time steps corresponding to multiplicities of the period of 0.125 (they are marked in order to reflect the sensor speeds). Furthermore, the same symbols corresponding to the sensors' positions at $t = 0$ are filled with black colour. In principle, the shapes of the trajectories obtained for different criteria are similar (the sensors follow the moving source). ☐

5.2 Sensor location for model discrimination

Process investigations often generate various mathematical models. Ideally, before any data have been analysed (or, better still, before collection of data), it would be desirable to know whether a proposed model is a suitable vehicle for unambiguous interpretation of observed behaviour, or alternatively, whether the data will constitute a discriminating test of the model's validity. If ambiguity remains, there is broad uncertainty in the model. This question is covered with the notion of the model's identifiability [19]. But sometimes several alternative and plausible models are proposed for the same physical situation and the results of the statistical analysis may depend heavily on the chosen model. Consequently, some criteria to identify a preferred model and assess its adequacy are desired.

An experiment especially designed for discrimination between the competing models is a good source of information about the model fit using minimum experimental effort. However, in experimental design theory the design problem for discrimination between models has drawn substantially less attention and has been developed for simple models only. Various criteria were considered in [9, 10, 36, 86, 87, 184, 217, 279]. The criterion called T-optimality, introduced by Atkinson and Fedorov [9, 10], constitutes a good choice as it has an interesting statistical interpretation as the power of a test for the fit of the second model when the first one is true. In what follows, we describe how to adapt it in the context of sensor location and to generate the corresponding solutions. The basic idea comes from the works [314,315] where it was applied for chemical kinetic modelling.

5.2.1   Competing models of a given distributed system

Identification of the correct dynamic model is a process consisting of several stages, which include data collection, fitting and diagnostic checking. The fitting and checking may not lead to a unique conclusion if the observations are taken in parts of the regression region and experimental conditions which suit more than one model. Box and Hill [26] illustrate this problem for a catalytic reaction which may go along four different mechanisms. They say "Unfortunately, it is easy to collect data that are well fitted by a large number of different models. Different research groups commonly claim widely varying mechanisms for the same mechanical system." Hence, a planning stage should precede the data collection. For each specific dynamic system the data collection needs to be especially designed (tailor-made). It is not really possible to find a common design which would suit a wide range of dynamic systems. However, procedures based on general theoretical results can be proposed for calculating the designs for model discrimination for some classes of models.

The results presented in the following are in a general form and they can be directly applied to various deterministic dynamic systems which constitute a primary motivation to study T-optimum designs. These systems are continuous-time models and we are specifically interested in their large class given in terms of PDEs. We consider two competing models defined in a spatial domain Ω and described by the following (possibly nonlinear) equations:

$$\mathcal{M}_\beta : \quad \frac{\partial y_\beta}{\partial t} = \mathcal{F}_\beta \left(x, t, y_\beta, \frac{\partial y_\beta}{\partial x_1}, \frac{\partial y_\beta}{\partial x_2}, \frac{\partial^2 y_\beta}{\partial x_1^2}, \frac{\partial^2 y_\beta}{\partial x_2^2}, \theta_\beta \right), \quad \beta = 1, 2, \quad (5.34)$$

where $x \in \Omega$, $t \in T$ stands for time, $T = (0, t_f)$, $y_\beta = y_\beta(x, t)$ denotes the state variable with values in $\mathbb{R}$ and $\mathcal{F}_\beta$ is some known function which may include terms accounting for given *a-priori* forcing inputs. For simplicity, we assume that both the equations are accompanied by the same boundary and initial conditions. Here $\theta_\beta \in \Theta_\beta$ denotes a vector of constant unknown parameters, Θ_β being a given set. Note that (5.34) implicitly defines mappings $\eta_\beta : \bar{\Omega} \times T \times \Theta_\beta \to \mathbb{R}$ such that $\eta_\beta(\,\cdot\,, \,\cdot\,, \theta_\beta)$ coincides with the solution $y_\beta(\,\cdot\,, \,\cdot\,)$ for any fixed θ_β, $\beta = 1, 2$.

For now, assume that the system is observed via N pointwise sensors placed at points $x^1, \ldots, x^\ell \subset \bar{\Omega}$, $\ell \leq N$ which take measurements at fixed time instants $t_1, \ldots, t_K \subset T$. The solutions of mechanistic models (5.34) give the expectations of the two competing statistical models, where random errors of observations are included. Discrimination between models $\mathcal{M}_1$ and $\mathcal{M}_2$, based on an experimental design is equivalent to discrimination between two statistical models:

$$\mathcal{S}_\beta: \quad z_{ij}^k = \eta_\beta(x^i, t_k, \theta_\beta^{(0)}) + \varepsilon_{ij},$$

$$i = 1, \ldots, \ell; \quad j = 1, \ldots, r_i; \quad k = 1, \ldots, K; \quad \beta = 1, 2, \quad (5.35)$$

where the observations z_{ij}^k correspond to noise-corrupted solutions of the PDEs (5.34) for some prior values $\theta_\beta^{(0)}$ of the parameters θ_β. This leads to the question of how to select an appropriate sensor location. A serious complication here is the presence of unknown parameter vectors θ_β. They have to be estimated (if not given *a priori*) and one possible approach is to design an experiment for both estimation of θ_β and discrimination between the models [7, 196]. In this section we are interested in model discrimination where one of the models is assumed to be known completely and the parameters of the other model are estimated in the procedure of finding an optimum design. Another complication is that the relationship η_β is defined implicitly through the solution of a PDE, but this obstacle is only of minor importance due to the availability of extremely efficient solvers performing numerical integration of such equations.

5.2.2 Theoretical problem setup

We consider a general nonlinear multiresponse statistical model where the observations $z_{ij}^k \in \mathbb{R}$ of process *responses* are described by

$$z_{ij}^k = \eta(x^i, t_k) + \varepsilon_{ij}, \quad i = 1, \ldots, \ell; \quad j = 1, \ldots, r_i; \quad k = 1, \ldots, K. \quad (5.36)$$

Here the nonlinear functional η is the *true* model of the process, $x^i \in X \subset \bar{\Omega}$ is the spatial coordinate of a measurement, $x_i \neq x_\kappa$ whenever $i \neq \kappa$, X is a known compact set, and the terms ε_{ij} represent random errors of measurements. The errors ε_{ij} are sampled from a distribution satisfying

$$\mathrm{E}[\varepsilon_{ij}] = 0, \quad \mathrm{E}[\varepsilon_{ij}\varepsilon_{\kappa\ell}] = \begin{cases} \sigma^2 & \text{if } i = \kappa \text{ and } j = \ell, \\ 0 & \text{otherwise,} \end{cases} \quad (5.37)$$

where σ^2 is a positive constant. The additional index j is necessary when $r_i > 1$ sensors are located at point x_i. Then the total number of sensors is $\sum_{i=1}^n r_i = N$.

Furthermore, let $\eta_1(x, t, \theta_1)$ and $\eta_2(x, t, \theta_2)$ be two competing model functions. The functions $\eta_1 : \mathbb{R}^{d+1+m_1} \to \mathbb{R}$ and $\eta_2 : \mathbb{R}^{d+1+m_2} \to \mathbb{R}$ are given *a priori*, where $\theta_1 \in \Theta_1 \subset \mathbb{R}^{m_1}$ and $\theta_2 \in \Theta_2 \subset \mathbb{R}^{m_2}$ are vectors of constant but unknown parameters (Θ_1 and Θ_2 denote some known compact sets). The purpose of the experiment is to determine which of the model functions, $\eta_1(x, t, \theta_1)$ or $\eta_2(x, t, \theta_2)$, is the true model $\eta(x, t)$. There is no loss of generality in assuming that it is the first model, i.e., $\eta(x, t) = \eta_1(x, t, \theta_1)$, where θ_1

is regarded as known prior to the experiment (this value could be obtained based on some preliminary experiment). To make the discrimination between the models η_1 and η_2 efficient means to select x_i's to maximize the sum of squares for the lack of fit of the second model defined as follows:

$$T_{12}^0(\xi_N) = \min_{\theta_2 \in \Theta_2} \mathcal{T}_{12}^0(\xi_N, \theta_2), \tag{5.38}$$

where

$$\mathcal{T}_{12}^0(\xi_N, \theta_2) = \sum_{i=1}^{\ell} p_i \sum_{k=1}^{K} \{\eta(x_i, t_k) - \eta_2(x_i, t_k, \theta_2)\}^2, \tag{5.39}$$

and the collection of variables

$$\xi_N = \begin{Bmatrix} x_1, & \ldots, & x_\ell \\ p_1, & \ldots, & p_\ell \end{Bmatrix} \tag{5.40}$$

is, as usual, called the normalized N-observation exact design of the experiment. The x_i's are the design support points and the p_i's are the weights representing the proportions of replications of each design point, i.e.,

$$p_i = \frac{r_i}{N}, \quad \sum_{i=1}^{\ell} p_i = 1,$$

Motivations behind the criterion (5.38) are intuitively clear, as it constitutes a measure of discrepancy between the responses of both the models: a good design for discriminating between the models will then provide a large lack of fit in terms of the sum of squares for the second model [7]. What is more, they are also confirmed by theory. Indeed, assume that the disturbances are Gaussian and independent and, for simplicity, that $\ell = N$ and $p_i = 1/N$, $i = 1, \ldots, N$, i.e.,

$$\xi_N = \begin{Bmatrix} x_1, & \ldots, & x_N \\ 1/N, & \ldots, & 1/N \end{Bmatrix}. \tag{5.41}$$

The ratio of the likelihoods of $\mathcal{M}_1$ and $\mathcal{M}_2$ has the form

$$L = \exp\left\{ \frac{1}{2\sigma^2} \left[\sum_{i=1}^{N}\sum_{k=1}^{K}(z_i^k - \eta_2(x^i, t_k, \theta_2))^2 - \sum_{i=1}^{N}\sum_{k=1}^{K}(z_i^k - \eta_1(x_i, t_k, \theta_1))^2 \right] \right\} \tag{5.42}$$

Now, let us fix estimates $\hat{\theta}_1$ and $\hat{\theta}_2$. It follows that

$$2\sigma^2 \, \mathrm{E}\,[\ln L]$$

$$= \sum_{i=1}^{N}\sum_{k=1}^{K} \mathrm{E}\left[(z_i^k - \eta_2(x^i, t_k, \hat{\theta}_2))^2 - (z_i^k - \eta_1(x^i, t_k, \hat{\theta}_1))^2 \right]$$

$$= \sum_{i=1}^{N}\sum_{k=1}^{K} \mathrm{E}\left[(z_i^k - \eta_2(x^i, t_k, \hat{\theta}_2))^2 \right] - \sum_{i=1}^{N}\sum_{k=1}^{K} \mathrm{E}\left[(z_i^k - \eta_1(x^i, t_k, \hat{\theta}_1))^2 \right]. \tag{5.43}$$

Expressing $z_i^k - \eta_\beta(x^i, t_k, \hat{\theta}_\beta)$ as $z_i^k - \eta(x^i, t_k) + \eta(x^i, t_k) - \eta_\beta(x^i, t_k, \hat{\theta}_\beta)$ for $\beta = 1, 2$, we have

$$
\mathrm{E}\Big[\big(z_i^k - \eta_\beta(x^i, t_k, \hat{\theta}_\beta)\big)^2\Big]
$$

$$
= \mathrm{E}\Big[\big(z_i^k - \eta(x^i, t_k)\big)^2\Big] + \big(\eta(x^i, t_k) - \eta_\beta(x^i, t_k, \hat{\theta}_\beta)\big)^2. \quad (5.44)
$$

Consequently, we obtain the following result:

$$
2\sigma^2 \, \mathrm{E}\,[\ln L]
$$

$$
= \sum_{i=1}^{N}\sum_{k=1}^{K} \mathrm{E}\Big[\big(z_i^k - \eta(x^i, t_k)\big)^2\Big] + \sum_{i=1}^{N}\sum_{k=1}^{K}\big(\eta(x^i, t_k) - \eta_2(x^i, t_k, \hat{\vartheta}_2)\big)^2
$$

$$
- \sum_{i=1}^{N}\sum_{k=1}^{K} \mathrm{E}\Big[\big(z_i^k - \eta(x^i, t_k)\big)^2\Big] - \sum_{i=1}^{N}\sum_{k=1}^{K}\big(\eta(x^i, t_k) - \eta_1(x^i, t_k, \hat{\vartheta}_1)\big)^2
$$

$$
= \sum_{i=1}^{N}\sum_{k=1}^{K}\big(\eta(x^i, t_k) - \eta_2(x^i, t_k, \hat{\vartheta}_2)\big)^2 - \sum_{i=1}^{N}\sum_{k=1}^{K}\big(\eta(x^i, t_k) - \eta_1(x^i, t_k, \hat{\vartheta}_1)\big)^2
$$

$$
= N\mathfrak{T}_{12}^0(\xi_N, \hat{\theta}_2) - \sum_{i=1}^{N}\sum_{k=1}^{K}\big(\eta(x^i, t_k) - \eta_1(x^i, t_k, \hat{\vartheta}_1)\big)^2.
$$

$$(5.45)$$

Note that the last term in the above sum tends to a constant in probability as $N \to \infty$ [104, 121, 265], so that it is stochastically bounded [43] and can be neglected when maximizing $2\,\mathrm{E}\,[\log L]$. In other words,

$$
\mathrm{E}\,[\ln L] \sim \frac{N}{2\sigma^2} T_{12}^0(\xi_N) \quad \text{as } N \to \infty, \qquad (5.46)
$$

i.e., for normally distributed errors ε_{ij} and large N, $T_{12}^0(\xi_N)$ is proportional (in mean) to the logarithm of the ratio of the likelihoods associated with both the models. Accordingly, $\mathrm{E}\,[\log L]$ will maximized if the criterion T_{12} is maximized, cf. also [77, 78].

Clearly, the solutions depend on the assumed choice of the true model, as well as on the values of its parameters θ_1. This implies that a construction of *a priori* solutions is impossible and we are faced with a problem analogous to the search for the so-called *locally optimum designs* while trying to estimate unknown parameters θ_2 of a nonlinear regression function η_2. The dependency of the optimal solution on the model parameters is an unappealing characteristic of nonlinear experimental designs. This predicament can be partially circumvented by relying on a nominal value of θ_1, the results of a preliminary experiment or a sequential design which consists of multiple alternating of experimentation and estimation steps.

Note that the design (5.40) defines a discrete probability distribution on its support points, i.e., on $x_1, \ldots, x_\ell$. This discrete nature of N-observation exact designs causes serious difficulties, as the resultant numerical analysis problem is not amenable to solution by standard optimization techniques, particularly when N is large. A commonly used device for this problem is to extend the definition of the design, cf. Chapter 3. When N is large, the p_i's can be considered as real numbers in the interval $[0,1]$, not necessarily integer multiples of $1/N$. One more step to widen the class of admissible designs further is to consider all probability measures ξ over X which are absolutely continuous with respect to the Lebesgue measure and satisfy by definition the condition

$$\int_X \xi(\mathrm{d}x) = 1. \tag{5.47}$$

The set of all such measures will be denoted by $\Xi(X)$. This relaxation of the original optimization problem makes searching for an optimum design much more tractable, but usually leads to approximate solutions. These, however, are generally accepted in the practice of optimum experimental design, as was already indicated in Chapter 3. This approach simplifies the optimization problem and allows us to derive the necessary conditions for optimality presented in the next section.

Moreover, for a large K and the points $t_1, \ldots, t_K$ densely covering the time interval T, the sum over time instants can be replaced by the integral over T. Accordingly, the continuous generalization of the optimality criterion (5.38) is of the form

$$T_{12}(\xi) = \min_{\vartheta_2 \in \Theta_2} \int_X \Delta_{12}(x, \theta_2)\, \xi(\mathrm{d}x), \tag{5.48}$$

where

$$\Delta_{12}(x, \theta_2) = \int_0^{t_f} \left[\eta(x,t) - \eta_2(x,t,\theta_2) \right]^2 \mathrm{d}t. \tag{5.49}$$

A design

$$\xi^\star = \arg \max_{\xi \in \Xi(X)} T_{12}(\xi) \tag{5.50}$$

is the *locally T_{12}-optimum design $\xi^\star$*.

5.2.3 T_{12}-optimality conditions

The optimization problem defined in (5.50) is far from being trivial and does not possess closed-form solutions in most interesting practical situations, which makes it necessary to look for appropriate numerical approximations.

In the remainder of this section, the following assumptions will be needed:

(A1) X and Θ_2 are compact sets,

(A2) η is a continuous function on $X \times T$,

(A3) η_2 is a continuous functions on $X \times T \times \Theta_2$.

The theorem characterizing T_{12}-optimum designs can be formulated as follows:

THEOREM 5.1

Assume that the minimization problem defined in (5.48) possesses a unique solution $\theta_2^* \in \Theta_2$ for a design ξ^*. Under Assumptions (A1)–(A3) a necessary and sufficient condition for ξ^* to be T_{12}-optimal is that for each $x \in X$,

$$\Delta_{12}(x, \theta_2^*) \le T_{12}(\xi^*). \tag{5.51}$$

The equality in (5.51) is attained at all support points of ξ^*. Furthermore, the set of all the corresponding optimal measures ξ^* is convex.

PROOF First we examine the one-sided directional derivative of T_{12}. For simplicity of notation, we let φ stand for

$$\varphi(\xi, \theta_2) = \int_X \Delta_{12}(x, \theta_2)\, \xi(\mathrm{d}x). \tag{5.52}$$

Then

$$T_{12}(\xi) = \min_{\theta_2 \in \Theta_2} \varphi(\xi, \theta_2). \tag{5.53}$$

The continuity of η and η_2, taken in conjunction with the Bounded Convergence Theorem [248, Cor. 6, p. 161], yields the continuity of the mappings

$$(\alpha, \theta_2) \longmapsto \varphi(\xi + \alpha\delta\xi, \theta_2) \tag{5.54}$$

and

$$(\alpha, \theta_2) \longmapsto \frac{\partial\varphi}{\partial\alpha}(\xi + \alpha\delta\xi, \theta_2). \tag{5.55}$$

Consequently, Theorem 3.3 of [223] implies

$$\delta_+ T_{12}(\xi; \delta\xi) = \min_{\theta_2 \in \Theta_2(\xi)} \delta_+\varphi(\xi, \theta_2; \delta\xi), \tag{5.56}$$

where

$$\Theta_2(\xi) = \left\{ \bar{\theta}_2 \in \Theta_2 : \bar{\theta}_2 = \arg\min_{\theta_2 \in \Theta_2} \varphi(\xi, \theta_2) \right\} \tag{5.57}$$

and $\delta_+\varphi(\xi, \theta_2; \delta\xi)$ stands for the one-sided differential of φ at ξ with increment $\delta\xi$, and θ_2 interpreted as a fixed parameter.

According to the main assumption of the theorem, for an optimal design ξ^*, i.e., the one which maximizes $T_{12}(\xi)$, the set $\Theta_2(\xi^*)$ consists of only one point θ_2^*, and therefore

$$\delta_+ T_{12}(\xi^*; \delta\xi) = \int_X \Delta_{12}(x, \theta_2^*)\, \delta\xi(\mathrm{d}x). \tag{5.58}$$

Putting $\delta\xi = \xi - \xi^*$, after some calculations, we get

$$\delta_+ T_{12}(\xi^*; \xi - \xi^*) = \int_X \psi(x, \xi^*)\,\xi(dx), \qquad (5.59)$$

where

$$\psi(x, \xi) = \Delta_{12}(x, \theta_2^*) - \int_X \Delta_{12}(x, \theta_2^*)\,\xi(dx). \qquad (5.60)$$

It follows that

$$\max_{\xi \in \Xi(X)} \delta_+ T_{12}(\xi^*; \xi - \xi^*) = \max_{\xi \in \Xi(X)} \int_X \psi(x, \xi^*)\,\xi(dx). \qquad (5.61)$$

The optimality of the design ξ^* implies that (5.61) must be nonpositive. Note, however, that we have

$$\int_X \psi(x, \xi^*)\,\xi^*(dx) = 0, \qquad (5.62)$$

which forces the nonnegativity of the maximum on the right-hand side of (5.61). From this we see that

$$\max_{\xi \in \Xi(X)} \int_X \psi(x, \xi^*)\,\xi(dx) = 0. \qquad (5.63)$$

Clearly, condition (5.63) is sufficient, too. In fact, for a fixed $\theta_2 \in \Theta_2$, φ in (5.52) is a linear function of ξ, and hence T_{12} becomes concave, cf. Theorem B.21, which means that the necessary condition (5.63) becomes sufficient as well.

We also claim that the mapping $x \mapsto \psi(x, \xi^*)$ attains its maximum value of zero at all the support points of ξ^*. Indeed, suppose that this were false. Then we could find a set $X' \subset \operatorname{supp}\xi^*$ and a scalar a such that

$$\int_{X'} \psi(x, \xi^*)\,\xi^*(dx) \le a < 0 \qquad (5.64)$$

and

$$\psi(x, \xi^*) = 0 \quad \text{for } x \in \operatorname{supp}\xi^* \setminus X'. \qquad (5.65)$$

But this would yield

$$\int_X \psi(x, \xi^*)\,\xi^*(dx) \le a < 0, \qquad (5.66)$$

which contradicts (5.62).

It remains for us to show that the set of all optimal measures ξ^* is convex. But this is immediate, since the mapping $\xi \mapsto T_{12}(\xi)$ is concave. This finishes the proof. ∎

Note that Theorem 5.1 might be generalized to a situation when the parameter estimate θ_2^* is not unique, which is related to parameter nonidentifiability (the proof proceeds on the same lines, but the resultant form of (5.51) is much more cumbersome). In practice, however, this problem will not arise if in the numerical construction of designs a regularization term is added to the least-squares equations for θ_2, i.e., if we restrict our attention to designs of the form $\xi = (1 - \alpha)\xi_p + \alpha\bar{\xi}$, where α is a small fixed number, $\xi_p \in \Xi(X)$ and $\bar{\xi}$ is a regular discrete design for which the least-squares equations possess a unique solution.

5.2.4 Numerical construction of T_{12}-optimum designs

Theorem 5.1 provides us with a test for the optimality and is used to verify that an intuitively sensible design measure is (or is not) T_{12}-optimal. However, to exploit its full potential, we require more than this, i.e., we need algorithms that would enable us to construct T_{12}-optimal design measures. One possibility is to try to adapt the iterative scheme suggested in [9,85], where each step consists in looking for a most informative point $x \in X$, including it in the set of support points of the current design and then updating the weights. Here we take a slightly different approach in which we treat the maximin problem as a nondifferentiable, or nonsmooth, optimization problem to which general numerical algorithms can be applied. For this purpose, we assume that an optimal design has the form (5.40) for a fixed ℓ which should be chosen as a sufficiently large number (in practice, we may only guess this value, as no theoretical results exist regarding the number of support points in T-optimal designs).

To shorten notation, let us introduce the vector

$$\gamma = (x_1, \ldots, x_\ell, p_1, \ldots, p_\ell) \in \Gamma, \tag{5.67}$$

where

$$\Gamma = \left\{ \gamma \in X^\ell \times [0,1]^\ell : \sum_{i=1}^{\ell} p_i = 1 \right\}, \tag{5.68}$$

and write

$$J(\gamma, \theta) = \sum_{i=1}^{\ell} p_i \int_0^{t_f} \left[\eta(x_i, t) - \eta_2(x_i, t, \theta) \right]^2 dt. \tag{5.69}$$

Thus our original optimum experimental design problem reduces to finding

$$\gamma^* = \arg\max_{\gamma \in \Gamma} \left\{ \min_{\theta_2 \in \Theta_2} J(\gamma, \theta_2) \right\}. \tag{5.70}$$

The operation of taking the pointwise minimum of an infinite number of functions (i.e., the elements of the family $\left\{ J(\cdot, \theta_2) \right\}_{\theta_2 \in \Theta_2}$) generates, in general, a nonsmooth function. Our approach is to try to eliminate the nondifferentiability by transforming the nonsmooth problem into a smooth one.

Let us observe first that the initial maximin optimization problem (5.70) can be viewed as the maximization of a scalar α, subject to the constraint

$$\min_{\theta_2 \in \Theta_2} J(\gamma, \theta_2) \geq \alpha \qquad (5.71)$$

which, in turn, is equivalent to the infinite set of constraints

$$\{ J(\gamma, \theta_2) \geq \alpha, \quad \theta_2 \in \Theta_2 \}. \qquad (5.72)$$

This task can be solved using some algorithms for inequality-constrained semi-inifinite optimization [61, 62, 113, 214, 215, 250, 257], but in practice the simple relaxation algorithm proposed in [266] turns out to perform well in the considered nonlinear experimental design problems, as is also suggested in [222, 349]. It consists in relaxing the problem by taking into account only a finite number of constraints (5.72). The relaxation procedure is represented by the following steps:

ALGORITHM 5.1 Finding T_{12}-optimum designs

Step 1. *Choose an initial parameter vector $\theta_2^{(1)} \in \Theta_2$ and define the first set of representative values $\mathbb{Z}^{(1)} = \{\theta_2^{(1)}\}$. Set $k = 1$.*

Step 2. *Solve the current relaxed problem*

$$\gamma^{(k)} = \arg \max_{\gamma \in \Gamma} \left\{ \min_{\theta_2 \in \mathbb{Z}^{(k)}} J(\gamma, \theta_2) \right\}.$$

Step 3. *Solve the minimization problem*

$$\theta_2^{(k+1)} = \arg \min_{\theta_2 \in \Theta_2} J(\gamma^{(k)}, \theta_2).$$

Step 4. *If*

$$J(\gamma^{(k)}, \theta_2^{(k+1)}) \geq (1 - \epsilon) \min_{\theta_2 \in \mathbb{Z}^{(k)}} J(\gamma^{(k)}, \theta_2),$$

where ϵ is a small predetermined positive constant, then $\gamma^{(k)}$ is a sought maximin solution, otherwise include $\theta_2^{(k+1)}$ into $\mathbb{Z}^{(k)}$, increment k, and go to Step 2.

It can be proved [266] that the above algorithm terminates in a finite number of iterations for any given $\epsilon > 0$. The technique can be regarded as a nonlinear cutting-plane method.

Computationally, Step 3 is crucial and also it is the most time-consuming step in the algorithm. Complications arise, among other things, due to the necessity of calculating a global minimum of $J(\gamma^{(k)}, \cdot)$ which is usually multimodal (getting stuck in one of local minima leads to premature termination of the algorithm). Therefore, while implementing this part of the computational procedure an effective global optimizer is essential. Based on numerous computer experiments it was found that the adaptive random search (ARS) strategy discussed in Section 3.1.3.1 is especially suited for that purpose if the admissible set Θ_2 is a hypercube, i.e., the admissible range for θ_{2i}, $i = 1, \ldots, p_2$ is in the form

$$\theta_{2i\,\min} \leq \theta_{2i} \leq \theta_{2i\,\max}. \qquad (5.73)$$

As to maximization in Step 2, local methods permitting nonlinear constraints perform well. Clearly, they may occasionally fail to find the global maximum; then a way to alleviate this problem is to run the corresponding maximization algorithm with a pattern of different starting points.

Less obvious is the need for scaling the objective function J whose values may turn out to be very small (e.g., of an order of 10^{-4}). Neglecting this factor will trigger termination criteria prematurely [202]. Without scaling, the objective function becomes substantially 'flat' at some iteration point, including the optimum. Scaling, that is multiplying the objective function by a constant, increases its sensitivity to changes in the variables or constraint values and eliminates the flatness.

Example 5.2
Consider the following simple model of an air pollutant proliferation:

$$\frac{\partial y(x,t)}{\partial t} - \Delta y(x,t) = 10\exp\left[-100((x_1 - 0.25)^2 + (x_2 - 0.25)^2)\right] \qquad (5.74)$$
$$+ 10\exp\left[-100((x_1 - 0.75)^2 + (x_2 - 0.75)^2)\right],$$

where $x \in \Omega = [0,1]^2$, $t \in T = [0,1]$, subject to the boundary and initial conditions

$$\begin{cases} y(x,0) = 0.1, & x \in \Omega, \\ y(x,t) = 0.1(1-t), & (x,t) \in \partial\Omega \times T. \end{cases} \qquad (5.75)$$

The state y can be interpreted as the pollutant concentration, and the forcing input on the right-hand side of (5.74) corresponds to two pollution sources modelled by Gaussian functions with the centres located at points $a_1 = (0.25, 0.25)$ and $a_2 = (0.75, 0.75)$.

The above description constitutes the reference model corresponding to a normal state of the system. The alternative model differs from the reference model only on the forcing input because it assumes the presence of only one source of pollution, i.e.,

$$\frac{\partial y(x,t)}{\partial t} - \Delta y(x,t) = \alpha\exp\left[-\beta\left((x_1 - \gamma)^2 + (x_2 - \rho)^2\right)\right] \qquad (5.76)$$

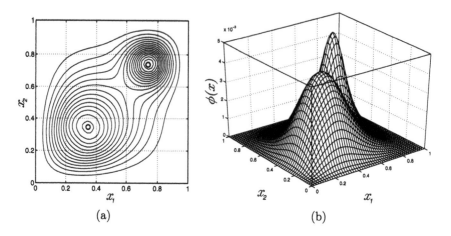

(a)   (b)

FIGURE 5.2
Contour (a) and surface (b) plots of the sensitivity function $\phi(x)$ in Example 5.2.

on $\Omega \times T$, subject to the same boundary and initial conditions as those in the reference model. Thus we obtain the vector of parameters $\theta_2 = \text{col}[\alpha, \beta, \gamma, \rho]$.

By appropriately selecting constraints imposed on the unknown parameters, we can make the alternative model correspond to situations of a significant increase or a decrease in the emission from one of the pollution sources. Thus the task of a proper recognition of the actual system state leads to discrimination between both the models. A T-optimum experimental design can be treated as a first step towards the detection and diagnosis of an abnormal mode of system functioning.

In order to determine a T-optimum design, a program was written completely in the Lahey-Fujitsu Fortran 95 v 5.7 environment supplemented with the IMSL numerical library. Solutions of all PDEs were obtained using the finite-element method. As for implementation of Algorithm 5.1, minimization with respect to the parameters was performed using the ARS algorithm coupled with the local search based on the Levenberg-Marquardt algorithm, while maximization with respect to the design was accomplished using the sequential-quadratic programming method.

In our computational study the following nominal ranges for the unknown parameters were used:

$$10 \le \alpha \le 30, \quad 250 \le \beta \le 350, \quad 0.15 \le \gamma \le 0.25, \quad 0.15 \le \rho \le 0.25. \quad (5.77)$$

In the optimization process, the maximum of the forcing input in the alternative model was found in the vicinity of the source a_1 in the reference model.

The ultimate approximation of the optimum design consists of two points:

$$\xi^\star = \left\{ \begin{matrix} (0.7377, 0.7377), & (0.3453, 0.3453) \\ 0.8889, & 0.1111 \end{matrix} \right\}. \qquad (5.78)$$

The values of the parameters corresponding to the largest values of the T-optimality criterion at $\xi^\star$ were $\theta_2^\star = \mathrm{col}[10, 250, 0.25, 0.25]$. The contour and surface plots of the sensitivity function

$$\phi(x) = \int_T |\eta(x, t) - \eta_2(x, t; \theta_2^\star)|^2 \, dt \qquad (5.79)$$

are presented in Fig. 5.2 together with the optimal locations of the design support points. It is easy to see that the sensors must be placed in the vicinity of both emission sources in the reference model. These locations simultaneously constitute the two global maxima of the sensitivity function, which conforms to Theorem 5.1.   □

5.3   Conclusions

This chapter has first considered the question of the achievable accuracy in response prediction and forecasting. In order to tackle this problem mathematically, three intuitively clear design criteria in output space were proposed as measures of the prediction accuracy. The nondifferentiability of two of them was also addressed and an approach to alleviate this difficulty was proposed. Such a formulation may constitute a potential solution to fill the gap existing between refined optimum experimental design theory which has been developed for years with emphasis on parameter space, and modern control theory where output space criteria are by all means more important than those which are primarily related to the accuracy of parameter values.

In turn, the discussion in the second part of the chapter has concentrated on the problem of optimum experimental design for model structure discrimination, where the T-optimality criterion proposed by Atkinson and Fedorov was tailored to the framework of continuous-time measurements. Its use involves an SDP problem, but such problems appear frequently in engineering, cf. [113, 214], and well-developed methods of their solution exist and are still being improved. The increasing power of modern computers makes such large-scale computations routine, so that the interest in using minimax criteria such as the T-optimality one is expected to be revived. This issue is going to be addressed once more in the next chapter on robust designs.

6

Robust designs for sensor location

As was already emphasized (cf. Section 2.4.4), one of the main difficulties associated with optimization of sensor locations is the dependence of optimal solutions on the true values θ_{true} of the parameters to be estimated. Since these values are unknown, an obvious and common approach is to use one of the locally optimal designs described in previous chapters for some prior estimate θ^0 of θ_{true} in lieu of θ_{true} itself (it can be, e.g., a nominal value for θ or a result of a preliminary experiment). But θ^0 may be far from θ_{true} and, simultaneously, properties of locally optimal designs can be very sensitive to changes in θ [92]. Such prior uncertainty on θ^0 is not taken into account by any optimization procedure to determine local designs and an experimental setting thus obtained may consequently be far from optimal. This has even raised some doubts among experimenters about the practical use of nonlinear experimental design at all [348].

Several more cautious approaches have been proposed so far in an attempt to surmount this difficulty, but none of them is flawless and the problem still remains a real challenge for researchers. The aim of this chapter is to briefly outline the existing methods of making experimental designs independent of the true parameter values and to discuss how these methods can be adopted in the framework of the sensor location problem where, to the best of our knowledge, robust approaches have not been applied yet.

6.1 Sequential designs

Since a good choice of design depends on true parameter values, a very natural idea is to alternate experimentation and estimation steps. Accordingly, the total time horizon is divided into several contiguous parts and each of them is related to the corresponding stage of the experiment. At each stage, in turn, a locally optimal sensor location is determined based on the available parameter estimates (nominal parameter values can be assumed as initial guesses for the first stage), measurements are taken at the newly calculated sensor positions, and the data obtained are then analysed and used to update the parameter estimates (see Fig. 6.1). In this general scheme, it is intuitively supposed that each estimation phase improves our knowledge about the parameters and this

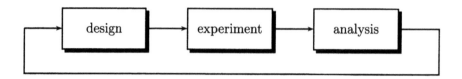

FIGURE 6.1
A general scheme of sequential design.

knowledge can then be used to improve the quality of the next experiment to be performed.

Owing to its simplicity, sequential design is commonly considered as a universal panacea for the shortcomings of local designs. Let us note, however, that the following important questions are to be faced [92] and the answers to them are by no means straightforward:

1. How many subintervals should be chosen?

2. How do the initial estimates of parameters influence the design?

3. What are the asymptotic properties of sequential procedures, i.e., does the generated design 'tend' in any sense to a design which would be optimal in terms of the true θ?

Some developments regarding a theoretical justification for the sequential approach and its convergence properties can be found, e.g., in [92, 348, 349]. But even though this technique can be warranted to a certain extent, it is often impractical because the required experimental time may be too long and the experimental cost may be too high.

In engineering practice it is sometimes known that θ_{true} belongs to a given compact set Θ_{ad}. In such a case, the following 'naïve' technique is often employed [85]: Θ_{ad} is covered with an appropriate grid $\bar{\Theta}_{\text{ad}}$ of a reasonable size and then the behaviour of $\Psi[M(s^\star(\theta^i), \theta^j)]$ is analysed for potential pairs $(\theta^i, \theta^j) \in \bar{\Theta}_{\text{ad}}^2$, where $M(s, \theta)$ stands for the FIM calculated at θ for a design s and $s^\star(\theta) = \arg\min_{s \in S_{\text{ad}}} \Psi[M(s, \theta)]$. Consequently, a 'compromise' design can be determined such that it is good enough for any θ from the discretized set Θ_{ad}. Clearly, this may involve tremendeous calculations if the number of grid nodes is large. Nevertheless, this approach constitutes an origin for more systematic methods of robust design which are delineated in what follows.

6.2 Optimal designs in the average sense

6.2.1 Problem statement

If it is known that the range of possible θ values reduces to a given compact set Θ_{ad}, then a more cautious approach to the control of the properties of the sensor location over Θ_{ad} consists in a probabilistic description of the prior uncertainty in θ, characterized by a prior distribution ς which may have been inferred, e.g., from previous observations collected on similar processes. The criterion to be minimized is then the expectation of the corresponding local optimality criterion J over Θ_{ad}, i.e.,

$$J_{\mathrm{E}}(s) = \mathrm{E}_{\theta}\{J(s, \theta)\} = \mathrm{E}_{\theta}\{\Psi[M(s, \theta)]\} = \int_{\Theta_{\mathrm{ad}}} \Psi[M(s, \theta)]\,\varsigma(\mathrm{d}\theta), \qquad (6.1)$$

Using ς makes it possible to remove the dependence of the FIM on θ.

As for possible choices of ς, it is customary to assume that

$$\varsigma(\mathrm{d}\theta) = p(\theta)\,\mathrm{d}\theta, \qquad (6.2)$$

where p signifies the prior probability density function for θ. Some examples of p are as follows [282]:

- If the true θ is known exactly as θ^0, then we have

$$p(\theta) = \delta(\theta - \theta^0),$$

  where θ^0 is the Dirac delta function.

- If θ is limited to a region Θ_{ad}, but we have no other information, then we can assume a uniform distribution on Θ_{ad}, i.e.,

$$p(\theta) = \begin{cases} 1/\operatorname{meas}(\Theta_{\mathrm{ad}}) & \text{if } \theta \in \Theta_{\mathrm{ad}}, \\ 0 & \text{otherwise}, \end{cases}$$

  where $\operatorname{meas}(\Theta_{\mathrm{ad}})$ stands for the Lebesgue measure of Θ_{ad}.

- If the expected value μ_{θ} and the covariance matrix V_{θ} of θ can be estimated, but we have no other information, we can assume that p is the probability density function for a multivariate normal distribution

$$p(\theta) = \frac{1}{(2\pi)^{m/2}(\det V_{\theta})^{1/2}} \exp\left[-\frac{1}{2}(\theta - \mu_{\theta})^{\mathsf{T}} V_{\theta}^{-1}(\theta - \mu_{\theta})\right],$$

  where a cutoff and an appropriate normalization should be additionally imposed if this $p(\theta)$ does not diminish to a negligibly small value before the limits of Θ_{ad} are reached.

A notable feature of the approach is that the number of expectation criteria is greater than the number of their counterparts in the local case. Indeed, for the most popular D-optimal design we have, e.g., the following choices [349]:

- ED-optimal design, which maximizes

$$J_{\text{ED}}(s) = \text{E}_\theta\{\det(M(s,\theta))\}, \qquad (6.3)$$

- EID-optimal design, which minimizes

$$J_{\text{EID}}(s) = \text{E}_\theta\{1/\det(M(s,\theta))\}, \qquad (6.4)$$

- ELD-optimal design, which maximizes

$$J_{\text{ELD}}(s) = \text{E}_\theta\{\ln\det(M(s,\theta))\}. \qquad (6.5)$$

It turns out that the above criteria usually yield different optimal solutions and hence care must be exercised while adopting a particular option (their advantages and drawbacks are discussed in [349]). The other cost functions could be handled in a similar manner.

The extension of the approach to continuous designs of Section 3.3 does not present a problem, as we may introduce

$$\Psi_{\text{E}}(\xi) = \int_{\Theta_{\text{ad}}} \Psi[M(\xi,\theta)]\,\varsigma(\text{d}\theta). \qquad (6.6)$$

Since integrating acts as a linear operator, Theorem 3.1 of p. 45 and Theorem 3.2 of p. 48 can be rewritten in this new framework, practically without any changes by introducing

$$\psi(x,\xi) = \int_{\Theta_{\text{ad}}} \hat{\psi}(x,\xi,\theta)\,\varsigma(\text{d}\theta), \qquad (6.7)$$

$$\phi(x,\xi) = \int_{\Theta_{\text{ad}}} \hat{\phi}(x,\xi,\theta)\,\varsigma(\text{d}\theta), \qquad (6.8)$$

$$c(\xi) = \int_{\Theta_{\text{ad}}} \hat{c}(\xi,\theta)\,\varsigma(\text{d}\theta), \qquad (6.9)$$

where $\hat{\psi}(x,\xi,\theta)$, $\hat{\phi}(x,\xi,\theta)$ and $\hat{c}(\xi,\theta)$ denote the respective quantities of Section 3.3 calculated for an indicated parameter vector θ. There are only two striking differences. First of all, Carathéodory's theorem cannot be directly applied since Ψ_{E} depends on different matrices $M(\xi,\theta)$ for different θ's, which implies in turn that the existence of an optimal design with no more than $m(m+1)/2$ support points is no longer guaranteed. Secondly, except for very special situations, an optimal design cannot be obtained analytically and algorithmic procedures are thus needed. Theorem 3.2 provides a basis for efficient numerical algorithms to determine approximations to locally optimal designs. Unfortunately, its counterpart for an expectation criterion is conctructive only if the prior distribution ς is discrete with a moderate number of support points. Clearly, this remark pertains to any numerical scheme described in previous chapters, since the main intricacy remains the same: in order to directly minimize (6.1) or (6.6), we have to evaluate expectations, i.e., multidimensional integrals, which is extremely time consuming. Luckily, it turns out that approximations to an optimal design can be determined without any evaluation of mathematical expectation. This constitutes the subject of the next subsection.

6.2.2 Stochastic-approximation algorithms

A direct minimization of (6.1) is highly complicated by the fact that an expected value of a local cost function has to be evaluated, which is plausible only when the prior distribution ς is discrete. Let us observe, however, that the situation is by no means hopeless, as this framework is typical for the application of stochastic-approximation techniques. Indeed, it is standard for a stochastic optimization problem that the objective function is not explicitly known [210, 274], i.e., that there is no computer programme which finds its exact value at each value of the decision variable in a reasonable time (otherwise, we would just have to solve a deterministic nonlinear optimization problem). Based on the validity of the law of large numbers, a stochastic optimization problem is approximated in such a way that the uncertain random quantities in the original problem are replaced by artificially generated random variables. If these random variables are produced in advance to construct an approximate empirical problem, then we deal with the so-called nonrecursive methods being part of the broad family of Monte Carlo methods. From a practical point of view, recursive methods are sometimes more interesting. In these methods random samples are drawn only at the moment when they are requested. The total number of such random draws does not have to be determined at the beginning, but it can be adaptively chosen during the progress of estimation [148, 210].

In the context of a nonlinear experimental design, the idea to employ algorithms of stochastic approximation was suggested and successfully applied to robust-design problems in [221, 347, 349]. Owing to evident similarities of that setting to the sensor location problem considered in our monograph, the same technique can be put into execution in this slightly modified framework. A simple classical Robbins-Monro algorithm, also known as the stochastic-gradient algorithm, corresponds to the following iterative procedure:

$$s^{(k+1)} = \Pi_{\mathcal{S}_{\mathrm{ad}}} \left(s^{(k)} - \gamma_k \left(\frac{\partial \Psi[M(s, \theta^{(k)})]}{\partial s} \right)^{\mathsf{T}}_{s=s^{(k)}} \right), \quad k = 0, 1, \ldots, \quad (6.10)$$

where $\theta^{(k)}$ is randomly generated according to the prior distribution ς and

$\Pi_{\mathcal{S}_{\mathrm{ad}}}$ denotes the orthogonal projection onto the set $\mathcal{S}_{\mathrm{ad}} = X^N$, where X signifies a spatial zone where the measurements are allowed (a compact subset of $\bar{\Omega}$). The sequence of decreasing scalar steps $\{\gamma_k\}$ must guarantee an implicit averaging of the outcomes of the simulation, which is attained if the following conditions are satisfied [148]:

$$\gamma_k \geq 0, \quad \sum_{k=0}^{\infty} \gamma_k = \infty, \quad \sum_{k=0}^{\infty} \gamma_k^2 < \infty. \tag{6.11}$$

The most common practice is to use the harmonic sequence

$$\gamma_k = \frac{b}{k+1}, \quad b > 0, \quad k = 0, 1, \dots. \tag{6.12}$$

The Robbins-Monro procedure requires the existence of unbiased estimates of the gradient $\partial \mathrm{E}_\theta \{\Psi[M(s,\theta)]\}/\partial s$, but it is a simple matter to check that

$$
\mathrm{E}_\theta \left\{ \frac{\partial \Psi[M(s,\theta)]}{\partial s} \right\} = \int_{\Theta_{\mathrm{ad}}} \frac{\partial \Psi[M(s,\theta)]}{\partial s} \varsigma(\mathrm{d}\theta)
$$
$$
= \frac{\partial}{\partial s} \int_{\Theta_{\mathrm{ad}}} \Psi[M(s,\theta)] \varsigma(\mathrm{d}\theta) = \frac{\partial}{\partial s} \mathrm{E}_\theta \{\Psi[M(s,\theta)]\}, \tag{6.13}
$$

provided that all derivatives $\partial^2 y/\partial x_i \partial \theta_j$ are continuous in $\bar{\Omega} \times T \times \Theta_{\mathrm{ad}}$. Consequently, the quantity $\partial \Psi[M(s, \theta^{(k)})]/\partial s$ in (6.10) constitutes an unbiased estimate of the gradient of the expectation criterion (6.1).

Under some classical assumptions [78, 148, 210] which are satisfied when the system state y is sufficiently smooth, convergence almost surely of the algorithm (6.10) is ensured. Note, however, that convergence to a global minimum is not guaranteed. As pointed out in [221], it can be accelerated by changing the value of $s^{(k)}$ only when the angle between the gradients at iterations $k-1$ and k is greater than $\pi/2$.

Projection $\Pi_{\mathcal{S}_{\mathrm{ad}}}(s)$ denotes the closest point in $\mathcal{S}_{\mathrm{ad}}$ to s and is introduced to avoid the situations where $s^{(k+1)}$ does not belong to $\mathcal{S}_{\mathrm{ad}}$. The uniqueness of such a mapping is guaranteed if $\mathcal{S}_{\mathrm{ad}}$ is convex. But if the closest point fails to be unique, a closest point should be selected such that the function $\Pi_{\mathcal{S}_{\mathrm{ad}}}(\cdot)$ is measurable [148, p. 100]. Let us recall that the projection can sometimes be performed without resorting to sophisticated optimization algorithms. For example, if $\mathcal{S}_{\mathrm{ad}}$ is a hyperrectangle, i.e., there are real numbers $a_i < b_i$, $i = 1, \dots, n$ such that $\mathcal{S}_{\mathrm{ad}} = \{s = (s_1, \dots, s_n) : a_i \leq s_i \leq b_i, \ i = 1, \dots, n\}$, then we have

$$
\left[\Pi_{\mathcal{S}_{\mathrm{ad}}}(s)\right]_i = \begin{cases} b_i & \text{if } s_i > b_i, \\ s_i & \text{if } a_i \leq s_i \leq b_i, \\ a_i & \text{if } s_i < a_i, \end{cases} \tag{6.14}
$$

where $[\cdot]_i$ is the i-th component of a vector, $i = 1, \dots, n$.

When solving sensor-location problems, it is occasionally necessary to include additional constraints regarding, e.g., the admissible distances between the sensors since, as pointed out in [347], robust designs based on expectaton criteria inherit many properties of local designs, including replication of measurements, which in our context means that sensor clusterization may be observed. Formally, the corresponding constraints can be taken into account by an appropriate redefinition of the admissible set S_{ad}, but this would essentially complicate the projection. Note, however, that the constraints on the distances are merely a guide in that they should not be violated by much, but they *can* be violated, i.e., we simply deal with the so-called *soft* constraints. Such constraints can be added to the Robbins-Monro algorithm directly by adding appropriate penalty-function terms to the performance index [148, p. 120]. The idea is more or less obvious and therefore the corresponding details are omitted.

The convergence result tells us that we will get the desired point if we let the procedure run for a sufficiently long time. Clearly, such a statement is unsatisfactory for practical purposes. What is really needed is a statement about the precision of $s^{(k)}$ for a finite k. This would allow us to make a decision about whether the procedure should be terminated. But this topic is also classical and the corresponding results regarding stopping criteria and confidence regions for the solutions can be found in [210, p. 297].

Another question is the optimum measurement problem for moving sensors. Of course, this case can be easily reduced to a static framework after a parameterization of the trajectories, but it turns out that the Robbins-Monro algorithm can also be generalized to minimizing noisy functionals in infinite-dimensional, real separable Hilbert spaces. This requires operating on H-valued random variables and the theoretical results are scattered in the literature, see, e.g., $[21, 70, 100, 147, 192, 253, 254, 267, 346, 354]$, but despite all that, such a generalization can still be done. In $[305, 306, 309, 325]$ based on the general description of the sensor motions,

$$\dot{s}(t) = f(s(t), u(t)) \quad \text{a.e. on } T, \quad s(0) = s_0, \tag{6.15}$$

the following general form of the performance index to be minimized was considered:

$$J_{\text{E}}(s_0, u) = \text{E}_\theta\big\{J(s_0, u, \theta)\big\}, \tag{6.16}$$

where $J(s_0, u, \theta) = \Psi[M(s, \theta)]$. Here u is assumed to be an element in the set

$$\mathcal{U} = \big\{u \in L^2(T; \mathbb{R}^r) : u_l \le u(t) \le u_u \ \text{a.e. on } T\big\} \tag{6.17}$$

and $s_0 \in S_{\text{ad}}$. This determines the set of feasible pairs (s_0, u) which will be denoted by $\mathcal{F} = S_{\text{ad}} \times \mathcal{U}$. Clearly, $\mathcal{U}$ is a closed convex set of $L^2(T; \mathbb{R}^r)$ and if S_{ad} is convex (by assumption it is closed), then so is $\mathcal{F}$ treated as a subset of the separable Hilbert space $\mathcal{H} = \mathbb{R}^n \times L^2(T; \mathbb{R}^r)$.

The corresponding version of the Robbins-Monro stochastic-gradient algorithm is

$$(s_0^{(k+1)}, u^{(k+1)}) = \Pi_{\mathcal{F}}\big((s_0^{(k)}, u^{(k)}) - \gamma_k \nabla J(s_0^{(k)}, u^{(k)}, \theta^{(k)})\big), \quad k = 0, 1, \ldots,$$
(6.18)

where $\nabla J(s_0^{(k)}, u^{(k)}, \theta^{(k)})$ stands for the gradient of $J(\cdot, \cdot, \theta^{(k)})$ calculated at $(s_0^{(k)}, u^{(k)})$ and $\Pi_{\mathcal{F}}$ denotes the orthogonal projection onto $\mathcal{F}$ in $\mathcal{H}$. It follows easily that $\Pi_{\mathcal{F}}(s_0, u) = (\Pi_{S_{ad}}(s_0), \Pi_{\mathcal{U}}(u))$ and

$$[\Pi_{\mathcal{U}}(u)]_i \, (t) = \begin{cases} u_{ui} & \text{if } u_i(t) > u_{ui}, \\ u_i(t) & \text{if } u_{li} \le u_i(t) \le u_{ui}, \\ u_{li} & \text{if } u_i(t) < u_{li}. \end{cases}$$
(6.19)

As regards computation of $\nabla J(s_0^{(k)}, u^{(k)}, \theta^{(k)}) \in \mathcal{H}$, it may be easily concluded that

$$\nabla J(s_0^{(k)}, u^{(k)}, \theta^{(k)}) = (\zeta(0), f_u^{\mathsf{T}}(\cdot)\zeta(\cdot)),$$
(6.20)

where the adjoint mapping ζ solves the Cauchy problem

$$\dot{\zeta}(t) + f_s^{\mathsf{T}}(t)\zeta(t) = -\sum_{i=1}^{m}\sum_{j=1}^{m} c_{ij}\left(\frac{\partial \chi_{ij}}{\partial s}\right)^{\mathsf{T}}_{\substack{s=s(t) \\ \theta=\theta^{(k)}}}, \quad \zeta(t_f) = 0,$$
(6.21)

$\langle \cdot, \cdot \rangle$ stands for the inner product in the appropriate Euclidean space, χ_{ij}'s are defined in (4.117), and c_{ij}'s are the components of the matrix

$$\overset{\circ}{\Psi}(s) = \{c_{ij}\}_{m \times m} = \left.\frac{\partial \Psi[M]}{\partial M}\right|_{M=M(s,\theta^{(k)})}.$$
(6.22)

The derivations are in principle the same as in Appendix H.1, but a thorough proof necessitates additional assumptions on f and sensitivity coefficients $\partial y/\partial \theta_i$ (they should be Lipschitz continuously differentiable on bounded sets), as well as an introduction of some supplementary notions and hence it is omitted. The interested reader is referred to Appendix 5.6 of [215], where analogous technicalities are exhaustively treated within the framework of a general optimal-control problem.

The above results suggest that solutions to robust sensor-location problems with minimax criteria can be obtained almost as simply as those for classical local design criteria, which constitutes a sound argument for the delineated approach. Note, however, that this assertion concerns only the manner in which the computations are organized. The approach itself sometimes raises the objection that it is not clear that values of Ψ are directly comparable for different values of θ. Moreover, the locally optimal values of Ψ may vary considerably with θ. Hence the resulting robust designs may tend to look like locally optimal designs for θ values with large associated variances. Consequently, it is a good idea to have alternative approaches in order to compare and analyse the obtained solutions. An additional option is offered by minimax criteria.

6.3 Optimal designs in the minimax sense

6.3.1 Problem statement and characterization

An experiment which is good on the average may prove very poor for some particular values of the parameters associated with very low probability densities [222]. If we do not accept such a situation, then a way out is to optimize the worst possible performance of the experiment over the prior admissible domain for the parameters Θ_{ad}, which leads to minimization of the criterion

$$J_{\mathrm{MM}}(s) = \max_{\theta \in \Theta_{\mathrm{ad}}} J(s, \theta) = \max_{\theta \in \Theta_{\mathrm{ad}}} \Psi[M(s, \theta)]. \tag{6.23}$$

In other words, we provide maximum information to a parameter vector θ which is the most difficult to identify in Θ_{ad}. For example, we may seek to maximize the MMD-optimality criterion [349]

$$J_{\mathrm{MMD}}(s) = \min_{\theta \in \Theta_{\mathrm{ad}}} \det(M(s, \theta)). \tag{6.24}$$

Thus the best experimental conditions in the worst circumstances are preferred to the best ones on the average.

Clearly, the same minimax approach can be taken in the case of continuous designs studied in Section 3.3, viz. we can consider minimization of the performance index

$$J_{\mathrm{MM}}(\xi) = \max_{\theta \in \Theta_{\mathrm{ad}}} \Psi[M(\xi, \theta)] \tag{6.25}$$

and we shall start on this framework, as some nonobvious characterizations of the corresponding optimum designs can be derived prior to resorting to numerical methods. The main idea is to observe that this setting can be treated in much the same way as that of nondifferentiable design criteria, where an optimal design

$$\xi_{\mathrm{M}}^{\star} = \arg\min_{\xi \in \Xi(X)} \max_{a \in A} \Psi[M(\xi), a] \tag{6.26}$$

is to be determined, A being a given compact set of parameters which cannot be controlled by the experimenter, see Section 3.2 for the relevant details.

Consider the problem of minimizing the criterion (6.25), where

$$M(\xi, \theta) = \int_X \Upsilon(x, \theta) \, \xi(\mathrm{d}x) \tag{6.27}$$

and

$$\Upsilon(x, \theta) = \frac{1}{t_f} \int_0^{t_f} f(x, t, \theta) f^{\mathsf{T}}(x, t, \theta) \, \mathrm{d}t,$$

for the linear regression (3.1). As regards the regularity of f, assume that it is continuous in $X \times T \times \Theta_{\mathrm{ad}}$. The resulting design problem can be handled in much the same way as (6.26). For example, in the case of the D-optimality criterion analysis analogous to that in the proof of Theorem 3.5 implies that a necessary and sufficient condition for $\xi^\star$ to be optimal is the existence of a probability measure $\omega^\star$ on $\Theta(\xi^\star)$ such that

$$\max_{x \in X} \frac{1}{t_f} \int_{\Theta(\xi^\star)} \left\{ \int_0^{t_f} f^{\mathsf{T}}(x, t, \theta) M^{-1}(\xi^\star, \theta) f(x, t, \theta) \, dt \right\} \omega^\star(d\theta) \le m, \quad (6.28)$$

where

$$\Theta(\xi) = \{\hat{\theta} \in \Theta_{\mathrm{ad}} : \det(M(\xi, \hat{\theta})) = \min_{\theta \in \Theta_{\mathrm{ad}}} \det(M(\xi, \theta))\}. \quad (6.29)$$

The Carathéodory theorem ensures the existence of $\omega^\star$ with no more than $m+1$ support points. Obviously, this conclusion remains the same if we replace $f(x, t, \theta)$ by the sensitivity coefficients $(\partial y(x, t; \theta)/\partial \theta)^{\mathsf{T}}$. Note, however, that implementation of the foregoing results in the form of a constructive computer algorithm is by no means straightforward.

Another drawback of minimax design is that criteria which are invariant with respect to transformations of θ for the calculation of locally optimal designs may no longer be invariant with respect to the minimax design criterion [92].

6.3.2 Numerical techniques for exact designs

As regards exact designs for stationary sensors, let us observe first that the initial minimax optimization problem (6.23) can be solved with the use of some algorithms for inequality-constrained, semi-infinite optimization [215, Sec. 3.5]. We wish to determine the value of the design variable s which maximizes $J(s) = \max_{\theta \in \Theta_{\mathrm{ad}}} \Psi[M(s, \theta)]$, but this task is complicated by the fact that each evaluation of $J(s)$ involves maximizing $\Psi[M(s, \theta)]$ with respect to $\theta \in \Theta_{\mathrm{ad}}$. Note that the problem so formulated can be cast as the following semi-inifinite programming one: Minimize a scalar α, subject to the constraint

$$\max_{\theta \in \Theta_{\mathrm{ad}}} \Psi[M(s, \theta)] \le \alpha \quad (6.30)$$

which is, in turn, equivalent to the infinite set of constraints

$$\{\Psi[M(s, \theta)] \le \alpha, \quad \theta \in \Theta_{\mathrm{ad}}\}. \quad (6.31)$$

The maximization above is to be performed with respect to s and the auxiliary variable $\alpha \in \mathbb{R}$. The term "semi-infinite programming" (SIP) derives from the property that we have finitely many decision variables (these are the elements of α and s) in infinitely many constraints (there are as many constraints as the number of elements in the set Θ_{ad}). Numerical approaches to solving SIP problems are characterized in [61, 62, 113, 214, 250, 257].

In practice, as was already the case while determining T-optimum designs in Section 5.2.4, the simple relaxation algorithm proposed by Shimizu and Aiyoshi [266] turns out to perform well in the considered nonlinear experimental design problems, as is also suggested in [222, 349]. It consists in relaxing the problem by taking into account only a finite number of constraints (6.31). The corresponding version of the relaxation procedure is represented by the following algorithm:

ALGORITHM 6.1 Finding minimax designs

Step 1. *Choose an initial parameter vector $\theta^{(1)} \in \Theta_{\mathrm{ad}}$ and define the first set of representative values $\mathbb{Z}^{(1)} = \{\theta^{(1)}\}$. Set $k = 1$.*

Step 2. *Solve the current relaxed problem*

$$s^{(k)} = \arg \min_{s \in S_{\mathrm{ad}}} \left\{ \max_{\theta \in \mathbb{Z}^{(k)}} \Psi[M(s, \theta)] \right\}.$$

Step 3. *Solve the maximization problem*

$$\theta^{(k+1)} = \arg \max_{\theta \in \Theta_{\mathrm{ad}}} \Psi[M(s^{(k)}, \theta)].$$

Step 4. *If*

$$\Psi[M(s^{(k)}, \theta^{(k+1)})] \leq \max_{\theta \in \mathbb{Z}^{(k)}} \Psi[M(s^{(k)}, \theta)] + \epsilon,$$

where ϵ is a small predetermined positive constant, then $s^{(k)}$ is a sought minimax solution, otherwise include $\theta^{(k+1)}$ into $\mathbb{Z}^{(k)}$, increment k, and go to Step 2.

The usefulness of the algorithm in finding best sensor locations was confirmed in [308].

As for the case of moving sensors whose movements are given by (6.15), we may consider minimization of the functional

$$J_{\mathrm{MM}}(s_0, u) = \max_{\theta \in \Theta_{\mathrm{ad}}} J(s_0, u, \theta). \tag{6.32}$$

Clearly, this can be treated by some numerical algorithms of optimal control, but a much simpler approach consists in making use of the dependence [13, p. 33]

$$\left\{ \frac{1}{\mathrm{meas}(\Theta_{\mathrm{ad}})} \int_{\Theta_{\mathrm{ad}}} J^{\mu}(s_0, u, \theta) \, d\theta \right\}^{1/\mu} \xrightarrow[\mu \to \infty]{} \max_{\theta \in \Theta_{\mathrm{ad}}} J(s_0, u, \theta), \tag{6.33}$$

where, without restriction of generality, it is assumed that J takes on only nonnegative values. Hence we may write

$$J_{\mathrm{MM}}(s_0, u) = \max_{\theta \in \Theta_{\mathrm{ad}}} J(s_0, u, \theta) \approx \left\{ \frac{1}{\mathrm{meas}(\Theta_{\mathrm{ad}})} \int_{\Theta_{\mathrm{ad}}} J^{\mu}(s_0, u, \theta) \, d\theta \right\}^{1/\mu} \tag{6.34}$$

and replace minimization of (6.32) by minimization of the 'smooth' functional

$$J_\mu(x_0, u) = \frac{1}{\text{meas}(\Theta_{\text{ad}})} \int_{\Theta_{\text{ad}}} J^\mu(s_0, u, \theta) \, d\theta = E_\theta \{ J^\mu(s_0, u, \theta) \} \qquad (6.35)$$

for a sufficiently large μ, where the expectation is calculated for the uniform distribution on Θ_{ad}. The latter can be solved in much the same way as in Section 6.2.2.

Example 6.1

In what follows, our aim is to apply the delineated numerical algorithms to the two-dimensional heat equation

$$\frac{\partial y(x,t)}{\partial t} = \frac{\partial}{\partial x_1} \left(\kappa(x) \frac{\partial y(x,t)}{\partial x_1} \right) + \frac{\partial}{\partial x_2} \left(\kappa(x) \frac{\partial y(x,t)}{\partial x_2} \right)$$
$$+ 10[x_1 + (1 - x_1)t], \quad (x,t) \in \Omega \times T = (0,1)^3$$

subject to the conditions

$$y(x,0) = 0, \quad x \in \Omega,$$
$$y(x,t) = 0, \quad (x,t) \in \partial\Omega \times T.$$

Let us assume the following form of the diffusion coefficient to be identified:

$$\kappa(x) = \theta_1 + \theta_2 x_1 + \theta_3 x_2.$$

The objective is to estimate κ (i.e., the parameters θ_1, θ_2 and θ_3) as accurately as possible, based on the measurements made by three moving sensors. For this purpose, the ED- and approximated MMD-optimum design procedures have been implemented in Fortran 95 programming language and run on a PC (Pentium IV, 2.40 GHz, 512 MB RAM). During simulations, maximization of the criterion (6.24) was replaced by minimization of the smoothed functional (6.35) for $J(s_0, u, \theta) = 1/\det(M(s,\theta))$ and $\mu = 5$. As regards the prior knowledge about the parameters, we were working under the assumption that

$$0.05 \leq \theta_1 \leq 0.15, \quad 0.0 \leq \theta_2 \leq 0.2, \quad 0.0 \leq \theta_3 \leq 0.2, \qquad (6.36)$$

and that they were characterized by uniform distributions on the corresponding intervals.

As for the sensor dynamics, we assumed that it was not of primary concern, so we adopted the simple model

$$\dot{s}(t) = u(t), \quad s(0) = s_0.$$

The sensor velocities were considered under the restrictions

$$|u_i(t)| \leq 0.7, \quad i = 1, \dots, 6$$

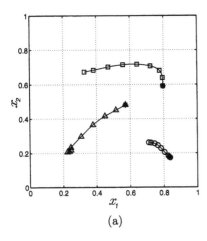
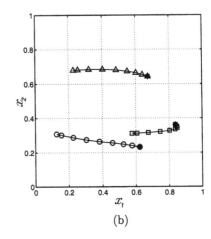

(a)   (b)

FIGURE 6.2
Optimum sensor trajectories: ED-design criterion (a) and approximated MMD-design criterion (b).

and approximated by piecewise linear polynomials (for 40 divisions of the interval $[0, t_f]$). In both the cases, the algorithm (6.18) started from

$$s_0^0 = (0.8, 0.2, 0.8, 0.4, 0.8, 0.6), \quad u^0 = 0.$$

Moreover, the harmonic sequence (6.12) was employed for $b = 10^{-2}$.

Figure 6.2 shows the sensor trajectories obtained in 200 iterations (on aggregate, approximately ten minutes of CPU time were used to run all the simulations). Circles, squares and triangles denote consecutive sensor positions. Furthermore, the sensors' positions at $t = 0$ are marked by asterisks. Note that the form of the forcing term in our PDE implies that the 'snapshots' of the corresponding solution (i.e., the state of our DPS) at consecutive time moments resemble those of a hat-shaped surface whose maximum moves from the right to the left boundary of Ω. In addition, the diffusion coefficient values in the upper right of Ω are on the average greater than those in the lower left. This means that the state changes during the system evolution are quicker when we move up and to the right (on the other hand, the system reaches the steady state there earlier). This fact explains in a sense the form of the obtained trajectories (the sensors tend to measure the state in the regions where the distributed system is more sensitive with respect to the unknown parameter κ, i.e., in the lower part of Ω). As can be seen, the results obtained for the ED- and approximated MMD-optimality criteria are to some extent similar. ∎

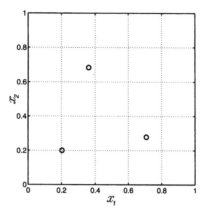

FIGURE 6.3
Optimum positions of stationary sensors for the MMD-design criterion.

Example 6.2
In the settings of Example 6.1 the problem of locating three stationary sensors was also solved through the direct maximization of the criterion (6.24) via the algorithm proposed by Shimizu and Aiyoshi, starting from $Z^{(1)} = \{(0.1, 0.1, 0.1)\}$. The nonlinear-programming subtasks were solved using the adaptive random search strategy of p. 54 and a sequential constrained quadratic programming (SQP) method [22, 178, 275, 276]. The convergence tolerance was set at $\epsilon = 5 \times 10^{-3}$.

Figure 6.3 shows the optimal sensor positions obtained in two iterations of the algorithm. They correspond to

$$s^\star = \begin{pmatrix} 0.203342, \ 0.200193, \ldots \\ 0.360944, \ 0.683536, \ldots \\ 0.707838, \ 0.278700 \end{pmatrix}$$

and $\theta_1 = 0.15$, $\theta_2 = 0.2$ $\theta_3 = 0.2$, i.e., the largest admissible values of the parameters. ▯

Note that the discussed minimax approach may lead to overly conservative designs. Much of the conservatism of the designs can be attributed to the worst-case nature of the associated performance index. Figure 6.4 shows how sensor configurations restraining the maxima of the criterion Ψ are preferred to those for which Ψ takes on low values in most cases. An analogous situation takes place in controller design as far as $\mathcal{H}^\infty$-norm minimization for achieving robust stabilization and μ-synthesis for achieving guaranteed performance and robust stabilization are concerned [343]. Consequently, it seems more reasonable to pay attention to designs which are efficient for most values of the unknown parameters. From Section 6.2.2 we know, however, that calcu-

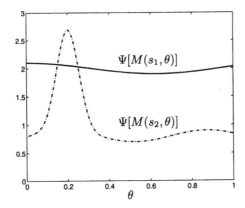

FIGURE 6.4
Illustration of the conservatism of the minimax approach. The design criterion Ψ is plotted for two different sensor configurations s_1 and s_2. The horizontal variable θ represents the unknown parameter. Configuration s_1 is preferred to s_2, although s_2 outperforms s_1 for most values of θ.

lation of optimal designs in the average sense is complicated by the necessity of properly estimating expectations. An attempt to overcome this obstacle via stochastic approximation does not eliminate the presence of many local minima.

A computer-intensive approach aimed at settling these problems to some extent is presented in the next section. It relies on randomized algorithms which have proved useful in robust controller synthesis.

6.4 Robust sensor location using randomized algorithms

In recent years, it has been shown that many problems in the robustness analysis and synthesis of control systems are either NP-complete or NP-hard, cf. the survey paper by Blondel and Tsitsiklis [24] or the tutorial introduction by Vidyasagar [342], which means that producing exact solutions may be computationally prohibitively demanding. Since the notions of computational complexity are not widely known among researchers concerned with both control theory and experimental design, we start by briefly recalling some relevant notions which strive to clarify this topic at an elementary level. This constitutes a prerequisite for understanding the idea of a randomized algorithm for sensor location to be presented in what follows.

6.4.1   A glance at complexity theory

Complexity theory constitutes part of the theory of computation dealing with the resources required during computation to solve a given problem. The most common resources are time (how many steps does it take to solve a problem) and space (how much memory does it take to solve a problem). If we define an algorithm as a procedure that can solve a problem, then the P class of problems is composed of the ones for which we can find an algorithm that runs in polynomial time (i.e., the maximum number of operations required by the algorithm on any problem instance of size n increases no faster than some polynomial in n). These are considered tractable problems, i.e., the ones which can realistically be solved even for reasonably large input sizes. As examples, sorting and searching are classical P problems. Intuitively, we think of the problems in P as those that can be solved reasonably fast. In turn, a problem is said to be NP if, given a candidate solution to this problem, it is possible to determine in polynomial time whether or not this candidate solution constitutes a true one, but for which no known polynomial algorithm to find the true solution exists. Solving a maze illustrates well this subtle definition: to find a way through a maze means to investigate a whole set of possible routes, tracing them to the end, before discovering a way out, but if we are given a maze with a route already drawn, it is trivial to check whether it is a correct route to the end. Clearly, a P problem (whose solution time is bounded by a polynomial) is automatically NP as well. The notation NP stands for 'nondeterministic polynomial' since originally NP problems were defined in terms of nondeterministic Turing machines.

Linear programming, long known to be NP and thought not to be P, was shown to be P by Khachiyan in 1979 [129]. But in general, the question whether the P and NP classes coincide still remains open, and inevitably the strong suspicion is that this is not the case. If N and NP are not equivalent (and probably this is so), then the solution of NP problems requires (in the worst case) an exhaustive search. Some famous NP problems are, e.g., the travelling salesman problem (given a set of locations and distances between them, find the shortest path that visits all of the locations once and only once, without ever visiting a location twice, and ends up at the same place where it began), the prime factoring problem (given an integer, find its prime factors), the graph colouring problem (given a graph with n items in it, where no two adjacent vertices can have the same colour, and an integer k, determine if there is a colouring of the graph that uses no more than k colours), or the packing problem (given a bag with some prespecified capacity and a collection of n items of various sizes and values, find the most valuable set of items you can fit into the bag).

There is also a special class of NP-hard problems. A problem is said to be NP-hard if every problem in NP can be reduced to it in polynomial time. Alternatively, a problem is NP-hard if solving it in polynomial time would make it possible to solve all problems in the NP class in polynomial time.

That is, a problem is NP-hard if an algorithm for solving it can be translated into one for solving any other NP problem (NP-hard thus means "at least as hard as any NP-problem," although it might, in fact, be harder, so another intuitive definition is "either the hardest problem in NP, or even harder than that"). Therefore, given a black box (often called an oracle) that solves an NP-hard problem in a single time step, it would be possible to solve every NP problem in polynomial time using a suitable algorithm which utilizes the black box (by first executing the reduction from this problem to the NP-hard problem and then executing the oracle). An example of an NP-hard problem is the decision problem SUBSET-SUM: Given a set of integers, do any nonempty subset of them add up to zero?

Note that not all of the NP-hard problems are necessarily NP problems themselves — they can be harder, e.g., the halting problem: Given a program, will it eventually halt when started with a particular input? A possible approach to solve it might be to run the program with this input and assert 'yes' if it halts. But if the program does not halt, it is not apparent how to detect this in finite time and to assert that the answer is 'no.' Therefore this problem is not in NP since all problems in NP are decidable, i.e., for each of them there exists an algorithm that always halts with the right binary answer. The halting problem cannot thus be solved in general given any amount of time.

A problem which is both NP and NP-hard is termed NP-complete. As a consequence, if we had a polynomial-time algorithm for an NP complete problem, we could solve all NP problems in polynomial time. Note that NP and NP-complete are both sets of decision problems, problems which have a 'yes' or 'no' answer. The NP-hard class includes both decision problems and other problems. For example, consider the problem FIND-SUBSET-SUM formulated as follows: Given a finite set of integers, return a subset of them that adds up to zero, or return the empty set if there is none. That is not a 'yes' or 'no' question, so it could not be NP-complete. It happens to be NP-hard. There are also decision problems that are NP-hard. For example, the halting problem is NP-hard, but it is not NP-complete. Informally, the NP-complete problems are the "toughest" problems in NP in the sense that they are the ones most likely not to be in P. However, it is not really correct to say that NP-complete problems are the hardest problems in NP. Assuming that P and NP are not equal, it is guaranteed that an infinite number of problems are in NP that are neither NP-complete nor in P. Some of these problems may actually have higher complexity than some of the NP-complete problems.

At present, all known algorithms for NP-complete problems require time which is exponential in the problem size. It is unknown whether there are any faster algorithms. Therefore, in order to solve an NP-complete problem for any nontrivial problem size, one of the following approaches is used:

- *Approximation:* An algorithm which quickly finds a suboptimal solution which is within a certain (known) range of the optimal one. Not all NP-complete problems have good approximation algorithms, and for some problems finding a good approximation algorithm is enough to solve the problem itself.

- *Probabilistic:* An algorithm which probably yields good average runtime behaviour for a given distribution of the problem instances — ideally, one that assigns low probability to "hard" inputs.

- *Special cases:* An algorithm which is probably fast if the problem instances belong to a certain special case.

- *Heuristic:* An algorithm which works "reasonably well" on many cases, but for which there is no proof that it is always fast.

6.4.2 NP-hard problems in control-system design

The continuous development of both high-performance hardware and various powerful software packages for matrix computations made it possible for control theorists to focus their attention on multi-input, multi-output systems and to attempt to apply theoretical advances to practical systems whose main characteristics were high orders or high dimensionalities of the underlying mathematical models. What is more, in response to requirements for controllers designed such that the performance of the controlled system be 'robust' to variations in the plant, the methodologies of $\mathcal{H}^\infty$ control and μ-synthesis were developed in the 1980s and successfully applied in a diverse set of difficult applications, including a tokamak plasma control system for use in nuclear fusion power generation, a flight control system for the space shuttle or an ultrahigh performance missile control system [32, 37, 69, 112]. This progress naturally highlighted the issue of how the computational difficulties of designing controllers 'scale up' as the size of the system (the number of inputs and/or outputs, the dimension of the state space, etc.) increases [342]. The investigations carried out down this route led to the astonishing conclusion that numerous control system analysis and controller design problems are in fact NP-hard. In [342] Vidyasagar gives a long list of such problems, among which we evidence here the following examples:

- *Robust stability:* Given matrices $A_l, A_u \in \mathbb{R}^{n \times n}$ determine whether every matrix A satisfying $A_l \geq A \geq A_u$ is stable in the sense that all its eigenvalues have negative real parts.

- *Robust nonsingularity:* Under the same conditions as above, determine whether every matrix A is nonsingular.

- *Robust nonnegative definiteness:* Given matrices $B_l, B_u \in \mathrm{Sym}(n)$ determine whether every matrix $B \in \mathrm{Sym}(n)$ satisfying $B_l \geq B \geq B_u$ is nonnegative definite.

- *Constant output feedback stabilization with constraints:* Given matrices $A, B, C, K_l, K_u \in \mathbb{R}^{n \times n}$ determine whether there exists a 'feedback' matrix K satisfying $K_l \geq K \geq K_u$ and such that $A + BKC$ is stable.

All these problems which are only the simplest instances of situations which must be confronted in practice turn out to be NP-hard or to belong to even higher-complexity classes. The NP-hardness implies that in general they constitute intractable problems when we attempt to solve them exactly. Other examples of NP-hard problems encountered, e.g., in the analysis and design of linear, nonlinear and hybrid systems, as well as in stochastic control, can be found [24, 207, 341–343]

A very natural question thus concerns what can be done to tackle such problems. If the conjecture P $\neq$ NP is true, NP-hardness eliminates the possibility of constructing algorithms which run in polynomial time and are correct for all problem instances. But this does not necessarily mean that those problems are intractable. Many NP-hard problems become considerably simpler if we decide to relax the notion of the solution. One of the possible options down this route is to apply polynomial-time randomized algorithms which provide solutions to arbitrarily high levels of accuracy and confidence [207, 340–343]. In what follows, we adapt the approach proposed by Vidyasagar to our sensor location problem.

6.4.3 Weakened definitions of minima

Let S_{ad} be a set of admissible N-sensor configurations $s = (x^1, \ldots, x^N)$ and Θ_{ad} the set of admissible parameters. Assume that the design criterion of interest $\Psi[M(s, \theta)]$ can be replaced by an equivalent criterion $G : S_{ad} \times \Theta_{ad} \rightarrow [0, 1]$ in the sense that $G(s, \theta) = h(\Psi[M(s, \theta)])$ for some strictly increasing function $h : \mathbb{R} \rightarrow [0, 1]$. Such a rescaling will be essential for the randomized algorithm outlined in what follows. In principle, finding such a transformation is rather easy, as, e.g., we can define the following counterparts of the common design criteria:

- D-optimum designs:

$$G(s, \theta) = \frac{1}{1 + \det(M(s, \theta))}, \tag{6.37}$$

- A-optimum designs:

$$G(s, \theta) = \frac{\operatorname{tr}(M^{-1}(s, \theta))}{1 + \operatorname{tr}(M^{-1}(s, \theta))}, \tag{6.38}$$

- Sensitivity criterion:

$$G(s, \theta) = \frac{1}{1 + \operatorname{tr}(M(s, \theta))}. \tag{6.39}$$

Given θ, minimization of $G(\,\cdot\,,\theta)$ is thus equivalent to minimization of the criterion $\Psi[M(\,\cdot\,,\theta)]$.

The above problem reformulation aimed at optimizing the expected or average performance of the estimation procedure involves choosing $s \in S_{\mathrm{ad}}$ to minimize the design criterion $J : S_{\mathrm{ad}} \to [0,1]$,

$$J(s) = \mathrm{E}_\theta[G(s,\theta)], \tag{6.40}$$

where the expectation is defined with respect to a probability distribution ς on the set Θ_{ad}. Such an approach does not prevent the resulting solution from occasionally performing poorly for some parameters θ which are associated with low probability values, but it involves much less conservatism than the minimax designs.

Obviously, attempts to find the exact value of a configuration $s^\star$ globally minimizing (6.40) are doomed to failure in practical settings, as this kind of problems are inherently NP-hard. More precisely, given a number J_0, it is NP-hard to determine whether or not $J_0 \geq J(s^\star)$ [343]. Given that an exact minimization of J is computationally excessive, we have to settle for approximate solutions.

Vidyasagar [343] thus considers in turn the following three definitions of 'near' minima.

DEFINITION 6.1   *Given $\epsilon > 0$, a real number J_0 is said to be a* Type 1 *near minimum of $J(\,\cdot\,)$ to accuracy ϵ, or an approximate near minimum of $J(\,\cdot\,)$ to accuracy ϵ, if*

$$\inf_{s \in S_{\mathrm{ad}}} J(s) - \epsilon \leq J_0 \leq \inf_{s \in S_{\mathrm{ad}}} J(s) + \epsilon, \tag{6.41}$$

or equivalently,

$$\left| J_0 - \inf_{s \in S_{\mathrm{ad}}} J(s) \right| \leq \epsilon. \tag{6.42}$$

This relaxation of the notion of the minimum is rather natural, but it still remains too strong for our purposes, as it may be NP-hard to compute even an approximation to J_0. This constitutes the motivation to look for other relaxations. One option is as follows:

DEFINITION 6.2   *Let ν be a given probability measure on S_{ad}. Given $\alpha > 0$, a real number J_0 is called a* Type 2 *near minimum of $J(\,\cdot\,)$ to level α, or a probable near minimum of $J(\,\cdot\,)$ to level α, if*

$$J_0 \geq \inf_{s \in S_{\mathrm{ad}}} J(s) \tag{6.43}$$

and

$$\nu\{s \in S_{\mathrm{ad}} : J(s) < J_0\} \leq \alpha. \tag{6.44}$$

This relaxed definition allows for the existence of a set Z in which there are better values of $J(s)$ than J_0, but the probability measure of this set $\nu(Z)$ is no greater than α. Additionally, we have

$$\inf_{s \in S_{\mathrm{ad}}} J(s) \le J_0 \le \inf_{s \in S_{\mathrm{ad}} \backslash Z} J(s), \tag{6.45}$$

i.e., J_0 is bracketed by the infimum of $J(\cdot)$ over all points of S_{ad} and the infimum of $J(\cdot)$ over 'nearly' all points of S_{ad}.

Some polynomial-time algorithms have been invented to find minima of such a type [288, 289, 341], yet they require that the exact values of $J(s)$ be computable for any given $s \in S_{\mathrm{ad}}$. In reference to (6.40) this requirement is too strong (the calculation of the precise values of an expectation involves computation of multivariate probability integrals by either numerical approximation, or running long-drawn Monte Carlo simulations [286]). This inspires us to introduce the following ultimate variant of the sought definition.

DEFINITION 6.3 *Let ν be a given probability measure on S_{ad}. Given $\epsilon, \alpha > 0$, we call a real number J_0 a Type 3 near minimum of $J(\cdot)$ to accuracy ϵ and level α, or a probably approximate near minimum of $J(\cdot)$ to accuracy ϵ and level α, if*

$$J_0 \ge \inf_{s \in S_{\mathrm{ad}}} J(s) - \epsilon \tag{6.46}$$

and

$$\nu\big\{ s \in S_{\mathrm{ad}} : J(s) < J_0 - \epsilon \big\} \le \alpha. \tag{6.47}$$

The above definition implies the existence of a set $Z \subset S_{\mathrm{ad}}$ whose probability measure $\nu(Z)$ does not exceed α and for which we have

$$\inf_{s \in S_{\mathrm{ad}}} J(s) - \epsilon \le J_0 \le \inf_{s \in S_{\mathrm{ad}} \backslash Z} J(s) + \epsilon, \tag{6.48}$$

This slightly weakened version of Definition 6.2 forms a basis for the construction of the universal randomized algorithm outlined in what follows.

6.4.4 Randomized algorithm for sensor placement

Observe that if we were able to calculate the precise values of $J(s)$ for any $s \in S_{\mathrm{ad}}$, then given a probability measure ν on S_{ad}, it would be rather a simple matter to determine a Type 2 near minimum of $J(\cdot)$ to level α. Indeed, taking an i.i.d. sample $s_1, \ldots, s_n$ from ν, we could set

$$\bar{J}(\tilde{s}) = \min_{1 \le i \le n} J(s_i), \tag{6.49}$$

where $\tilde{s} = (s_1, \ldots, s_n) \in S_{\mathrm{ad}}^n$. Lemma B.2 yields

$$\nu^n \Big\{ \tilde{s} \in S_{\mathrm{ad}}^n : \nu \big\{ s \in S_{\mathrm{ad}} : J(s) < \bar{J}(\tilde{s}) \big\} > \alpha \Big\} \leq (1-\alpha)^n, \qquad (6.50)$$

where ν^n denotes the product probability measure.

This bound can be used to manipulate the accuracy of the solution. Namely, given level and confidence parameters $\alpha, \delta \in (0,1)$, we could choose an integer n such that

$$(1-\alpha)^n \leq \delta \qquad (6.51)$$

or equivalently,

$$n \geq \frac{\ln(1/\delta)}{\ln(1/(1-\alpha))}. \qquad (6.52)$$

From (6.50) it would then follow that

$$\nu \big\{ s \in S_{\mathrm{ad}} : J(s) < \bar{J}(\tilde{s}) \big\} > \alpha \qquad (6.53)$$

with probability no greater than δ, i.e.,

$$\nu \big\{ s \in S_{\mathrm{ad}} : J(s) < \bar{J}(\tilde{s}) \big\} \leq \alpha \qquad (6.54)$$

at least with probability (or confidence coefficient) $1-\delta$. In other words, $\bar{J}$ so generated would be a probable near minimum of $J(\cdot)$ to level α with confidence coefficient at least $1-\delta$. Unfortunately, this straightforward idea cannot be implemented since the determination of each value of $J(s)$ necessitates the computation of the expected value of $G(s, \cdot)$.

Since the precise values of

$$J(s) = \mathrm{E}_\theta \big[G(s, \theta) \big] = \int_{\Theta_{\mathrm{ad}}} G(s, \theta)\, \varsigma(\mathrm{d}\theta) \qquad (6.55)$$

cannot be computed, a very first idea is to take an i.i.d. sample $\theta_1, \ldots, \theta_r$ from ς and to estimate (6.55) by the sample average

$$\widehat{\mathrm{E}} \big[G(s, \theta); \tilde{\theta} \big] = \frac{1}{r} \sum_{j=1}^r G(s, \theta_j), \qquad (6.56)$$

where $\tilde{\theta}$ stands for the multisample $(\theta_1, \ldots, \theta_r) \in \Theta_{\mathrm{ad}}^r$.

Obviously, $\widehat{\mathrm{E}}[G; \tilde{\theta}]$ is itself a random variable on the product space Θ_{ad}^r equipped with the product probability measure ς^r. The law of large numbers implies that these Monte Carlo estimates converge to $\mathrm{E}_\theta[G]$ as r tends to ∞. As for a measure of the closeness of $\widehat{\mathrm{E}}[G; \tilde{\theta}]$ to $\mathrm{E}_\theta[G]$ for a finite r, the bound given by Hoeffding's inequality (cf. Appendix B.11.1) can be used:

$$\varsigma^r \Big\{ \tilde{\theta} \in \Theta_{\mathrm{ad}}^r : \big| \widehat{\mathrm{E}} \big[G(s, \theta); \tilde{\theta} \big] - \mathrm{E}_\theta[G(s, \theta)] \big| > \epsilon \Big\} \leq 2 \exp(-2r\epsilon^2). \qquad (6.57)$$

This means that once an r-element i.i.d. sample has been drawn from ς, we may assert that $\widehat{E}[G; \tilde\theta]$ is at a maximum distance of ϵ to $E_\theta[G]$ with confidence coefficient $1 - 2\exp(-2r\epsilon^2)$. Consequently, if we wish to estimate (6.55) with a predefined accuracy ϵ and a confidence coefficient $1 - \delta$, it is suffcient to draw an r-element i.i.d. sample $\tilde\theta$ from ς, where

$$r \geq \frac{1}{2\epsilon^2} \ln\left(\frac{2}{\delta}\right), \tag{6.58}$$

and to determine the corresponding arithmetic mean $\widehat{E}[G; \tilde\theta]$.

Having clarified how to steer the accuracy of approximating expectations, we can return to the key idea of the algorithm for finding near minima of J outlined at the beginning of this section, and to make the necessary modifications so as to replace the exact values of J by the corresponding arithmetic means. The final form of this scheme is due to Vidyasagar [343].

ALGORITHM 6.2 Finding a probably approximate near minimum

Step 1: *Select desired values of α, ϵ and δ.*

Step 2: *Choose integers*

$$n \geq \frac{\ln(2/\delta)}{\ln(1/(1-\alpha))}, \quad r \geq \frac{1}{2\epsilon^2} \ln\left(\frac{4n}{\delta}\right). \tag{6.59}$$

Step 3: *Take i.i.d. samples $s_1, \ldots, s_n$ and $\theta_1, \ldots, \theta_r$ from ν and ς, respectively. Define*

$$\hat{J}_i = \frac{1}{r} \sum_{j=1}^{r} G(s_i, \theta_j), \quad i = 1, \ldots, n. \tag{6.60}$$

Step 4: *Determine*

$$\hat{J}_0 = \min_{1 \leq i \leq n} \hat{J}_i, \tag{6.61}$$

which is automatically a Type 3 near minimum of J to accuracy ϵ and level α with confidence coefficient $1 - \delta$.

The proof of the correctness of this algorithm is rather easy. From Hoeffding's inequality and (6.59) we have

$$\varsigma^r\left\{\tilde\theta \in \Theta_{ad}^r : |\hat{J}_1(\tilde\theta) - J(s_1)| > \epsilon \text{ or } \ldots \text{ or } |\hat{J}_n(\tilde\theta) - J(s_n)| > \epsilon\right\}$$

$$\leq \sum_{i=1}^{n} \varsigma^r\left\{\tilde\theta \in \Theta_{ad}^r : |\hat{J}_i(\tilde\theta) - J(s_1)| > \epsilon\right\} \tag{6.62}$$

$$\leq 2n\exp(-2r\epsilon^2) \leq 2n\exp\left(-2\frac{1}{2\epsilon^2}\ln\left(\frac{4n}{\delta}\right)\epsilon^2\right) = \frac{\delta}{2},$$

so that it is legitimate to say that

$$|\hat{J}_i - J(s_i)| \leq \epsilon, \quad i = 1, \ldots, n \tag{6.63}$$

with confidence coefficient $1 - \delta/2$.

Define

$$\bar{J} = \min_{1 \leq i \leq n} J(s_i). \tag{6.64}$$

Choosing any $k \in \{1, \ldots, n\}$ such that $J(s_k) = \bar{J}$, we also see that

$$\hat{J}_0 - \bar{J} = \hat{J}_0 - J(s_k) \leq \hat{J}_k - J(s_k) \leq \epsilon. \tag{6.65}$$

Similarly, choosing any $p \in \{1, \ldots, n\}$ such that $\hat{J}_0 = \hat{J}_p$, it follows that

$$\bar{J} - \hat{J}_0 = \bar{J} - \hat{J}_p \leq J(s_p) - \hat{J}_p \leq \epsilon. \tag{6.66}$$

Consequently, we get

$$|\hat{J}_0 - \bar{J}| \leq \epsilon \tag{6.67}$$

with confidence coefficient $1 - \delta/2$.

In order for $\hat{J}_0$ to be a Type 3 near minimum of J with confidence $1 - \delta$, we have to show that

$$\hat{J}_0 \geq \inf_{s \in S_{\text{ad}}} J(s) - \epsilon \tag{6.68}$$

and

$$\nu\{s \in S_{\text{ad}} : J(s) < \hat{J}_0 - \epsilon\} \leq \alpha, \tag{6.69}$$

both with confidence $1 - \delta$.

The relationship (6.68) follows from (6.67), as we have

$$\nu^n\left\{\tilde{s} \in S_{\text{ad}}^n : \hat{J}_0(\tilde{s}) \geq \inf_{s \in S_{\text{ad}}} J(s) - \epsilon\right\}$$
$$= \nu^n\left\{\tilde{s} \in S_{\text{ad}}^n : \inf_{s \in S_{\text{ad}}} J(s) - \hat{J}_0(\tilde{s}) \leq \epsilon\right\} \tag{6.70}$$
$$\geq \nu^n\left\{\tilde{s} \in S_{\text{ad}}^n : \bar{J}(\tilde{s}) - \hat{J}_0(\tilde{s}) \leq \epsilon\right\} > 1 - \frac{\delta}{2} > 1 - \delta.$$

As for (6.69), we must show that

$$\nu^n\left\{\tilde{s} \in S_{\text{ad}}^n : \nu\{s \in S_{\text{ad}} : J(s) < \hat{J}_0(\tilde{s}) - \epsilon\} > \alpha\right\} \leq \delta. \tag{6.71}$$

To this end, it suffices to observe that the left-hand side of (6.71) can be decomposed as follows:

$$\nu^n \Big\{ \tilde{s} \in S^n_{ad} : \nu \{ s \in S_{ad} : J(s) < \hat{J}_0(\tilde{s}) - \epsilon \} > \alpha \Big\}$$

$$= \nu^n \Big\{ \tilde{s} \in S^n_{ad} : \nu \{ s \in S_{ad} : J(s) < \hat{J}_0(\tilde{s}) - \epsilon \} > \alpha$$

$$\text{and } \bar{J}(\tilde{s}) < \hat{J}_0(\tilde{s}) - \epsilon \Big\}$$

$$+ \nu^n \Big\{ \tilde{s} \in S^n_{ad} : \nu \{ s \in S_{ad} : J(s) < \hat{J}_0(\tilde{s}) - \epsilon \} > \alpha$$

$$\text{and } \bar{J}(\tilde{s}) \geq \hat{J}_0(\tilde{s}) - \epsilon \Big\}, \tag{6.72}$$

and each term in the sum is no greater than $\delta/2$. Indeed, from (6.67) we have

$$\nu^n \Big\{ \tilde{s} \in S^n_{ad} : \nu \{ s \in S_{ad} : J(s) < \hat{J}_0(\tilde{s}) - \epsilon \} > \alpha \text{ and } \bar{J}(\tilde{s}) < \hat{J}_0(\tilde{s}) - \epsilon \Big\}$$

$$\leq \nu^n \Big\{ \tilde{s} \in S^n_{ad} : \bar{J}(\tilde{s}) < \hat{J}_0(\tilde{s}) - \epsilon \Big\} \leq \frac{\delta}{2}, \tag{6.73}$$

and from (6.50) we conclude that

$$\nu^n \Big\{ \tilde{s} \in S^n_{ad} : \nu \{ s \in S_{ad} : J(s) < \hat{J}_0(\tilde{s}) - \epsilon \} > \alpha \text{ and } \bar{J}(\tilde{s}) \geq \hat{J}_0(\tilde{s}) - \epsilon \Big\}$$

$$\leq \nu^n \Big\{ \tilde{s} \in S^n_{ad} : \nu \{ s \in S_{ad} : J(s) < \bar{J}(\tilde{s}) \} > \alpha \text{ and } \bar{J}(\tilde{s}) \geq \hat{J}_0(\tilde{s}) - \epsilon \Big\}$$

$$\leq \nu^n \Big\{ \tilde{s} \in S^n_{ad} : \nu \{ s \in S_{ad} : J(s) < \bar{J}(\tilde{s}) \} > \alpha \Big\} \leq \frac{\delta}{2}. \tag{6.74}$$

Example 6.3

As a simple test of the proposed approach, the DPS characterized in Example 2.4 was considered. It was assumed that estimation of the parameter θ was to be performed based on measurements of one sensor located at point x^1. Since there is only one parameter to be identified, the FIM is actually a scalar value and hence we may identify the values of $\Psi[M(x^1, \theta)]$ with those of $M(x^1, \theta)$ themselves. The closed form of $M(x^1, \theta)$ is given in Example 2.4.

Assuming that θ is characterized by a uniform distribution on the interval $(0, 1)$, we introduce the criterion

$$J(x^1) = \mathrm{E}_\theta \big[(1 + M(x^1, \theta))^{-1} \big] = \int_0^1 (1 + M(x^1, \theta))^{-1} \, d\theta.$$

Its minimization was accomplished using Algorithm 6.2 after setting $\delta = 0.05$ and $\epsilon = \alpha = 0.01$. These parameters implied the sample sizes

$$n = \left\lceil \frac{\ln(2/\delta)}{\ln(1/(1-\alpha))} \right\rceil = 368, \quad r = \left\lceil \frac{\ln(4n/\delta)}{2\epsilon^2} \right\rceil = 51451.$$

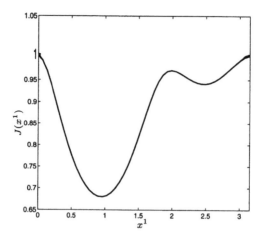

FIGURE 6.5

Results of the numerical evaluation of the expectations to be minimized in Example 6.3.

Setting $G(x^1, \theta) = (1 + M(x^1, \theta))^{-1}$, the appropriate script was written in MATLAB making use of the built-in routine **rand** to generate the samples of x^1 and θ in intervals $[0, \pi]$ and $[0, 1]$, respectively. Running it on a PC (Pentium IV, 2.40 GHz, 512 MB RAM) equipped with Windows 2000 and MATLAB 6.5 Rel. 13, the value of $\hat{J}_0 = 0.6804$ was obtained after some eight minutes. It corresponds to a probably approximate near minimum at $x^{1\star} = 0.9552$.

In order to verify this result, which seems reasonable anyway taking account of Fig. 2.3, the values of $J(x^1)$ were determined via numerical integration performed by **quadl** (MATLAB's routine implementing an adaptive high-order Lobatto quadrature). The corresponding results are shown in Fig. 6.5, from which it is clear that the minimum produced by Algorithm 6.2 is correct.   ⬜

6.5   Concluding remarks

We have shown that the difficulties created by the dependence of the solutions to the sensor-location problem on the parameters to be identified can be cleared up to a certain extent by the introduction of robust designs, on the analogy of the common procedure in optimum experimental design for nonlinear regression models. To the best of our knowledge, this issue has received no attention as yet in the context of parameter estimation for DPSs, in spite of the fact that it is of utmost importance while trying to find the best sensor positions. In particular, optimal designs in the average and minimax senses were introduced, appropriate optimality conditions for continuous designs were given and efficient algorithms to numerically obtain exact designs were discussed. Each choice between the two robust approaches has some advantages and some disadvantages. A final selection has to be made on the basis of secondary considerations. For example, if we are anxious to optimize the worst possible performances of the experiment designed, then optimal designs in the minimax sense are preferred. On the other hand, if the prior uncertainty on the parameters to be identified is characterized by some prior distribution, then the average experimental design is particularly suited for making the designs independent of the parameters.

The algorithms outlined in the first part of this chapter permit the application of any of the above policies at a reasonable computational cost. In particular, the use of the framework of stochastic approximation opens up the possibility of developing robust design strategies for moving sensors, since the corresponding solutions can be obtained almost as simply as the locally optimal designs discussed in the previous chapter.

The ever-increasing power of modern computers generates interest in using computer-intensive methods to determine optimum experimental designs. An example of such an approach was given in the second part of this chapter, where an attempt was made to apply a randomized algorithm to find optimal sensor locations in an average sense. Note, however, that those results are only introductory in character and much has still to be done towards this line of research. At first sight, algorithms of this type can be thought of as implementations of the simple rule: "Just try lots of sensor configurations and pick the best one." But as is evidenced by Vidyasagar, a sound mathematical justification exists for the approach and the results can be used to direct the search for better sensor configurations. Obviously, although sample sizes needed by randomized algorithms are theoretically finite, they may be much too large to extensively use this approach in practice. This constitutes the main weakness of this approach at present. Nevertheless, the reported results of applying randomized designs in controller design indicate that this may be a promising emerging technique which, after a sufficient amount of work to be done, will make the problems of robust design more tractable.

7

Towards even more challenging problems

So far in this monograph, we have surveyed some developments in the field of the optimum experimental design for DPSs related to sensor location and motivated by practical applications. Clearly, these considerations by no means cover all aspects of the reality and many difficult problems still remain open, challenging researchers concerned with spatial measurements. In this chapter, we investigate two possibilities of extending or applying the foregoing results to other settings, which are expected to become fields of laborious research in the near future. Specifically, spatial correlations are considered in Section 7.1. In turn, Section 7.2 deals with the application of our approach originating in the statistical literature to the output selection in control system design.

7.1 Measurement strategies in the presence of correlated observations

One of the characteristic properties of collecting spatial data is the presence of spatial correlations between observations made at different sites [31,57,85, 181]. This topic is, however, neglected in most works on sensor location for parameter estimation in dynamic DPSs. On one hand, such a simplification is very convenient as it leads to elegant theoretical results, but on the other hand, it is rarely justified in a large number of applications. Environmental monitoring, meteorology, surveillance, some industrial experiments and seismology are the most typical areas where the necessity for taking account of this factor may emerge. For example, amongst air quality data the following kinds of correlation exist [186]: (i) time correlations (autocorrelations) in series of measurements of a single pollutant at one staton, (ii) cross-correlations in measurements of several pollutants at one station, (iii) space correlations in concurrent measurements of a single pollutant at two stations, (iv) space correlations in measurements of a single pollutant at two stations, the times of observation differing by a fixed amount (lagged correlations). It is obvious that in many cases (but not always) several measurement devices situated close to one another do not give much more information than a single sensor.

It goes without saying that even simple correlation structures may substantially complicate the solution of the sensor-location problem. This is because the corresponding information matrix in this case is no longer a sum of information matrices corresponding to individual sensors and therefore the results of the convex design theory discussed in Chapter 3 cannot be directly used. To deal with this problem, two main approaches can be distinguished. The first option is to exploit some well-known constrained optimization algorithms which deal with suitably defined additional constraints accounting for admissible distances between observations in time and space. Direct application of optimization techniques limits the clusterization effect, but when the number of sensors becomes high, the problem complexity substantially increases. In terms of the interpretation and applicability, such an approach is very attractive in the context of mobile sensors, cf. Chapter 4. The second approach consists in taking account of the mutual correlations between measurements by appropriately modifying the information matrix and then attempting to imitate some known iterative methods for noncorrelated observations.

As in Chapter 3, consider the problem of finding the best locations $x^1, \ldots, x^N \in X$ of N stationary sensors whose outputs are modelled by the following output equations:

$$z_m^i(t) = y(x^i, t; \theta) + \varepsilon(x^i, t), \quad t \in [0, t_f], \quad i = 1, \ldots, N, \qquad (7.1)$$

where X is a compact subset of a given spatial domain Ω (the set of feasible sensor locations), $y(x, t) \in \mathbb{R}$ denotes the system state at location x^i and time instant t, and $\varepsilon(x, t)$ is the Gaussianly distributed measurement disturbance satisfying

$$\mathrm{E}\big[\varepsilon(x, t)\big] = 0, \qquad (7.2)$$

$$\mathrm{E}\big[\varepsilon(x, t)\varepsilon(\chi, \tau)\big] = q(x, \chi, t)\delta(t - \tau), \qquad (7.3)$$

$q(\,\cdot\,, \,\cdot\,, t)$ being a known continuous spatial covariance kernel and δ the Dirac delta function.

The best sensor positions are then sought so as to minimize a convex criterion Ψ defined on the FIM

$$M(x^1, \ldots, x^N) = \int_0^{t_f} F(t)C^{-1}(t)F^{\mathsf{T}}(t)\,\mathrm{d}t \qquad (7.4)$$

whose derivation for the appropriate best linear unbiased estimator is contained in Appendix C.2. F is given by

$$F(\cdot) = \Big[\, f(x^1, \cdot) \quad \cdots \quad f(x^N, \cdot) \,\Big] \qquad (7.5)$$

where

$$f(x^i, \cdot) = \left(\frac{\partial y(x^i, \cdot\,; \theta)}{\partial \theta}\right)^{\mathsf{T}}_{\theta = \theta^0}, \qquad (7.6)$$

θ^0 being a prior estimate to the unknown parameter vector θ. The quantity $C(\cdot)$ is the $N \times N$ covariance matrix with elements

$$c_{ij}(t) = q(x^i, x^j, t), \quad i, j = 1, \ldots, N. \tag{7.7}$$

Introducing the matrix $W(t) = [w_{ij}(t)] = C^{-1}(t)$, we can rewrite the FIM as follows:

$$M(x^1, \ldots, x^N) = \sum_{i=1}^{N} \sum_{j=1}^{N} \int_0^{t_f} w_{ij}(t) f(x^i, t) f^{\mathsf{T}}(x^j, t) \, dt. \tag{7.8}$$

For notational convenience, in what follows, we shall use the following compact notation to denote a solution to our problem:

$$\xi = \{x^1, \ldots, x^N\}. \tag{7.9}$$

Each ξ will then be called a *discrete* (or *exact*) *design* and its elements *support points*. Since all measurements are to be taken on one run of the process (7.1), so that replications are not allowed, the admissible designs must satisfy the condition $x^i \neq x^j$ for $i, j = 1, \ldots, N$ and $i \neq j$. Thus, from among all discrete designs we wish to select the one which minimizes $J(\xi) = \Psi[M(\xi)]$. We call such a design a Ψ-*optimum design*.

This design problem differs from the much studied classical design problem because the matrix W here may be nondiagonal, due to the correlations between different observations. This constitutes a major obstacle when trying to directly apply numerical algorithms existing in the optimum experimental design literature, which were constructed for experiments with uncorrelated observations. The technical point is that the algorithms make explicit use of the fact that, for uncorrelated observations, the FIM can be expressed as the sum of the FIMs of the individual points forming the design. This is no longer true when the observations are correlated.

7.1.1   Exchange algorithm for Ψ-optimum designs

Scheme of the algorithm

In order to calculate exact designs, we generalized here the numerical algorithm set forth in the monograph by Brimkulov *et al.* [31], cf. also the monograph by Näther [189] and the paper by Uciński and Atkinson [313], where it had been employed to determine D-optimum sampling points while estimating linear-in-parameters expectations of random fields. It is an exchange algorithms, which means that starting from an arbitrary initial N-point design $\xi^{(0)}$, at each iteration one support point is deleted from the current design and a new point from the domain X is included in its place in such a way as to maximize the resulting decrease in the value of the criterion Ψ.

The general scheme of the algorithm is outlined below. For iteration k, we use the following notation: $F^{(k)}$ is the current matrix of sensitivity coefficients in (7.5), $W^{(k)}$ stands for the current weighting matrix, $M^{(k)}$ means the resulting FIM and $D^{(k)} = \{M^{(k)}\}^{-1}$. Moreover, $\mathfrak{J} = \{1, \ldots, N\}$ and $\xi_{x^i \leftrightarrows x}^{(k)}$ means the design in which the support point x^i was replaced by x.

ALGORITHM 7.1 Exchange algorithm for correlated observations

Step 1. *Select an initial design*

$$\xi^{(0)} = \{x^{1(0)}, \ldots, x^{N(0)}\}$$

such that $x^{i(0)} \neq x^{j(0)}$ for $i, j \in \mathfrak{J}$ and $i \neq j$. Calculate the matrices $F^{(0)}$, $W^{(0)}$ and $M^{(0)}$, and then the value of $\Psi[M^{(0)}]$. If $\Psi[M^{(0)}] = \infty$, select a new initial design and repeat this step. Set $k = 0$.

Step 2. *Determine*

$$(i^\star, x^\star) = \arg \max_{(i,x) \in \mathfrak{J} \times X} \Delta(x^i, x),$$

where

$$\Delta(x^i, x) = \{\Psi[M(\xi^{(k)})] - \Psi[M(\xi_{x^i \leftrightarrows x}^{(k)})]\} / \Psi[M(\xi^{(k)})].$$

Step 3. *If $\Delta(x^{i^\star}, x^\star) \leq \delta$, where δ is some given positive tolerance, then STOP. Otherwise, set*

$$\xi^{(k+1)} = \xi_{x^{i^\star} \leftrightarrows x^\star}^{(k)}$$

and determine $F^{(k+1)}$, $W^{(k+1)}$ and $M^{(k+1)}$ corresponding to $\xi^{(k+1)}$. Set $k \leftarrow k + 1$ and go to Step 2.

From the point of view of nonlinear programming, if we treat the design problem as an optimization problem with decision variables $x^1, \ldots, x^N$ and the performance index $\Psi[M(\xi)]$, the algorithm outlined above constitutes a Gauss-Seidel-like algorithm. In the classical Gauss-Seidel relaxation scheme (also called the block coordinate ascent method, cf. [22]), each iteration consists of N one-dimensional search steps with respect to variables $x^1, \ldots, x^N$ taken in cyclic order. If a decrease in $\Psi[M(\xi)]$ is attained for some i after performing the search, then the corresponding variable x^i is immediately updated. The design algorithm outlined above differs from the Gauss-Seidel scheme in the update, as it takes place only for the coordinate x^i for which the resulting decrease in $\Psi[M(\xi)]$ is the largest. As a result, the delineated design algorithm possesses convergence properties similar to those of the Gauss-Seidel one, i.e., only convergence to a local maximum can be guaranteed. This implies that several restarts must often be performed from different initial designs so as to obtain an approximation to a global minimum. Owing to optimized updates of the matrices in Steps 2 and 3 through elimination of matrix inverses (see the following comments), the algorithm is fast enough in spite of this inconvenience.

Implementation details

The speed of Algorithm 7.1 can be highly improved by reduction of time-consuming operations, such as inversions of matrices in Step 2. We here give some details, generalizing some derivations from [31, 313]. With no loss of generality, we assume that we are always to replace point x^N by a point $x \in X$. Indeed, deletion of a point x^i for $i < N$ can be converted to this situation by interchanging positions of x^i and x^N in $\xi^{(k)}$, which then should be followed by swapping the i-th and N-th columns in $F^{(k)}$, as well as interchanging the i-th and N-th rows and the i-th and N-th columns in $W^{(k)}$. Such a replacement greatly simplifies the resulting formulae and makes the implementation easier.

Assume that we are to pass from $\xi^{(k)}$ to $\xi^{(k)}_{x^N \leftrightarrows x}$. This requires two stages: the removal of x^N from $\xi^{(k)}$ and the augmentation of the resulting design by x. Clearly, both stages imply changes in all the matrices corresponding to the current design. Applying the Frobenius theorem, cf. Appendix B.1, we can derive the forms of the necessary updates which guarantee the extreme efficiency of the calculations.

Stage 1: Deletion of x^N from $\xi^{(k)}$. Write $F^{(k)}(t)$ as

$$F^{(k)}(t) = \left[\ F_r(t)\ \vdots\ f_r(t)\ \right], \tag{7.10}$$

where $F_r(t) \in \mathbb{R}^{m \times (N-1)}$ and $f_r(t) \in \mathbb{R}^m$. Deletion of x^N from $\xi^{(k)}$ implies removing $f_r(t)$ from $F^{(k)}(t)$, so that we then have $F_r(\cdot)$ instead of $F^{(k)}(\cdot)$. But some changes are also necessary in matrices $W^{(k)}(\cdot)$ and $M^{(k)}$.

Namely, decomposing the symmetric matrix $W^{(k)}(t)$ into

$$W^{(k)}(t) = \left[\begin{array}{c|c} V_r(t) & b_r(t) \\ \hline b_r^{\mathsf{T}}(t) & \gamma_r(t) \end{array}\right], \tag{7.11}$$

where $V_r(t) \in \mathrm{Sym}(N-1)$, $b_r(t) \in \mathbb{R}^{N-1}$, $\gamma_r(t) \in \mathbb{R}$, we set

$$g_r(t) = \frac{1}{\sqrt{\gamma_r(t)}} F_r(t) b_r(t) + \sqrt{\gamma_r(t)} f_r(t), \tag{7.12}$$

and then calculate, respectively, the following counterparts of the matrices

$W^{(k)}(\cdot)$ and $M^{(k)}$:

$$W_r(t) = V_r(t) - \frac{1}{\gamma_r(t)} b_r(t) b_r^{\mathsf{T}}(t), \tag{7.13}$$

$$M_r = M^{(k)} - \int_0^{t_f} g_r(t) g_r^{\mathsf{T}}(t)\, \mathrm{d}t. \tag{7.14}$$

Stage 2: Inclusion of x into the design resulting from Stage 1. We construct

$$F_a(t) = \left[\, F_r(t) \,\vdots\, f_a(t) \,\right], \tag{7.15}$$

where $f_a(t) = f(x, t)$. Such an extension of the matrix of sensitivity coefficients influences the form of matrices $W_r(\cdot)$ and M_r obtained at Stage 1. In order to determine their respective updated versions $W_a(\cdot)$ and M_a, we define

$$v_a(t) = \mathrm{col}\left[q(x, x^1, t), \dots, q(x, x^N, t)\right], \tag{7.16}$$

$$\gamma_a(t) = 1/[q(x, x, t) - v_a^{\mathsf{T}}(t) W_r(t) v_a(t)], \tag{7.17}$$

$$b_a(t) = -\gamma_a(t) W_r(t) v_a(t), \tag{7.18}$$

$$g_a(t) = \frac{1}{\sqrt{\gamma_a(t)}} F_r(t) b_a(t) + \sqrt{\gamma_a(t)} f_a(t). \tag{7.19}$$

Then we have

$$W_a(t) = \left[\begin{array}{c:c} W_r(t) + \dfrac{1}{\gamma_a(t)} b_a(t) b_a^{\mathsf{T}}(t) & b_a(t) \\ \hdashline b_a^{\mathsf{T}}(t) & \gamma_a(t) \end{array} \right], \tag{7.20}$$

$$M_a = M_r + \int_0^{t_f} g_a(t) g_a^{\mathsf{T}}(t)\, \mathrm{d}t. \tag{7.21}$$

Completion of Step 2 means that points $x^{i^*} \in \xi^{(k)}$ (identified with x^N) and $x^* \in X$ which guarantee the largest decrease in the Ψ-optimality criterion for the current iteration have been found. At this juncture, we may update the matrices as follows:

$$F^{(k+1)} = F_a^*, \quad W^{(k+1)} = W_a^*, \quad M^{(k+1)} = M_a^*, \tag{7.22}$$

where the quantities on the right-hand sides are those of (7.15), (7.20) and (7.21), respectively, evaluated at x^*. The relatively simple form of the above formulae guarantees the efficiency of the algorithm which thus becomes very similar to Fedorov's exchange algorithm for determining exact designs, cf. [82].

The efficiency of the computations at both the stages of Algorithm 7.1 results from application of Theorem B.3 which provides convenient formulae for the inverse of block-partitioned matrices.

When deleting point x^N at Stage 1, the covariance matrix $C^{(k)}(t)$ can be partitioned as follows:

$$C^{(k)}(t) = \left[\begin{array}{c:c} C_r(t) & v_r(t) \\ \hdashline v_r^\mathsf{T}(t) & \alpha_r(t) \end{array} \right], \tag{7.23}$$

where $C_r(t) \in \mathrm{Sym}(N-1)$, $v_r(t) \in \mathbb{R}^{N-1}$, $\alpha_r(t) \in \mathbb{R}$. On the other hand, $C^{(k)}(t)$ constitutes the inverse of $W^{(k)}(t)$ and hence

$$C^{(k)}(t) = \left\{ W^{(k)}(t) \right\}^{-1} = \left[\begin{array}{c:c} V_r(t) & b_r(t) \\ \hdashline b_r^\mathsf{T}(t) & \gamma_r(t) \end{array} \right]^{-1}$$

$$= \left[\begin{array}{c:c} W_r^{-1}(t) & -\dfrac{1}{\gamma_r(t)} W_r^{-1}(t) b_r(t) \\ \hdashline -\dfrac{1}{\gamma_r(t)} b_r^\mathsf{T}(t) W_r^{-1}(t) & \dfrac{1}{\gamma_r(t)} + \dfrac{1}{\gamma_r^2(t)} b_r^\mathsf{T}(t) W_r^{-1}(t) b_r(t) \end{array} \right]. \tag{7.24}$$

Equating the corresponding blocks of (7.23) and (7.24), we see that $C_r(t) = W_r^{-1}(t)$, which implies $W_r(t) = C_r^{-1}(t)$, i.e., $W_r^{-1}(t)$ is the inverse of the covariance matrix for the measurements made at the reduced set of points $x^1, \ldots, x^{N-1}$.

By the above, making use of the partitions (7.10) and (7.11), we can write

$$F^{(k)}(t) W^{(k)}(t) \left\{ F^{(k)}(t) \right\}^\mathsf{T}$$

$$= \left[\begin{array}{c:c} F_r(t) & f_r(t) \end{array} \right] \left[\begin{array}{c:c} V_r(t) & b_r(t) \\ \hdashline b_r^\mathsf{T}(t) & \gamma_r(t) \end{array} \right] \left[\begin{array}{c} F_r^\mathsf{T}(t) \\ \hdashline f_r^\mathsf{T}(t) \end{array} \right]$$

$$= \left[\begin{array}{c:c} F_r(t) & f_r(t) \end{array} \right] \left[\begin{array}{c:c} W_r(t) + \dfrac{1}{\gamma_r(t)} b_r(t) b_r^\mathsf{T}(t) & b_r(t) \\ \hdashline b_r^\mathsf{T}(t) & \gamma_r(t) \end{array} \right] \left[\begin{array}{c} F_r^\mathsf{T}(t) \\ \hdashline f_r^\mathsf{T}(t) \end{array} \right] \tag{7.25}$$

$$= F_r(t) W_r(t) F_r^\mathsf{T}(t) + \dfrac{1}{\gamma_r(t)} F_r(t) b_r(t) b_r^\mathsf{T}(t) F_r^\mathsf{T}(t)$$

$$\quad + F_r(t) b_r(t) f_r^\mathsf{T}(t) + f_r(t) b_r^\mathsf{T}(t) F_r^\mathsf{T}(t) + \gamma_r(t) f_r(t) f_r^\mathsf{T}(t)$$

$$= F_r(t) W_r(t) F_r^\mathsf{T}(t) + g_r(t) g_r^\mathsf{T}(t),$$

which clearly forces (7.14).

In much the same way, when including a point x into the design resulting from Stage 1 so as to form an augmented design $\xi_a = \{x^{1(k)}, \ldots, x^{N-1(k)}, x\}$, the respective covariance matrix is naturally partitioned as follows:

$$C_a(t) = \left[\begin{array}{c|c} C_r(t) & v_a(t) \\ \hline v_a^{\mathsf{T}}(t) & \alpha_a(t) \end{array} \right], \qquad (7.26)$$

where $\alpha_a(t) = q(x, x, t)$. Defining

$$\gamma_a(t) = \frac{1}{\alpha_a(t) - v_a^{\mathsf{T}}(t)C_r^{-1}(t)v_a(t)} = \frac{1}{q(x, x, t) - v_a^{\mathsf{T}}(t)W_r(t)v_a(t)}, \qquad (7.27)$$

from Theorem B.3 we see that

$$W_a(t) = C_a^{-1}(t)$$

$$= \left[\begin{array}{c|c} C_r^{-1}(t) + \gamma_a(t)C_r^{-1}(t)v_a(t)v_a^{\mathsf{T}}(t)C_r^{-1}(t) & -\gamma_a(t)C_r^{-1}(t)v_a(t) \\ \hline -\gamma_a(t)v_a^{\mathsf{T}}(t)C_r^{-1}(t) & \gamma_a(t) \end{array} \right]$$

$$= \left[\begin{array}{c|c} W_r(t) + \dfrac{1}{\gamma_a(t)}b_a(t)b_a^{\mathsf{T}}(t) & b_a(t) \\ \hline b_a^{\mathsf{T}}(t) & \gamma_a(t) \end{array} \right].$$

$$(7.28)$$

Consequently,

$$F_a(t)W_a(t)F_a^{\mathsf{T}}(t)$$

$$= \left[\begin{array}{c|c} F_r(t) & f_a(t) \end{array} \right] \left[\begin{array}{c|c} W_r(t) + \dfrac{1}{\gamma_a(t)}b_a(t)b_a^{\mathsf{T}}(t) & b_a(t) \\ \hline b_a^{\mathsf{T}}(t) & \gamma_a(t) \end{array} \right] \left[\begin{array}{c} F_r^{\mathsf{T}}(t) \\ \hline f_a^{\mathsf{T}}(t) \end{array} \right]$$

$$(7.29)$$

$$= F_r(t)W_r(t)F_r^{\mathsf{T}}(t) + \frac{1}{\gamma_a(t)}F_r(t)b_a(t)b_a^{\mathsf{T}}(t)F_r^{\mathsf{T}}(t)$$

$$\quad + F_r(t)b_a(t)f_a^{\mathsf{T}}(t) + f_a(t)b_a^{\mathsf{T}}(t)F_r^{\mathsf{T}}(t) + \gamma_a(t)f_a(t)f_a^{\mathsf{T}}(t)$$

$$= F_r(t)W_r(t)F_r^{\mathsf{T}}(t) + g_a(t)g_a^{\mathsf{T}}(t),$$

which gives (7.21).

As a by-product of the above derivations, we deduce that the consecutive designs $\xi^{(k)}$ will never contain repeated support points. Indeed, it suffices to consider Stage 2 after k iterations of Algorithm 7.1, i.e., when the design

$\xi^{(k)} = \{x^{1(k)}, \ldots, x^{N-1(k)}, x^{N(k)}\}$ is to be improved by replacing $x^{N(k)}$ by a point $x \in X$. If $x = x^{i(k)}$ for some $i \in \{1, \ldots, N-1\}$, then $v_a(t)$ and $f_a(t)$ coincide with the i-th columns of $C_r(t)$ and $F_r(t)$. The result is

$$W_r(t)v_a(t) = C_r^{-1}(t)v_a(t) = \text{col}\big[\overset{1}{0}, \ldots, \overset{i-1}{0}, \overset{i}{1}, \overset{i+1}{0}, \ldots, \overset{N-1}{0} \big]. \quad (7.30)$$

Consequently, writing

$$
\begin{aligned}
g_a(t) &= \frac{1}{\sqrt{\gamma_a(t)}} F_r(t)b_a(t) + \sqrt{\gamma_a(t)}f_a(t) \\
&= \frac{f_a(t) - F_r(t)W_r(t)v_a(t)}{\sqrt{\alpha_a(t) - v_a(t)W_r(t)v_a(t)}},
\end{aligned}
\quad (7.31)
$$

we see that both the numerator and the denominator tend to zero as $x \to x^i$. However, it can be shown that the convergence of the latter is slow enough to yield $g_a(t) \to 0$ as $x \to x^i$. It follows then that $M_a = M_r$, i.e., inclusion of a point coinciding with a point which already exists in the current design does not alter the FIM. In other words, additional measurements at a sensor location do not provide additional information about the estimated parameters, and hence the optimal design will automatically be clusterization-free.

The above numerical algorithm for the construction of Ψ-optimum designs involves the iterative improvement of an initial N-point design. In general, Algorithm 7.1 is attempting to find the minimum of a hypersurface with many local extrema, so it cannot be guaranteed to discover anything more than a local minimum. The probability of finding the global minimum can be increased by repeating the search several times from different starting designs, the generation of which often includes a random component. The probability of finding the global optimum can also be increased by the use of a more thorough search algorithm for improvement of the initial design, cf. [8].

7.2 Maximization of an observability measure

This section seeks to study a problem which has relatively often been addressed in the control literature: How should one choose the locations of measurement sensors for a given DPS so as to increase its degree of observability quantified by a suitable observability measure? Certainly the selecton of these sensor positions may have such a dramatic effect on the performance possibilities as to far outweigh the optimal "tuning" of the control signals that takes place after sensor locations are selected. Specific features of this problem and past approaches are surveyed, e.g., in [145, 330]. However, the results communicated by most authors are rather limited to the selection of stationary sensor positions. A generalization which imposes itself is to apply sensors which are capable of tracking points providing at a given time the best information about the system state. In particular, it happens frequently that the observation system comprises multiple sensors whose positions are already specified and it is desired to activate only a subset of them during a given time interval while the other sensors remain dormant [65]. Such a scanning strategy of taking measurements was already discussed in Section 4.1. Likewise, the key idea here is to operate on the density of sensors per unit area instead of the positions of individual sensors. Such conditions allow us to relax the discrete optimization problem in context and to replace it by its continuous approximation. Mathematically, this procedure involves looking for a family of 'optimal' probability measures defined on subsets of the set of feasible measurement points. In the sequel we report a nontrivial generalization of the results of Section 4.1 to output selection in control system design, where the approach based on sensor densities has not yet been employed, apart from [316] where the presented idea was outlined first.

7.2.1   Observability in a quantitative sense

Given a linear DPS described by a PDE model, consider its finite-dimensional approximation (e.g., obtained via the finite-element method) in the form of the following system of linear ODEs:

$$\frac{dy(t)}{dt} = A(t)y(t), \quad t \in T = [t_0, t_f],$$ (7.32)

$$y(t_0) = y_0,$$ (7.33)

such that $y(t) \in \mathbb{R}^n$ and $A(t) \in \mathbb{R}^{n \times n}$, which is augmented by the sensor location parameterized counterpart of the output equation

$$z(t) = C(t; \zeta(t))y(t),$$ (7.34)

where $z(t) \in \mathbb{R}^N$, $C(t; \zeta(t)) \in \mathbb{R}^{N \times n}$, and the notation emphasizes the dependence of the output matrix C on the current spatial configuration of sensors $\zeta(t)$ (to be determined in what follows).

As regards a quantitative measure for state observability, consider the observability Gramian [69, 330]

$$W(\zeta) = \int_{t_0}^{t_f} \Phi^{\mathsf{T}}(t, t_0) C^{\mathsf{T}}(t; \zeta(t)) C(t; \zeta(t)) \Phi(t, t_0) \, dt,$$ (7.35)

where the fundamental (or transition) matrix $\Phi(t, t_0)$ obeys

$$\frac{d\Phi(t, t_0)}{dt} = A(t)\Phi(t, t_0), \quad \Phi(t_0, t_0) = I,$$ (7.36)

I being the identity matrix.

An optimal sensor configuration strategy ζ^* can be found by minimizing some convex function Ψ defined on $W(\zeta)$ [330]. Common choices include the following:

1. $\Psi(W) = -\ln\det(W)$,

2. $\Psi(W) = \operatorname{tr}(W^{-1})$,

3. $\Psi(W) = -\operatorname{tr}(W)$,

4. $\Psi(W) = \lambda_{\max}(W^{-1})$,

where $\lambda_{\max}(\,\cdot\,)$ stands for the largest eigenvalue of its matrix argument. Since the last criterion is nondifferentiable when there are repeated eigenvalues, its use will not be considered here.

7.2.2 Scanning problem for optimal observability

Let us form an arbitrary partition of the time interval $T = [t_0, t_f]$ by choosing points $t_0 < t_1 < \cdots < t_R = t_f$ defining subintervals $T_r = [t_{r-1}, t_r)$, $r = 1, \ldots, R$. We then consider N scanning sensors which will possibly change their locations at the beginning of every time subinterval, but will remain stationary for the duration of each of the subintervals. Thus the sensor configuration ζ can be viewed as follows:

$$\zeta(t) = (x_r^1, \ldots, x_r^N) \quad \text{for } t \in T_r, \quad r = 1, \ldots, R, \qquad (7.37)$$

where $x_r^j \in X \subset \mathbb{R}^d$ stands for the location of the j-th sensor on the subinterval T_r, X being the part of the spatial domain where the measurements can be taken (mathematically, it should be a compact set).

Assume that the consecutive rows of the matrix $C(t; \zeta(t))$ in (7.34) correspond to contributions from different sensors, i.e.,

$$C(t, \zeta(t)) = \begin{bmatrix} \gamma^{\mathsf{T}}(x_r^1, t) \\ \vdots \\ \gamma^{\mathsf{T}}(x_r^N, t) \end{bmatrix} \quad \text{for } t \in T_r, \quad r = 1, \ldots, R, \qquad (7.38)$$

where $\gamma : X \times T \to \mathbb{R}^n$ is a given function. Then we can decompose the Gramian as follows:

$$W(\zeta) = \sum_{r=1}^{R} \sum_{j=1}^{N} \Upsilon_r(x_r^j), \qquad (7.39)$$

where

$$\Upsilon_r(x) = \int_{t_{r-1}}^{t_r} g(x, t) g^{\mathsf{T}}(x, t)\, \mathrm{d}t, \qquad (7.40)$$

$$g(x, t) = \Phi^{\mathsf{T}}(t, t_0) \gamma(x, t). \qquad (7.41)$$

We have thus arrived at the crucial point for the presented approach, as the proposed algorithm of finding the best sensor locations may only be employed on the condition that the Gramian constitutes the sum of some matrices, each of them being completely defined by the position of only one scanning sensor on one subinterval T_r, cf. (7.39).

7.2.3   Conversion to finding optimal sensor densities

As is known from Section 4.1, when the number of sensors N is large, the optimal sensor-location problem becomes extremely difficult from a computational point of view. Consequently, we propose to operate on the spatial density of sensors, rather than on the sensor locations.

Performing such a conversion, we approximate the density of sensors over the subinterval T_r by a probability measure $\xi_r(\mathrm{d}x)$ on the space $(X, \mathcal{B})$, where $\mathcal{B}$ is the σ-algebra of all Borel subsets of X. As regards the practical interpretation of the so-produced solutions, one possibility is to partition X into nonoverlapping subdomains ΔX_i of relatively small areas and then, on the subinterval T_r, to allocate to each of them the number

$$N_r(\Delta X_i) = \left\lceil N \int_{\Delta X_i} \xi_r(\mathrm{d}x) \right\rceil \tag{7.42}$$

of sensors ($\lceil \rho \rceil$ is the smallest integer greater than or equal to ρ).

Thus our aim is to find probability measures ξ_r, $r = 1, \ldots, R$ over X. For notational convenience, in what follows we shall briefly write $\xi = (\xi_1, \ldots, \xi_R)$ and call ξ a *design*.

Such an extension of the concept of the sensor configuration allows us to replace (7.39) by

$$W(\xi) = \sum_{r=1}^{R} \int_X \Upsilon_r(x)\, \xi_r(\mathrm{d}x). \tag{7.43}$$

A rather natural additional assumption is that the density $N_r(\Delta X_i)/N$ in a given part ΔX_i must not exceed some prescribed level. In terms of the probability measures, this amounts to imposing the conditions

$$\xi_r(\mathrm{d}x) \leq \omega(\mathrm{d}x), \quad r = 1, \ldots, R, \tag{7.44}$$

where $\omega(\mathrm{d}x)$ is a given measure satisfying $\int_X \omega(\mathrm{d}x) \geq 1$.

Defining $J(\xi) = \Psi[W(\xi)]$, we may phrase the scanning sensor location problem as the selection of

$$\xi^\star = \arg \min_{\xi \in \Xi(X)} J(\xi), \tag{7.45}$$

where $\Xi(X)$ denotes the set of all competing designs whose components satisfy (7.44) (note that $\Xi(X)$ is nonempty and convex). We call $\xi^\star$ the (Ψ, ω)-*optimal solution*.

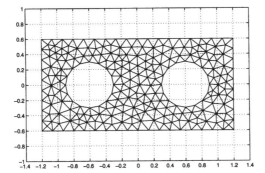

FIGURE 7.1
Spatial domain Ω and an unstructured mesh on it.

It is easy to see that the above design problem perfectly fits the framework introduced in Section 4.1 if we interpret the Gramian as the Fisher information matrix, as a precise way of determining the values of the matrices $\Upsilon_r(x)$ is unimportant for application of Algorithm 4.2. In order to warrant its use, it suffices to retain Assumptions (A1)–(A7) while replacing solely Assumption (A2) by the assumption that

$$\Upsilon_r \in C(X; \mathbb{R}^{n \times n}). \tag{7.46}$$

The only difference from the deliberations of Section 4.1 is the large dimensionality of the Gramian ($n \times n$, where n is typically the number of nodes in a finite-element mesh) compared with the usual sizes of the FIMs ($m \times m$, where m rarely exceeds 10). This may constitute a serious inconvenience while implementing the above idea and require care when selecting numerical algorithms for the solution of the computational tasks involved.

The following numerical example serves as a vehicle for the display of the practical performance of the solution technique which has just been briefly outlined.

Example 7.1
Consider the two-dimensional heat equation

$$\frac{\partial y}{\partial t} - \mu \Delta y = 0 \quad \text{in } \Omega \times T, \tag{7.47}$$

which describes the diffusion of heat over the time interval $T = [0, 1]$ in a body represented geometrically by the spatial domain Ω, where $y = y(x, t)$ is the state (temperature) and $\mu = 0.1$ is the diffusion coefficient. The form of Ω, which can represent, e.g., a metal block with two circular cracks or cavities, is given in Fig. 7.1.

Equation (7.47) is supplemented with the zero Dirichlet boundary conditions on the outer boundaries of Ω and the zero Neumann conditions on both the inner circular boundaries. Using the PDE Toolbox which provides a powerful and flexible environment for the study and solution of partial differential equations in MATLAB, cf. Appendix I.2, the triangular mesh of 243 nodes shown in Fig. 7.1 was built on the domain Ω using the graphical user interface implemented in the routine `pdetool`. The mesh nodes which do not lie on the outer boundary (there were 189 such nodes) were treated as candidates for locating $N = 90$ pointwise sensors, i.e., they formed the set $\tilde{X}$ in Algorithm 4.2. The observation horizon T was partitioned into four subintervals

$$T_r = \left[\frac{r-1}{4}, \frac{r}{4}\right), \quad r = 1, \ldots, 4. \tag{7.48}$$

The matrices $\Upsilon_r(x^j)$ for $r = 1, \ldots, 4$ and $j = 1, \ldots, 189$ were then computed in accordance with (7.40). In particular, approximation of the integrals was performed by dividing the interval T_r into seven equal subintervals and then using the trapezoidal rule. The ODE (7.36) was integrated by fixing the time step $\Delta t = 1/28$ and employing the backward difference method [162].

At this step, the stiffness and mass matrices resulting from applying the method of lines semidiscretization were needed. They were therefore assembled using the procedure `assempde`. Computation of the $\Upsilon_r(x^j)$'s took approximately 90 s on a Pentium IV 2.40 GHz computer equipped with 524 MB RAM and running Windows 2000. The initial selection of sensor configurations was performed at random several times so as to avoid getting stuck in a local minimum. Algorithm 4.2 was then run for each such starting configuration for the performance indices $\Psi_1[W] = -\ln\det(W)$ and $\Psi_2[W] = -\operatorname{tr}(W)$. For comparison, optimal locations of stationary sensors were also determined using the same technique (i.e., by setting $R = 1$). Convergence of the algorithm is rapid, as in the most time-consuming case of the criterion Ψ_1 one run took no more than three minutes. The results are shown in Figs. 7.2 and 7.3.

As was expected, major improvements are observed when scanning sensors are applied, since the ratios of the best absolute values for the scanning case to those for the stationary case equal 1.242 and 1.116 for the criteria $\Psi_1[W]$ and $\Psi_2[W]$, respectively. In applications, $\Psi_2[W]$ is often preferred, as its linearity with respect to the contributions from different sensors leads to extremely simple computations. Note, however, that this apparent advantage is illusory, since the resulting Gramian is close to singularity, which may cause some problems with the system observability. But this phenomenon is rather obvious, since a larger trace of a matrix does not necessarily imply that the matrix is nonsingular. This drawback is eliminated through the use of the determinant of the Gramian as the performance index.

As for the interpretation of the produced solution, the scanning sensors are to occupy positions which give the best information about the initial state. As time elapses, the candidate points close to the outer boundary provide less relevant information since their state is mostly determined by the Dirichlet boundary conditions. Thus, intuitively, the sensors should tend toward the centre of Ω, Such behaviour is exhibited by the solution for the

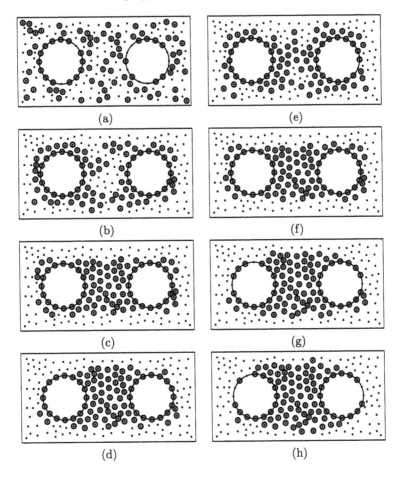

FIGURE 7.2
Optimal selection of consecutive sensor configurations for the criteria $\Psi_1[W] = -\ln\det(W)$ (panels (a)–(d)) and $\Psi_2[W] = -\operatorname{tr}(W)$ (panels (e)–(h)).

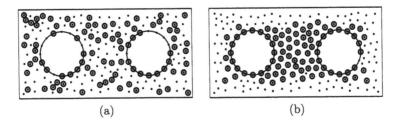

FIGURE 7.3
Optimal selection of stationary sensor locations for the criteria $\Psi_1[W] = -\ln\det(W)$ (panel (a)) and $\Psi_2[W] = -\operatorname{tr}(W)$ (panel (b)).

criterion $\Psi_1[W]$, which constitutes an additional argument for its superiority over $\Psi_2[W]$.   ☐

7.3 Summary

The main contribution of this chapter is to provide a numerical method for the construction of optimum designs in the presence of correlations. The crucial step is the algebra of Section 7.1.1 which leads to a rapid exchange algorithm that is very simple to implement. It constitites an extension of the algorithm proposed in [31], cf. also [189, 313], which can be further modified to take account of more complicated correlation structures [206].

In the second part of the chapter, a new approach to the problem of scanning sensor location for linear DPSs based on various criteria defined on the observability Gramian was demonstrated. A close connection was established between this problem and modern optimum experimental design theory. As in Section 4.1, the main idea here is to operate on the density of sensors per unit area instead of the positions of individual sensors. Likewise, it can be shown that the optimal solutions obey certain minimax properties that lead to a rapidly convergent algorithm which is nothing but Algorithm 4.2 with the Gramian in place of the Fisher information matrix. Note that, with minor changes, the technique can be extended to Kalman filtering and robust control.

8

Applications from engineering

The present chapter includes several examples which indicate potential applications of the theory and methods proposed in the monograph. They have been selected so as to be intuitively clear, and if possible, to possess solutions whose form can be justified to a great extent so as to validate the results produced by the applied algorithms. On the other hand, the examples are already far from triviality and may constitute prototypes for more sophisticated frameworks.

8.1 Electrolytic reactor

The first example concerns an application of the proposed methodology in the computer-assisted tomography of an electrolysis process [118, 268]. Consider a cylindrical chamber of a reactor filled with an electrolyte. In the interior of the chamber, two metallic conductors are placed on the plane of a conductive medium characterized by a spatially varying conductivity coefficient. The domain of interest $\Omega \subset \mathbb{R}^2$ which represents a planar cross-section of the reactor chamber is shown in Fig. 8.1. Its boundary $\Gamma = \bigcup_{i=1}^{3} \Gamma_i$ is split into three disjoint subsets:

$$\Gamma_1 = \left\{ (x_1, x_2) : x_1^2 + x_2^2 = 1 \right\}, \tag{8.1}$$

$$\Gamma_2 = \left\{ (-0.5 \pm 0.05, x_2) : -0.2 \le x_2 \le 0.2 \right\} \\ \cup \left\{ (x_1, \pm 0.2) : -0.55 \le x_1 \le -0.45 \right\}, \tag{8.2}$$

$$\Gamma_3 = \left\{ (0.5 \pm 0.05, x_2) : -0.2 \le x_2 \le 0.2 \right\} \\ \cup \left\{ (x_1, \pm 0.2) : -0.55 \le x_1 \le -0.45 \right\}. \tag{8.3}$$

A steady current flow through the metallic electrodes, taken in conjunction with the absence of external current sources, results in the physical model for this problem expressed in the form of the Laplace equation

$$\nabla \cdot \left(\sigma(x) \nabla y(x) \right) = 0, \quad x \in \Omega, \tag{8.4}$$

where y is the potential of the electric field in the conductive medium. The

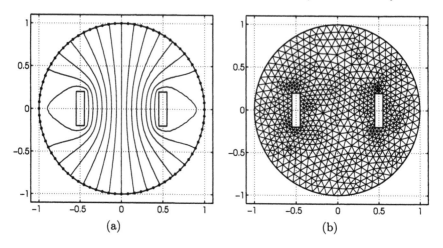

FIGURE 8.1
Domain Ω with admissible locations of discrete sensors versus the contour
plot of the electric potential (a), and the unstructured mesh for the FEM
discretization (749 nodes and 1392 triangles) (b).

respective boundary conditions are

$$
\begin{cases}
\dfrac{\partial y(x)}{\partial n} = 0 & \text{if } x \in \Gamma_1 \quad \text{(the natural boundary condition} \\
& \qquad\qquad\qquad \text{on the outer boundaries),} \\[6pt]
y(x) = -10 & \text{if } x \in \Gamma_2 \quad \text{(for the left metallic conductor),} \\
y(x) = 10 & \text{if } x \in \Gamma_3 \quad \text{(for the right metallic conductor).}
\end{cases}
\tag{8.5}
$$

where $\partial y/\partial n$ denotes the derivative in the direction of the unit outward normal
to Γ.

The purpose here is to find a D-optimum design to reconstruct the spatial
distribution of the conductivity coefficient modelled in the following form:

$$
\sigma(x) = 0.2 + \theta_1(x_1 - \theta_2)^2(x_1 - \theta_3)^2 + \theta_4 x_2^2,
\tag{8.6}
$$

which leads to large conductivity values near the metallic electrodes and low
ones at the outer boundary.

The aim of the parameter estimation is to optimally estimate the vector
$\theta = \mathrm{col}[\theta_1, \theta_2, \theta_3, \theta_4]$. The state equation (8.4) is thus supplemented by four
sensitivity equations, which results in a system of elliptic PDEs which have to
be solved numerically and the solution is to be stored in memory for further
design. To this end, procedures from the MATLAB PDE Toolbox [53], cf. also
Appendix I.2, based on a finite-element method solver (function **assempde**)
were employed with the initial estimate of $\theta^0 = \mathrm{col}[-0.2, 0.5, -0.5, -0.1]$ taken

as a nominal one. The distribution of the electric potential is illustrated in Fig. 8.1(a).

The requirement of a nondestructive experiment in the computer-assisted tomography forces the placement of the measurement electrodes only on the outer boundary Γ_1, which thus becomes the design space X. For this reason it is very convenient to derive its suitable parameterization. The simplest way to do so is to use the length $\lambda \in [0, 2\pi]$ of Γ_1, i.e.,

$$x_1(\lambda) = \cos(\lambda), \quad x_2(\lambda) = \sin(\lambda). \tag{8.7}$$

8.1.1 Optimization of experimental effort

In order to obtain the desired approximation to the D-optimal design, Algorithms 3.1 and 3.2 were implemented in a computer program using the Lahey/Fujitsu Fortran 95 compiler v.5.7 and a PC running Windows 2000 (Pentium 4, 2.40 GHz, 512 MB RAM).

In the case of Algorithm 3.2 the set X_d of possible locations for measurement electrodes was selected as a uniform grid along the outer boundary Γ_1, cf. Fig. 8.1(a),

$$X_d = \left\{ x(\lambda_i) : \lambda_i = \frac{1}{40}(i-1)\pi, \ i = 1, \ldots, 80 \right\}. \tag{8.8}$$

Those support points with assigned equal weights $p_i^{(0)} = 1/80$, $i = 1, \ldots, 80$ formed the starting design for weight-optimization procedures.

Both the algorithms produced very similar results, which proves the efficiency of the multiplicative weight-optimization procedure. Given the accuracy of $\eta = 10^{-5}$, Algorithm 3.2 converged after 523 iterations (5 seconds of operational time) to the following approximation of the D-optimal design:

$$\xi^\star = \left\{ \begin{array}{cccc} (0.9949, 0.0783), & (0.9220, 0.3819), & (0.4530, 0.8892), & (0.0783, 0.9949) \\ 0.2482, & 0.2529, & 0.2469, & 0.2520 \end{array} \right\}. \tag{8.9}$$

In turn, Algorithm 3.1 started from the design

$$\xi^{(0)} = \left\{ \begin{array}{cccc} (0.0, 1.0), & (0.5, \sqrt{3}/2), & (\sqrt{3}/2, 0.5), & (1.0, 0.0) \\ 0.25, & 0.25, & 0.25, & 0.25 \end{array} \right\}, \tag{8.10}$$

and attained the optimal solution

$$\xi^\star = \left\{ \begin{array}{cccc} (0.9941, 0.0877), & (0.9279, 0.3674), & (0.4530, 0.8915), & (0.0548, 0.9965) \\ 0.25 & 0.25 & 0.25 & 0.25 \end{array} \right\}. \tag{8.11}$$

using only five additions of new support points (with $\eta = 10^{-5}$), which took about 10 seconds.

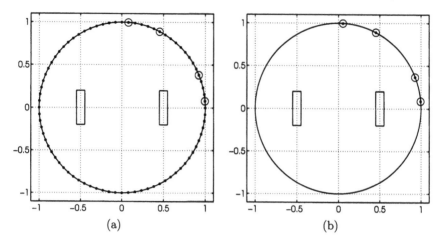

FIGURE 8.2
D-optimal design obtained via Algorithm 3.2 (open circles indicate the D-optimum sensor locations) (a), and Algorithm 3.1 (b).

The results are compared in Fig. 8.2. They turn out to be practically the same as far as the value of the D-optimality criterion is considered as a measure of the difference between them. This means that the discretization of the boundary was adequate to obtain a reasonably high quality of the solution.

An important observation is that due to the symmetry and antisymmetry of the problem with respect to the x_1- and x_2-axes, respectively, every new design created by a symmetrical mirroring of any support point with respect to those axes (with a suitable repartiton of the weights) constitutes an equivalent D-optimal solution.

8.1.2 Clusterization-free designs

As a last approach applied in our experiment, the clusterization-free algorithm (i.e., Algorithm 3.3) was used in the simplest form of a sequential one-point exchange procedure. In this case, the aim was to choose 30 locations of measurement electrodes from the discrete set X_d defined by (8.8). An initial design was created by randomly selecting support points. The algorithm calculated the solution very quickly (8 iterations for $\eta = 10^{-5}$). The initial and final distributions of the optimal support points are shown in Fig. 8.3.

Clearly, the quality of the solution is lower in the sense of the D-optimality criterion values because of the constraints imposed on the design measures. However, the clusterization effect is avoided and the D-optimal observational strategy tends to distribute the measurements in the vicinity of the best support points of the replicated designs obtained earlier (the symmetry of the

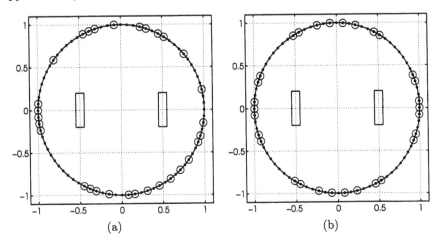

FIGURE 8.3
Initial (a) and D-optimal (b) designs for the clusterization-free algorithm.

problem is retained).

Note that the optimal measurement problem may become extremely diffi-cult in the case of the considered tomography process, since the most informa-tive measurements can be taken only inside the object (this is a most common situation in practical applications), and thus we may obtain ill-conditioned FIMs (they may be very close to singularity).

8.2 Calibration of smog-prediction models

As the next example, we consider the mass balance of chemical species and transport of tracers including advection, diffusion and reaction processes. One of the most tangible and interesting practical instances of this phenomenon is monitoring an air pollutant in order to produce a proper forecast of pollutant concentrations [64, 119, 180]. It is quite clear that a model in the form of a system of PDEs should be accurately calibrated as it will directly influence the quality of the forecast.

In our computer simulation, the transport and chemistry processes of the air pollutant over a given urban area in a mesoscale are considered. In reality, the contaminant interacts with components of the local atmosphere including other pollutants. However, to slightly simplify the situation, those are not included in our analysis. The entire process can thus be described by the

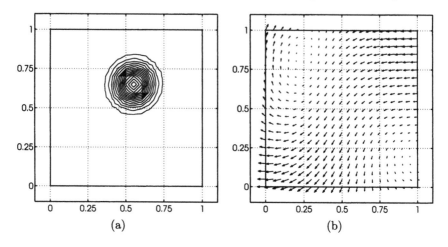

FIGURE 8.4
An urban area with the initial concentration of the pollutant (a) and the measured wind-velocity field (b).

advection-diffusion-reaction equation

$$\frac{\partial y(x,t)}{\partial t} + \nabla \cdot \big(v(x,t)y(x,t)\big) = \nabla \cdot \big(d(x)\nabla y(x,t)\big) - \kappa y(x,t), \qquad (8.12)$$

where $y(x,t)$ is the concentration observed in the normalized time interval $T = [0,1]$ and for a suitably rescaled spatial variable x belonging to the unit square Ω with the boundary Γ, cf. Fig. 8.4. The last term on the right-hand side with coefficient κ is an absorption term responsible for the decay of the pollutant due to the gas-phase kinetic reaction [119].

At the initial time instant, the source of the pollutant located at point $a = (0.55, 0.65)$ emits a chemical substance to the atmosphere, which spreads all over the domain Ω due to a complicated combination of diffusion and advection processes. The velocity vector v varies in space according to the model reflected in Fig. 8.4(b).

The phenomena mentioned above can be expressed by the following boundary and initial conditions:

$$\begin{cases} \dfrac{\partial y(x,t)}{\partial n} = 0 & \text{if } (x,t) \in \Gamma \times T, \\ y(x,0) = 30\exp(-80\|x-a\|^2) & \text{if } x \in \Omega. \end{cases} \qquad (8.13)$$

The distributed diffusion coefficient was assumed to be in the following form:

$$d(x) = \theta_1 + \theta_2(x_1 - \theta_4)^2 + \theta_3(x_2 - \theta_5)^2, \qquad (8.14)$$

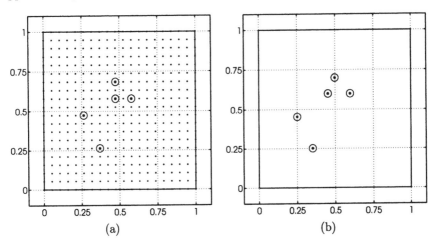

(a)　　　　　　　　　　　　(b)

FIGURE 8.5
D-optimal locations of air-pollution monitoring sensors determined by Algorithm 3.2 (a) and Algorithm 3.1 (b).

where the nominal vector of parameter values $\theta = \mathrm{col}[0.02, 0.01, 0.01, 0.5, 0.5]$ was assumed to be accessible from previous experiments in similar situations and the decay rate was set arbitrarily as $\kappa = 0.2$. The purpose was to find a D-optimal sensor allocation strategy for determining the most accurate estimates of the true parameters θ_i, $i = 1, \ldots, 5$.

The experiment consisted in a comparison of the strategies employing stationary and scanning sensors. In much the same way as in the previous example, the solutions of the PDEs were obtained using MATLAB's PDE Toolbox (20 divisions of the time interval and a regular (41×41)-node mesh over the rectangular domain. All design optimization algorithms were implemented using the Lahey/Fujitsu Fortran 95 compiler v.5.7 and ran on a PC equipped with a Pentium IV 1.7 GHz processor, 768 MB RAM and Windows 2000.

In the case of stationary sensors, Algorithms 3.1 and 3.2 were considered. As for the latter, a 20×20-point uniform grid was used and the algorithm started from a randomly generated design (in the sense of weights). It converged after 611 iterations for an accuracy of $\eta = 10^{-5}$. The obtained approximation of the optimal design contains five support points

$$
\xi^\star = \begin{cases} (0.3684, 0.2632), & (0.2631, 0.4737), & (0.4737, 0.5789), \\ \quad 0.2160, & \quad 0.2140, & \quad 0.1889, \end{cases} \cdots
$$
$$
\begin{array}{cc} (0.5789, 0.5789), & (0.4737, 0.6842) \\ \quad 0.2012, & \quad 0.1798 \end{array} \Bigg\}
\tag{8.15}
$$

and corresponds to the criterion value $\det(M(\xi^\star)) = 42.399$. In turn, Algorithm 3.1 was started from an initial six-point design generated randomly. For

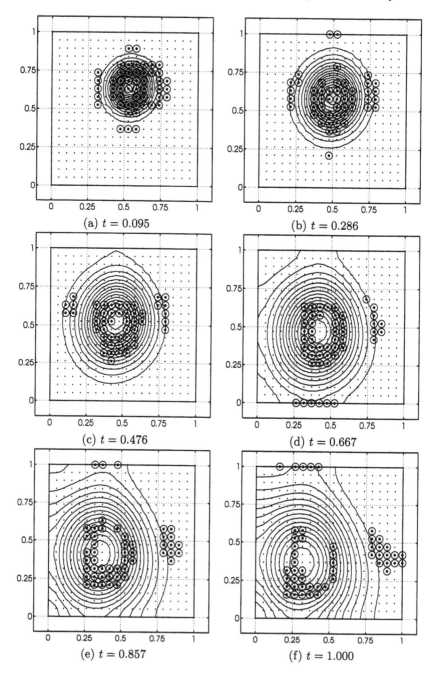

FIGURE 8.6

D-optimal scanning sensor activations for selected time moments in the air-quality monitoring example.

an accuracy of $\eta = 10^{-5}$ it converged after 5 iterations to the ultimate design

$$\xi^\star = \left\{ \begin{array}{ccc} (0.3518, 0.2513), & (0.2513, 0.4523), & (0.4523, 0.5990), \\ 0.2200, & 0.2085, & 0.1935, \end{array} \right. \cdots$$

$$(8.16)$$

$$\left. \begin{array}{cc} (0.5990, 0.5990), & (0.4975, 0.6985) \\ 0.2145, & 0.1635 \end{array} \right\},$$

which corresponds to $\det(M(\xi^\star)) = 47.212$. Both the solutions are shown in Fig. 8.5.

Finally, for the same task the scanning strategy was investigated with the assumption that the sensors may be placed at points of the same uniform grid as in the case of Algorithm 3.2. The partition of T was defined by the switching points $t_r = r/20$, $r = 0, \ldots, 20$. The task was to choose the best subset of 50 points from among 400 admissible sites over any resulting time subinterval. The solution was produced by Algorithm 4.2 in only 291 iterations with accuracy $\eta = 10^{-5}$ (it took less than 5 seconds). The results are shown in Fig. 8.6, where points stand for the sensor locations and open circles indicate the currently active sensors. In general, the locations of the activated sensors in successive time subintervals are not easy to explain. Obviously, the tendency to make observations at the positions where the greatest changes in the concentration occur is rather clear, since the advection is the dominant process and sensors tends to follow the wind field. However, the behaviours of some group of sensors are difficult to predict, which can be seen in Fig. 8.6, because it results from a sophisticated combination of advection, diffusion and reaction processes.

8.3 Monitoring of groundwater resources quality

Groundwater modelling is another interesting application which can be considered in the context of the practical usefulness of the developed methodology. Groundwater is one of the natural resources which sustained extensive damage in the past decades due to man's industrial activities [139, 283], e.g., due to solid and liquid waste dumps, storing and processing chemicals, etc. The pollutants spread through a covering layer and after some time they reach and pollute the groundwater [139, 282, 283]. Nowadays, the problem in question comprises not only determination of wells with a given amount of water discharge, but also the fulfilment of many quality requirements [283].

Consider a confined aquifer bounded by a river bank on one side and a discharge water channel on the opposite side. The rest of the boundary is assumed to be of an impermeable type or the flow through it is negligible. In this area, six possible locations for the observation wells, O_1 to O_6, were assumed. The whole situation is depicted in Fig. 8.7(a).

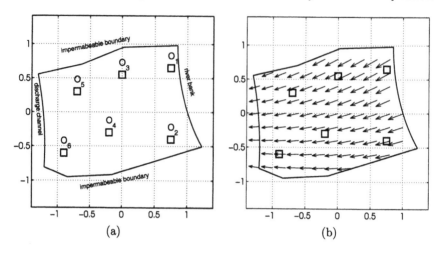

FIGURE 8.7
The general overview of a confined aquifer (a) and the velocity field with
locations of observation wells.

The river constitutes a source of pollution, which spreads over the aquifer
due to the hydrodynamic transport and dispersion. The contamination process can be described by a simplified two-dimensional model. In contrast to
the atmospheric pollution, the transport of substances in porous media takes
place in a much larger scale of time. The duration of such processes may be
equal to months or even years, so the pollution effects are also more perma-
nent. The model for changes in the pollutant concentration $y(x,t)$ over the
domain Ω in a normalized unit time interval T is given by [139, 283]

$$\frac{\partial y(x,t)}{\partial t} + \nabla \cdot \big(v(x)y(x,t)\big)$$

$$= \nabla \cdot \big(d(x)\nabla y(x,t)\big), \quad x \in \Omega, \quad t \in T = [0,1]. \quad (8.17)$$

The bank of the river is assumed to be contaminated all along its length
adjacent to the domain Ω with the same constant rate. Moreover, an initial
contamination of the aquifer with the considered substance can be neglected.
The left boundary of the aquifer is of a permeable type, but the interaction
between the water level in the channel and the groundwater surface is not as
ideal as on the river bank. This situation can be handled by the generalized
Neumann conditions. Finally, the pollutant is assumed not to affect the density of groundwater, i.e., it is approximately constant in time. Thus the initial
condition for (8.17) takes the following form:

$$y(x,0) = 0, \quad x \in \Omega, \quad (8.18)$$

whereas the boundary conditions are defined for individual parts of the boundary of Ω in the following manner:

River bank:

$$y(x,t) = 10, \quad (x,t) \in \Gamma_1 \times T, \tag{8.19}$$

Channel boundary:

$$\frac{\partial y(x,t)}{\partial n} + y(x,t) = 0, \quad (x,t) \in \Gamma_2 \times T, \tag{8.20}$$

Impermeable boundary:

$$\frac{\partial y(x,t)}{\partial n} = 0, \quad (x,t) \in (\Gamma_3 \cup \Gamma_4) \times T. \tag{8.21}$$

Optimization of the process of monitoring water quality in order to reduce the number of data sources and proper use of the available data is of great importance, since exploratory wells of observation and pumping types are very expensive and, additionally, not every location is possible to make a suitable drill. Because the possible number of observation wells is rather small, the application of the scanning strategy with an optimal switching schedule seems to be the most appropriate approach.

The parametric form of the spatially varying hydrodynamic dispersion was assumed as

$$d(x) = \theta_1 + \theta_2 x_1 + \theta_3 x_2, \tag{8.22}$$

where the elements of the vector $\theta^0 = \text{col}[0.4, -0.05, -0.03]$ were taken as the nominal parameter values for those of θ. The velocity field v of the transport medium established from the accompanying kinetic and hydrodynamic equations is illustrated in Fig. 8.7(b).

The main aim of the experiment was to find an optimal sensor activation policy for determining the most accurate estimates of the true parameters θ_1 to θ_3. As the number of the exploited wells should be minimal, a reasonable choice is to use at every time instant only two from among all the available six locations, cf. Fig. 8.7. The set of fifteen combinations of active wells, which were coded as successive integers being the levels of the input control signal u_c, is listed in Table 8.1.

The implementation of all routines for this example was performed in MATLAB 6.5. The PDEs were solved using exactly the same routines as in the previous application example with a FEM approximation of the domain (an unstructured triangle mesh with 430 nodes and 788 triangles) and 30 divisions of the time interval. Finally, the procedure based on the CPET approach with

TABLE 8.1

Combinations of the activated
observation wells

Active wells	Control u_c
$\{O_1, O_2\}$	0
$\{O_1, O_3\}$	1
$\{O_1, O_4\}$	2
$\{O_1, O_5\}$	3
$\{O_1, O_6\}$	4
$\{O_2, O_3\}$	5
$\{O_2, O_4\}$	6
$\{O_2, O_5\}$	7
$\{O_2, O_6\}$	8
$\{O_3, O_4\}$	9
$\{O_3, O_5\}$	10
$\{O_3, O_6\}$	11
$\{O_4, O_5\}$	12
$\{O_4, O_6\}$	13
$\{O_5, O_6\}$	14

the `fmincon` function in the role of the optimizer produced the following control signal describing the changes in the active well locations:

$$
u_c(t) = \begin{cases}
0 & \text{if } 0.000 \leq t < 0.226, \\
5 & \text{if } 0.226 \leq t < 0.387, \\
9 & \text{if } 0.387 \leq t < 0.645, \\
12 & \text{if } 0.645 \leq t < 0.935, \\
14 & \text{if } 0.935 \leq t \leq 1.000.
\end{cases}
\tag{8.23}
$$

The maximal number of switchings was assumed to be equal to two and the time required for computations was about 20 minutes (Pentium IV 1.7 GHz processor, 768 MB RAM, Windows 2000).

Figure 8.8 illustrates the optimal sensor activation policy versus contour plots of the pollutant concentration, where open circles indicate the actually exploited observation wells. The sensor activation strategy can be interpreted based on the fact that the measurements are taken possibly close to the forehead of the 'pollutant wave' moving from the river bank in the direction of the discharge channel of the aquifer. This is reflected by the sensor-activation schedule as the measurements are taken at the locations with the greatest changes in the concentration with respect to the parameters.

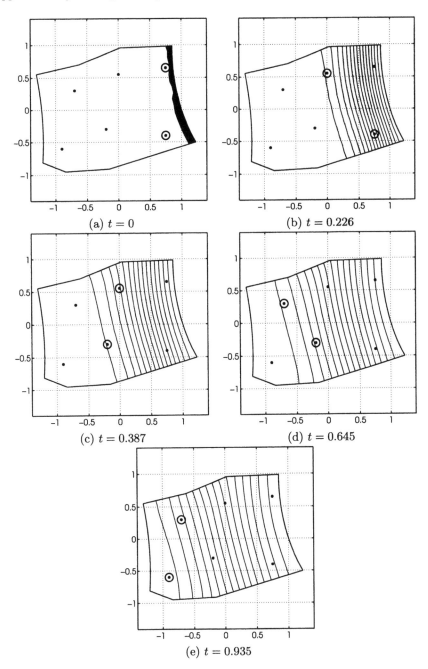

(a) $t = 0$

(b) $t = 0.226$

(c) $t = 0.387$

(d) $t = 0.645$

(e) $t = 0.935$

FIGURE 8.8
Consecutive switchings of the activated observation wells versus contour plots
of the pollutant concentration.

8.4 Diffusion process with correlated observational errors

We consider once more the atmospheric pollution process over a given urban area, but this time we take into account two active sources of pollution, which leads to changes in the pollutant concentration $y(x, t)$. In contrast to Section 8.2, in order to slightly simplify the analysis, assume that the velocity of the transport medium is zero everywhere. In such a way the entire process over the normalized observation interval $T = [0, 1]$ can be modelled with the following diffusion equation:

$$\frac{\partial y(x, t)}{\partial t} = \theta_1 \nabla^2 y(x, t) + f(x), \quad x \in \Omega = [0, 1]^2, \tag{8.24}$$

where the second term on the right-hand side imitates sources of contamination over Ω. Specifically, the function f takes the form

$$f(x) = \theta_2 \exp(-100\|x - a_1\|^2) + \theta_3 \exp(-100\|x - a_2\|^2), \tag{8.25}$$

where $a_1 = (0.4, 0.4)$ and $a_2 = (0.6, 0.6)$ are the locations of the sources.

Equation (8.24) is supplemented by homogeneous boundary and initial conditions:

$$y(x, t) = 0, \quad (x, t) \in \partial\Omega \times T, \tag{8.26}$$
$$y(x, 0) = 0, \quad x \in \Omega. \tag{8.27}$$

This time the task comprises determination of the locations of an arbitrary number of stationary sensors for estimation of the diffusion coefficient in the simplest scalar form θ_1 and the unknown amplitudes of the sources θ_2 and θ_3. The measurements are assumed to be corrupted by the noise which is zero-mean, uncorrelated in time and correlated in space with the covariance kernel in the simple form $k(x, \chi) = \exp(-\rho_x \|x - \chi\|)$.

The problem in (8.24)–(8.27) and the corresponding sensitivity equations were solved numerically with the use of the routines from MATLAB's PDE Toolbox. The nominal parameter values $\theta_1^0 = 0.02$, $\theta_2^0 = 10.0$ and $\theta_3^0 = 5.0$ were employed for that purpose.

In order to find the D-optimum locations for taking measurements, a program based on Algorithm 7.1 was written in Matlab v. 6.5, Rel. 12.1 and run on a PC equipped with Pentium IV, 1.7 MHz CPU, and Windows 2000. The two-dimensional search for a candidate to include into the current design was implemented using the routine fmincon from the Optimization Toolbox.

The influence of the mutual correlations of observations on the sensor allocation was tested by varying the coefficient ρ_x. The results obtained for different numbers of measurements are presented in Fig. 8.9.

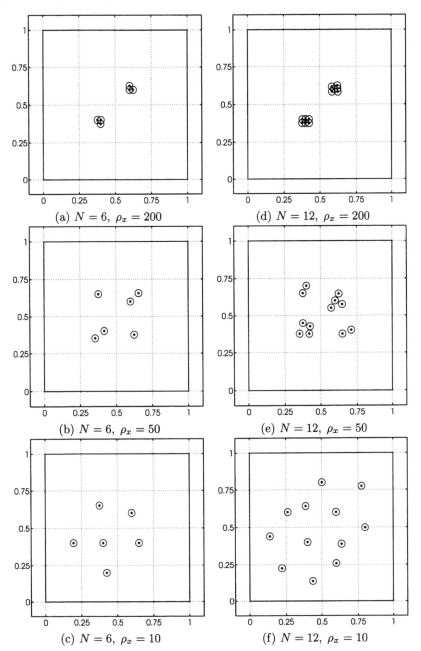

(a) $N = 6$, $\rho_x = 200$

(d) $N = 12$, $\rho_x = 200$

(b) $N = 6$, $\rho_x = 50$

(e) $N = 12$, $\rho_x = 50$

(c) $N = 6$, $\rho_x = 10$

(f) $N = 12$, $\rho_x = 10$

FIGURE 8.9
D-optimum sensor allocation for $N = 6$ and 12 measurements for small ($\rho_x = 200$), medium ($\rho_x = 50$) and considerable ($\rho_x = 10$) spatial correlations, respectively.

It can be easily seen that the level of the correlation strength ρ_x directly affects the distances between the sensors which are increasing when the correlation is more intense. If the correlation is small, then the measurements tend to cluster in the vicinity of the pollutant sources which coincide with the locations of the optimal design support for the case of independent measurements.

The problem is symmetric with respect to the axis $x_1 = x_2$, but the solutions only approximately retain the symmetry, which is an unavoidable effect of a finite precision of numerical computations. Nevertheless, the fact that the sensitivities $\partial y/\partial \theta_i$, $i = 2, 3$ attain their maxima at points a_1 and a_2, respectively, suggests that a two-point design

$$\xi^\star = \{(0.4, 0.4), (0.6, 0.6)\} \tag{8.28}$$

is D-optimal for the stationary sensor strategy on the assumption of independent measurements. This fact is empirically confirmed by the results for the small correlations. Indeed, points cluster in the close vicinity of the pollutant sources. Another important remark is that the higher the correlation, the lower the criterion value. This effect results from a more global character of the mutual influence of random-error realizations in the case of higher correlation values. Due to the interference with the estimation of the unknown parameters, the valuable information in observations is reduced.

8.5   Vibrating H-shaped membrane

In this application example, consider an H-shaped membrane vibrating under the internal elasticity forces. The exact shape of the membrane which defines the spatial domain Ω is given in Fig. 8.10(a).

The membrane is fixed on the top and bottom boundaries, and is free elsewhere. The amplitude $y(x, t)$ of the transverse vibrations over a given time interval $T = [0, 1]$ is described by the hyperbolic equation

$$\frac{\partial^2 y(x, t)}{\partial t^2} = \nabla \cdot (\gamma(x)\nabla y(x, t)) \quad \text{in } \Omega, \tag{8.29}$$

subject to the boundary and initial conditions

$$y(x, t) = 0 \quad \text{on } \{\bigcup_{i=1}^{4} \Gamma_i\} \times T, \quad \text{(top and bottom boundaries)} \tag{8.30}$$

$$\frac{\partial y(x, t)}{\partial n} = 0 \quad \text{on } \{\bigcup_{i=5}^{8} \Gamma_i\} \times T, \quad \text{(left and right boundaries, cutouts)} \tag{8.31}$$

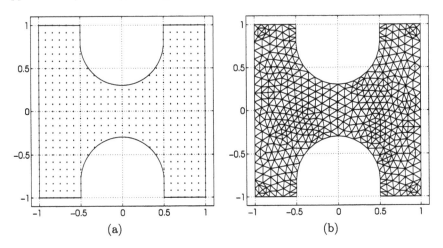

FIGURE 8.10
Membrane with potential sites where the measurements can be taken (a) and
an unstructured mesh for the problem (521 points, 928 triangles) (b).

and

$$y(x, 0) = \exp(-50\|x\|^2) \quad \text{in } \Omega, \quad \text{(initial disturbance)} \tag{8.32}$$

$$\frac{\partial y(x, 0)}{\partial t} = 0 \qquad \text{in } \Omega. \tag{8.33}$$

The coefficient of the transverse elasticity has the spatially varying form

$$\gamma(x) = \theta_1 + \theta_2 \cosh(x_1) + \theta_3 \sinh(x_2^2), \tag{8.34}$$

where the parameter values $\theta_1 = 1.0$, $\theta_2 = -0.1$ and $\theta_3 = -0.1$ were assumed
to be known prior to the experiment. Our purpose is to construct D- and
A-optimal scanning strategies for determining most accurate estimates of the
true parameters θ_1, θ_2 and θ_3 when applying $N = 60$ scanning sensors and
the partition of T defined a *priori* by the switching instants $t_r = r/20$, $r = 0, \ldots, 20$. The resulting optimal solutions are shown in Figs. 8.11 and 8.12,
where open circles indicate the positions of the activated sensors.

The initial design was generated by randomly selecting its support points.
Algorithm 4.2 employed in this example produced the solutions for $\eta = 10^{-5}$
after 61 and 64 iterations, respectively, practically within 10 seconds on a PC
(Pentium IV, 1.7 GHz, 768 MB RAM) using the Lahey/Fujitsu Fortran 95
compiler combined with the MATLAB PDE Toolbox.

At the initial time instant the external force generates a deformation at
the centre of the membrane, which starts to vibrate due to the elasticity
forces. During the selected time interval the initial impulse spreads over the

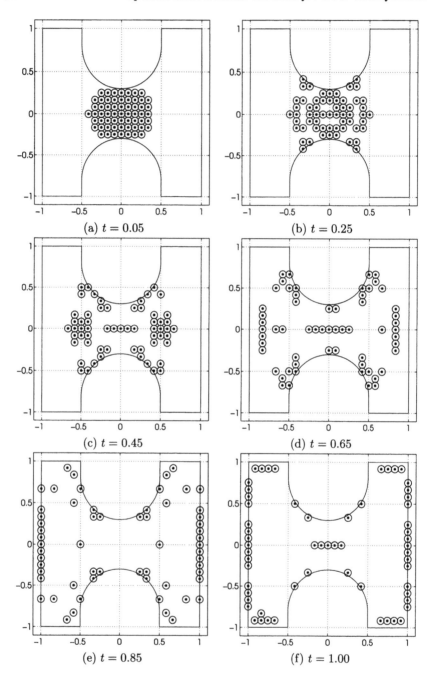

FIGURE 8.11
Selected successive D-optimal scanning sensor configurations for the membrane example.

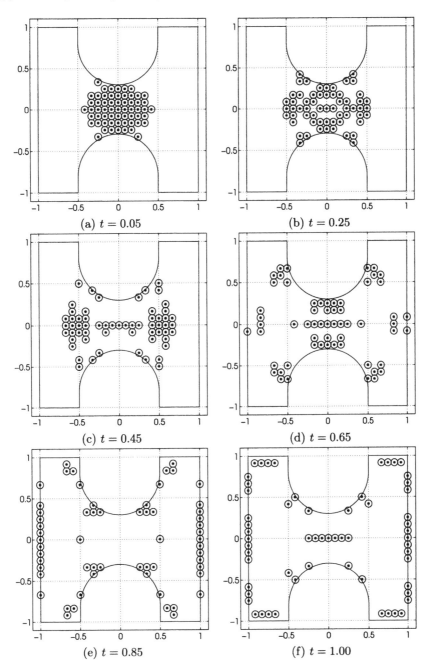

FIGURE 8.12
Selected successive A-optimal scanning sensor configurations for the membrane example.

entire domain in the form of a circular wave moving from the centre to the boundaries. Both the D- and A-optimal strategies in this computer simulation clearly tend to take most of the measurements near the forehead of the displaced vibrations.

9

Conclusions and future research directions

From an engineering point of view it is clear that sensor placement is an integral part of control design, particularly in control of DPSs, e.g., flexible structures, air-pollution processes, oil and gas production from deposits, etc. Its choice is fundamental in the sense that it determines the accuracy of the system characteristics which are identified from an identification experiment. On the other hand, an engineering judgement and trial-and-error analysis are quite often used to determine spatial arrangements of measurement transducers, in spite of the fact that the problem has been attacked from various angles by many authors and a number of relevant results have already been reported in the literature. What is more, although it is commonly known that this area of research is difficult, since the nonlinearity inherent in the sensor-location problem precludes simple solution techniques, some systematic attempts at obtaining optimal sensor positions are still made and the progression is towards more general models, more realistic criteria and better understanding of the nature of the optimal locations. Logically, the number of applications should proliferate, yet this is not the case. It seems that two main reasons explain why strong formal methods are not accepted in engineering practice. First, with the use of the existing approaches, only relatively simple engineering problems can be solved without resorting to numerical methods. Second, the complexity of most sensor location algorithms does not encourage engineers to apply them in practice.

The benefits from using systematic approaches to the sensor-location problem can by far outweigh the effort put forth to implement them, as is evidenced by the following example.

Example 9.1
Consider a diffusion process described by the parabolic equation

$$\frac{\partial y(x,t)}{\partial t} = \frac{\partial}{\partial x_1}\left(\kappa(x)\frac{\partial y(x,t)}{\partial x_1}\right) + \frac{\partial}{\partial x_2}\left(\kappa(x)\frac{\partial y(x,t)}{\partial x_2}\right)$$
$$+ 20\exp\left(-50(x_1 - t)^2\right), \quad x \in \Omega = (0,1)^2, \quad t \in (0,1), \tag{9.1}$$

subject to zero initial and boundary condtions:

$$y(x,0) = 0, \quad x \in \Omega, \tag{9.2}$$
$$y(x,t) = 0, \quad (x,t) \in \partial\Omega \times T. \tag{9.3}$$

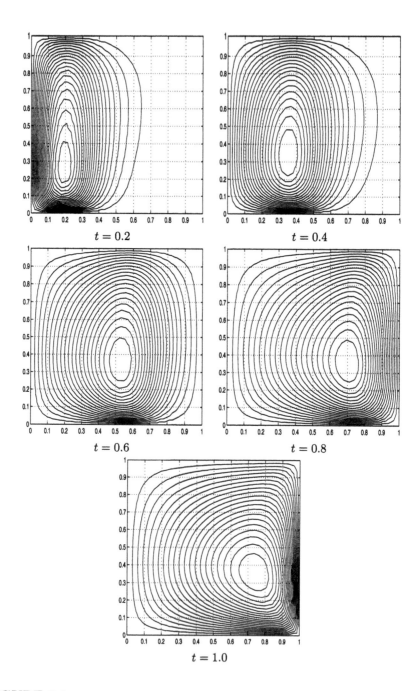

FIGURE 9.1
Contour plots of the solution for (9.1)–(9.3) at consecutive time moments.

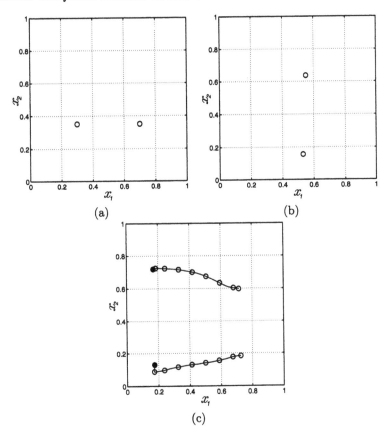

FIGURE 9.2
Sensor configurations investigated in Example 9.1: (a) An arbitrary location of stationary sensors; (b) A D-optimum placement of stationary sensors; (c) D-optimum trajectories of moving sensors.

The estimated diffusion coefficient is approximated by

$$\kappa(x) = \theta_1 + \theta_2 x_2, \qquad \theta_1 = 0.1, \quad \theta_2 = 0.2, \qquad (9.4)$$

and the given values of θ_1 i θ_2 are also treated as nominal and known prior to the estimation (this is in order to determine the sensitivity coefficients, and thereby the elements of the FIM).

The forcing input on the right-hand side of (9.1) imitates the action of a source whose support is the unit-length line segment constantly oriented parallelly to the x_2-axis and moving at a constant pace from the left to the right border of the domain Ω. It makes the system state to evolve such that its three-dimensional plot forms a surface which resembles a hill whose pick

TABLE 9.1

Results of parameter estimation in Example 9.1

	Arbitrary configuration	Stationary D-optimum	Moving D-optimum
Mean value of $\hat{\theta}_1$	.10275	.09998	.09991
Mean value of $\hat{\theta}_2$	.19641	.20036	.20022
Stand. dev. of $\hat{\theta}_1$	.04514	.00260	.00192
Stand. dev. of $\hat{\theta}_2$	.14042	.00793	.00647
95% Confidence interval for $\hat{\theta}_1$	$[.00972, .19028]$	$[.09479, .10521]$	$[.09616, .10383]$
95% Confidence interval for $\hat{\theta}_2$	$[-.08084, .48084]$	$[.18413, .21587]$	$[.18706, .21294]$
$\det(\mathrm{cov}\,\hat{\theta})$	$.93049 \times 10^{-7}$	$.15571 \times 10^{-9}$	$.61293 \times 10^{-10}$

follows the movement of the moving source, see also Fig. 9.1 where contour plots at consecutive time moments are shown.

In the simulation experiment three sensor configurations were tested:

1. An arbitrary location of stationary sensors at points $(0.30, 0.35)$ and $(0.70, 0.35)$, selected such that the crest of the moving wave representing the system state in three dimensions passes through them. Intuitively, this design seems justified by the fact that the measured state takes on maximal values there and the implied influence of the measurement errors should be less than anywhere else.

2. A D-optimum placement of stationary sensors at points $(0.53, 0.16)$ and $(0.55, 0.64)$.

3. D-optimum trajectories of moving sensors with the dynamics

$$\dot{s}(t) = u(t), \quad s(0) = s_0, \tag{9.5}$$

and the constraints

$$|u_i(t)| \le 0.7, \quad \forall t \in T, \quad i = 1, \ldots, 4 \tag{9.6}$$

imposed on the motion.

The output measurements of the system responses at the corresponding spatial points were disturbed by a Gaussianly distributed noise with zero mean and a standard deviation of 0.1 (for comparison, the maximal value of the state equalled about 2.1). It was also assumed that the measurements were performed at forty time moments evenly distributed in the time interval T.

Estimation of the parameters θ_1 i θ_2 was performed using the Levenberg-Marquardt algorithm. For each sensor configuration, the generation of simulated measurement data and parameter estimation were repeated 300 times, which made it possible to summarize the simulation results via Table 9.1.

The obtained results indicate that the parameter estimates can be considered unbiased. Also observe a marked difference between the standard deviations, or equivalently, confidence intervals, of the estimates obtained when passing from the arbitrary to the D-optimum configurations of stationary sensors (the average error becomes about twenty times smaller), as well as from stationary to moving sensors for the D-optimum case (a 20% increase in the accuracy should be noted).   □

Bearing the above remarks in mind, the original goal of the research reported in this monograph was simply to develop computationally efficient methods to solve practical sensor-location problems for a wide class of DPSs. In the process of executing this task, we have developed a theoretical foundation for the adopted approach and constructed several new algorithms for various types of computation. The following is a concise summary of the contributions provided by this work to the state of the art in optimal sensor-location for parameter estimation in DPSs:

- Systematizes characteristic features of the problem and analyses the existing approaches based on the notion of the Fisher information matrix.

- Develops an effective method for computing sensitivity coefficients which are indispensable when determining the elements of the FIM. This scheme based on the direct-differentiation approach and tricubic spline interpolation enables us to implement the algorithms for finding optimal sensor positions in an extremely efficient manner.

- Provides characterizations of continuous (or approximated) designs for stationary sensors, which allows an easy testing of any given sensor setting for optimality, and then clarifies how to adapt well-known algorithms of optimum experimental design for finding numerical approximations to the solutions.

- Presents the concept of clusterization-free locations along with a practical algorithm; this is a modified version of the effective method proposed by Fedorov in the context of linear regression models.

- Develops efficient methods of activating scanning sensors by generalizing the concept of clusterization-free locations and looking for a family of 'optimal' probability measures defined on subsets of the set of feasible measurement points. The resulting exchange algorithm is very easy to implement and extremely efficient. Since it can be used only when the number of sensors is relatively high, an approach based on some recent results of discrete-valued optimal control (the CPET technique) is suggested for situations when the number of scanning sensors is small.

- Extends Rafajłowicz's approach to constructing optimal trajectories of moving sensors and derives an alternative approach consisting of parameterization of sensor trajectories.

- Formulates and solves the problem of trajectory planning for moving sensors based on dynamic models of sensor motions and various constraints imposed on the movements. The line of development given here is original in that, to the best of our knowledge, there have been no approaches so far within such a framework in the context of parameter identification of DPSs. Specifically, it is shown how to reduce the problem to a state-constrained optimal-control problem. Then an effective method of successive linearizations is employed to solve it numerically. It is demonstrated that the proposed approach can tackle various challenging problems of vital importance, including motion planning along specified paths, utilization of minimax criteria, or selection of a minimal number of sensors. Simulation experiments validate the fact that making use of moving sensors may lead to dramatic gains in the values of the adopted performance indices, and hence to a much greater accuracy in the resulting parameter estimates.

- Discusses alternative optimality criteria such as that quantifying the achievable accuracy in response prediction and model structure discrimination. Such a formulation may constitute a potential solution to fill the gap existing between refined optimum experimental design theory which has been developed for years with emphasis on parameter space, and modern control theory where output space criteria are by all means more important than those which are primarily related to the accuracy of parameter values themselves. In turn, model discrimination can be extremely useful for fault detection and diagnosis in industrial processes.

- Introduces the notion of robust design of sensor locations, characterizes the corresponding solutions for continuous designs, and indicates how to find numerical solutions for exact designs efficiently. The advantage of the suggested schemes based on stochastic approximation is that the solutions are obtained almost as simply as for local-optimality criteria. As a consequence, robust trajectories for moving sensors can be determined at a reasonable computational cost. Minimax designs are also discussed and it is shown that they might not be as difficult as they seem at first sight. In addition to that, an attempt is made to apply randomized algorithms to find optimal sensor locations in an average sense.

- Provides a numerical method for the construction of optimum designs in the presence of spatial correlations. It leads to a rapid exchange algorithm being additionally very simple to implement.

- Develops a new approach to the problem of scanning sensor location for linear DPSs based on various criteria defined on the observability Gramian and establishes a close connection between this problem and modern optimum experimental design theory. With minor changes, the technique can be extended to Kalman filtering and robust control. A version suitable for online control architectures has been developed.

The approach suggested here has the advantage that it is independent of a particular form of the partial-differential equation describing the distributed system under consideration. The only requirement is the existence of sufficiently regular solutions to the state and sensitivity equations, and consequently nonlinear systems can also be treated within the same framework practically without any changes. Moreover, it can easily be generalized to three spatial dimensions and the only limitation is the amount of required computations.

We believe that our approach has significant advantages which will make it, with sufficient development, a leading approach to solving sensor-location problems facing engineers involved in applications. However, there still remain open problems regarding some important areas. What follows is a discussion of the areas for further investigation, besides applications.

Further development of robust approaches. Formally, robust sensor location is somewhat reminiscent of the problem of robustness analysis and design for uncertain control systems. In this respect, some ideas of statistical learning theory and randomized algorithms, which had already proved to be useful in robust controller design [289, 340, 342] were exploited here in the framework of the robust sensor location. A main drawback of this approach is its excessive conservatism, which results in sample sizes which are prohibitively higher than needed. Less conservative estimates of the sample sizes are thus demanded, and this question seems to be related to the property of the uniform convergence of empirical probabilities (UCEP) [340] whose possession should be investigated for particular classes of DPSs. Potential results can be of significance in attempts at finding sensor positions which would guarantee an extended identifiability of a given DPS (this concept was introduced and thoroughly studied by Sun [282]).

Coupled input and measurement system design. The problem of simultaneous optimum choice of the controls influencing a given DPS (including both the actuator location and the form of the input signals) and the locations in which to place the sensors was suggested in the survey paper [174]. Unfortunately, owing to the great complexity of the corresponding optimization problem, communications on this subject are rather limited. Some interesting preliminary results were reported in some early works of

Rafajłowicz [232–234] who then focused his attention solely on input optimization [235–237, 242, 243, 247]. The maturity of the optimal control theory of DPSs and the availability of more and more powerful computers encourage us to resume the research related to this idea.

Coupled parameter identification and experimental design. In the experiments discussed in this monographs, sensor allocation strategies are implemented offline, before collecting the measurement data (though some kind of feedback is encountered in sequential designs). On the other hand, it would be interesting to investigate the problem of simultaneously taking measurements, identifying system parameters and updating locations of moving sensors. Unfortunately, the number of publications related to online parameter estimation for infinite-dimensional dynamical systems is very limited, cf. e.g., [1, 18, 67], and distributed measurements are only considered. In spite of that, owing to potential applications, further development of this line of research is desirable.

Bridging the gap between spatial statistics and sensor-location techniques used in engineering. As was already mentioned in the Introduction, quite similar problems are treated while designing monitoring networks in the field of spatial statistics. Many of those results, e.g., the so-called approximate information matrices aimed at coping with the problem of correlated observations [181], seem to be perfectly suited for the tasks considered here. This topic deserves close attention, as numerous interesting interrelations between both the approaches are expected.

Model discrimination for fault detection and diagnosis in industrial systems. In Chapter 5 we treated discrimination between competing models of DPSs. With a little turning up, this may constitute a good basis for the development of methods for fault detection, i.e., determining whether a fault has occured in the considered process, and fault diagnosis, i.e., determining which fault has occured [51]. Process monitoring which implements both the tasks becomes extremely important in modern complex industrial systems. In the context of DPSs, the theory lags behind the needs created by applications. Promising results in this respect are reported in [206].

Close links to the methods of input/output selection in control system design. In Chapter 7 it was shown that some algorithms of sensor location for parameter estimation can be successfully employed in a seemingly different problem of maximizing observability measures arising in control-system design. The approach can be generalized to stochastic systems where Kalman filters are in common use. This suggests that solutions to the problem of sensor and controller placement, which have been elaborated in both fields for many years, can possibly find applications in a completely different setting of parameter estimation, and vice versa. This aspect of the sensor-location problem is one of the most promising research directions for the near future.

Appendices

A

List of symbols

$\mathbb{N}$	set of natural numbers
A^n	given a set A, it is the Cartesian product $\underbrace{A \times \cdots \times A}_{n \text{ times}}$
$\mathbb{R}^n$	finite-dimensional Euclidean space
t_f	total time of the observation process
t	time
Ω	simply connected, open and bounded subset of $\mathbb{R}^n$
$x = (x_1, \ldots, x_n)$	point of Ω
$\Gamma \equiv \partial\Omega$	boundary of Ω
$\bar{\Omega}$	$\Omega \cup \partial\Omega$
Q	space-time domain $\Omega \times]0, t_f[$
$F \circ G$	composition of mappings F and G
$F\|A$	restriction of a mapping $F : X \to Y$ to a subset $A \subset X$, i.e., the mapping $F \circ j$, where $j : A \ni x \mapsto x \in X$ is the canonical injection of A into X
ℓ_n^p	$\mathbb{R}^n$ endowed with the norm $\|x\|_{\ell_n^p} = \left(\sum_{i=1}^n |x_i|^p \right)^{1/p}$
ℓ_n^∞	$\mathbb{R}^n$ endowed with the norm $\|x\|_{\ell_n^\infty} = \max\limits_{1 \leq i \leq n} |x_i|$
$C(\bar{\Omega})$	class of all continuous real-valued functions on $\bar{\Omega}$
$C^1(\bar{\Omega})$	class of all continuously differentiable functions on $\bar{\Omega}$
$C^{2,1}(\bar{Q})$	class of all the functions f continuous on $\bar{Q}$ together with the derivatives $\partial f/\partial t$, $\partial f/\partial x_i$, $\partial^2 f/\partial x_i \partial x_j$, $1 \leq i, j \leq n$

$L^p(\Omega)$ space of classes of measurable functions f on Ω such that $\int_\Omega |f(x)|^p \, dx < \infty$; this is a Banach space for the norm

$$\|f\|_{L^p(\Omega)} = \left\{ \int_\Omega |f(x)|^p \, dx \right\}^{1/p}$$

$L^\infty(\Omega)$ space of classes of measurable functions f on Ω such that $x \mapsto |f(x)|$ is essentially bounded, i.e., $\exists C < \infty, |f(x)| \le C$ a.e. in Ω; this is a Banach space for the norm

$$\|f\|_{L^\infty(\Omega)} = \operatorname*{ess\,sup}_{x \in \Omega} |f(x)| = \inf \left\{ C : |f(x)| \le C \text{ a.e. in } \Omega \right\}$$

$\mathcal{L}(X, Y)$ set of all bounded linear operators from X to Y; if X and Y are both Banach spaces, then so is $\mathcal{L}(X, Y)$, endowed with norm

$$\|A\|_{\mathcal{L}(X,Y)} = \sup_{x \ne 0} \frac{\|Ax\|_Y}{\|x\|_X}$$

$\mathcal{L}(X)$ $\mathcal{L}(X, X)$

X' dual space of X, i.e., $\mathcal{L}(X, \mathbb{R})$

supp f support of f

$\partial^\alpha f$ for a multi-index $\alpha = (\alpha_1, \dots, \alpha_n) \in \mathbb{N}^n$ it is equal to $\partial^{|\alpha|} f / \partial x_1^{\alpha_1} \dots \partial x_n^{\alpha_n}$, where $|\alpha| = \alpha_1 + \dots + \alpha_n$

$\mathcal{D}(\Omega)$ space of smooth (i.e., infinitely differentiable) functions which have compact support in Ω; a sequence $\{\varphi_m\}$ of $\mathcal{D}(\Omega)$ tends to φ in $\mathcal{D}(\Omega)$ if

         (i) $\bigcup_m \operatorname{supp} \varphi_m \subset K \subset \Omega$, where K is compact

         (ii) for each $\alpha \in \mathbb{N}^n$, $\partial^\alpha \varphi_m$ tends uniformly to $\partial^\alpha \varphi$

$\mathcal{D}'(\Omega)$ space of distributions on Ω, i.e., the set of 'continuous' linear forms on $\mathcal{D}(\Omega)$: if T is a distribution on Ω and if $\langle T, \varphi \rangle$ stands for the pairing between $\mathcal{D}'(\Omega)$ and $\mathcal{D}(\Omega)$, then for every sequence $\{\varphi_m\}_{m=1}^\infty$ converging to φ in $\mathcal{D}(\Omega)$, $\langle T, \varphi_m \rangle$ tends to $\langle T, \varphi \rangle$

$H^1(\Omega)$	first-order Sobolev space, i.e., the set $\{f \in L^2(\Omega) : \partial f / \partial x_i \in L^2(\Omega), 1 \leq i \leq n\}$; this is a Hilbert space with the scalar product

$$(f,g)_{H^1(\Omega)} = \int_\Omega \left(fg + \sum_{i=1}^n \frac{\partial f}{\partial x_i} \frac{\partial g}{\partial x_i} \right) dx$$

$H^2(\Omega)$	second-order Sobolev space, i.e., the set $\{f \in L^2(\Omega) : \partial f / \partial x_i, \partial^2 f / \partial x_i \partial x_j \in L^2(\Omega), 1 \leq i,j \leq n\}$; this is a Hilbert space with the scalar product

$$(f,g)_{H^2(\Omega)} = \int_\Omega \left\{ \sum_{|\alpha| \leq 2} \partial^\alpha f \, \partial^\alpha g \right\} dx$$

$H_0^1(\Omega)$	closure of $\mathcal{D}(\Omega)$ in the space $H^1(\Omega)$; for domains with Lipschitz boundaries it may be concluded that $H_0^1(\Omega) = \{v \in H^1(\Omega) : v\vert_\Gamma = 0\}$
$H^{-1}(\Omega)$	dual space of $H_0^1(\Omega)$, i.e., $H^{-1}(\Omega) = \mathcal{L}(H_0^1(\Omega), \mathbb{R})$; this is a Hilbert space with the norm

$$\|F\|_{H^{-1}(\Omega)} = \sup_{0 \neq f \in H_0^1(\Omega)} \frac{|\langle F, f \rangle|}{\|f\|}$$

$L^2(0, t_f; X)$	space of classes of functions f from $]0, t_f[$ to a Banach space X such that f is measurable with respect to the Lebesgue measure and

$$\|f\|_{L^2(0,t_f;X)} = \left(\int_0^{t_f} \|f(t)\|_X^2 \, dt \right)^{1/2} < \infty$$

$C([0, t_f]; X)$	space of all those functions which are continuous from $[0, t_f]$ to a Banach space X; this is a Banach space endowed with norm

$$\|f\|_{C([0,t_f];X)} = \sup_{t \in [0, t_f]} \|f(t)\|_X$$

ran $\mathcal{T}$	range of an operator $\mathcal{T}$
ker $\mathcal{T}$	kernel of an operator $\mathcal{T}$
$\Pr(A)$	probability of an event A

Abbreviations

LPS	lumped parameter system
DPS	distributed parameter system
ODE	ordinary differential equation
PDE	partial differential equation
FIM	Fisher information matrix
OSL	optimal sensor location
SLI	OSL for system identification
SLE	OSL for state estimation
i.i.d.	independent, identically distributed

B

Mathematical background

B.1 Matrix algebra

Throughout the monograph, we adhere to the convention that $\mathbb{R}^{n \times k}$ denotes the linear space of real matrices with n rows and k columns.

A nonsingular matrix $A \in \mathbb{R}^{n \times n}$ is termed *unitary* if its inverse equals its transpose, i.e.,

$$A^{-1} = A^{\mathsf{T}}. \tag{B.1}$$

The *trace* of a square matrix $A = [a_{ij}] \in \mathbb{R}^{n \times n}$, denoted $\mathrm{tr}(A)$, is the sum if its diagonal elements:

$$\mathrm{tr}(A) = \sum_{i=1}^{n} a_{ii}. \tag{B.2}$$

Some easily proven properties are

(P1) $\mathrm{tr}(\alpha A) = \alpha \, \mathrm{tr}(A), \quad \forall \alpha \in \mathbb{R}, \quad \forall A \in \mathbb{R}^{n \times n}$.

(P2) $\mathrm{tr}(A + B) = \mathrm{tr}(A) + \mathrm{tr}(B), \quad \forall A, B \in \mathbb{R}^{n \times n}$.

(P3) $\mathrm{tr}(A^{\mathsf{T}}) = \mathrm{tr}(A), \quad \forall A \in \mathbb{R}^{n \times n}$.

(P4) $\mathrm{tr}(AB) = \mathrm{tr}(BA), \quad \forall A \in \mathbb{R}^{m \times n}, \quad \forall B \in \mathbb{R}^{n \times m}$.

(P5) $\mathrm{tr}(A^{\mathsf{T}} B) = \sum_{i=1}^{m} \sum_{j=1}^{n} a_{ij} b_{ij}, \quad \forall A, B \in \mathbb{R}^{m \times n}$.

The permutation rule (P4), can be extended to any cyclic permutation in a product, e.g.,

$$\mathrm{tr}(ABCD) = \mathrm{tr}(BCDA) = \mathrm{tr}(CDAB) = \mathrm{tr}(DABC) \tag{B.3}$$

provided that the resulting products are square matrices.

The Euclidean matrix scalar product on $\mathbb{R}^{n \times k}$ is defined as

$$\langle A, B \rangle = \mathrm{tr}(A^{\mathsf{T}} B) = \sum_{i=1}^{n} \sum_{j=1}^{k} a_{ij} b_{ij}. \tag{B.4}$$

This makes it possible to identify $\mathbb{R}^{n \times k}$ with a Euclidean space of dimension nk. It is a simple matter to show that

$$\langle A, B \rangle = \langle B, A \rangle = \langle B^\mathsf{T}, A^\mathsf{T} \rangle = \langle A^\mathsf{T}, B^\mathsf{T} \rangle. \tag{B.5}$$

The *determinant* of a square matrix $A = [a_{ij}] \in \mathbb{R}^{n \times n}$, denoted $\det(A)$, can be introduced by the following induction:

If $n = 1$, then

$$\det(A) = \det([a_{11}]) = a_{11}. \tag{B.6}$$

If $n > 1$, then

$$\det(A) = \sum_{j=1}^{n} (-1)^{i+j} a_{ij} \det(A_{ij}) \tag{B.7}$$

for any fixed i, $1 \leq i \leq n$, or

$$\det(A) = \sum_{i=1}^{n} (-1)^{i+j} a_{ij} \det(A_{ij}) \tag{B.8}$$

for any fixed j, $1 \leq j \leq n$, where $A_{ij} \in \mathbb{R}^{(n-1) \times (n-1)}$ is the submatrix of A obtained by deleting the i-th row and j-th column from A. Formulae (B.7) and (B.8) are called the Laplace expansions of $\det(A)$ along the i-th row and j-th column, respectively. (The value of $\det(A)$ so defined depends on neither i, nor j.)

Alternatively,

$$\det(A) = \sum_{\sigma} \mathrm{sgn}(\sigma) \prod_{i=1}^{n} a_{i\sigma(i)}, \tag{B.9}$$

where σ in the first summation runs through the set of all permutations on the set of numbers $\{1, \ldots, n\}$, $\sigma(i)$ stands for the number occupying the i-th position in σ, and $\mathrm{sgn}(\sigma)$ equals $+1$ or -1 depending on whether the permutation is even or odd, respectively. Recall that a permutation is said to be even if it can be expressed as a product of an even number of transpositions (i.e., swaps of two elements) which are necessary to pass from the permutation $(1, \ldots, n)$ to σ. Otherwise, it is called odd, cf. [98, p. 101] for details.

The definition of the determinant implies the following properties:

(P1) $\det(A) = \det(A^\mathsf{T})$, $\quad \forall A \in \mathbb{R}^{n \times n}$.

(P2) $\det(AB) = \det(A) \det(B) = \det(BA)$, $\quad \forall A, B \in \mathbb{R}^{n \times n}$.

(P3) $\det(I_n) = 1$ and $\det(0_n) = 0$, where I_n is the identity matrix of order n and 0_n is the $n \times n$ matrix with zero elements.

(P4) (*Multilinearity*) Write $A = [a_1, \ldots, a_n]$, where a_i is the i-th column of A. If $a_i = b + c$, then

$$\det\left([a_1, \ldots, a_{i-1}, b + c, a_{i+1}, \ldots, a_n]\right)$$
$$= \det\left([a_1, \ldots, a_{i-1}, b, a_{i+1}, \ldots, a_n]\right)$$
$$+ \det\left([a_1, \ldots, a_{i-1}, c, a_{i+1}, \ldots, a_n]\right).$$

Moreover, if β is a scalar, then

$$\det\left([a_1, \ldots, a_{i-1}, \beta a_i, a_{i+1}, \ldots, a_n]\right)$$
$$= \beta \det\left([a_1, \ldots, a_{i-1}, a_i, a_{i+1}, \ldots, a_n]\right).$$

(P5) If two columns or two rows of a matrix $A \in \mathbb{R}^{n \times n}$ are equal, then we have $\det(A) = 0$.

(P6) If $A = [a_{ij}] \in \mathbb{R}^{n \times n}$ is a triangular matrix, then its determinant is the product of the diagonal elements

$$\det(A) = \prod_{i=1}^{n} a_{ii}.$$

In particular, the determinant of a diagonal matrix equals the product of its diagonal elements.

(P7) The determinant of a matrix is nonzero if and only if it has full rank (see below).

(P8) The determinant of a unitary matrix equals -1 or $+1$.

Given $A \in \mathbb{R}^{n \times n}$, the equation $Av = \lambda v$, where λ is a scalar, will have nontrivial solutions only for particular values of λ that are roots of the polynomial $\det(A - \lambda I)$ of degree n. These roots λ_i, $i = 1, \ldots, n$ (possibly repeated and complex) and the corresponding solutions $v_i \neq 0$ are called the eigenvalues and the eigenvectors of A, respectively. The eigenvalues depend continuously on the elements of A.

If a matrix $A \in \mathbb{R}^{n \times n}$ is triangular, then its eigenvalues coincide with its diagonal elements. In particular, the eigenvalues of a diagonal matrix tally with its diagonal elements. For any matrix $A \in \mathbb{R}^{n \times n}$ the eigenvalues of $cI + A$ are equal to $c + \lambda_1, \ldots, c + \lambda_n$.

We can view a matrix $A \in \mathbb{R}^{m \times n}$ as a set of consecutive column vectors $a_1, \ldots, a_n$. The *column rank* of A is the dimension of the vector space which is spanned by $a_1, \ldots, a_n$, or equivalently, the largest number of linearly independent column vectors A contains.

Considering, instead, the set of vectors obtained by using the rows of A instead of the columns, we define the *row rank* of A as the largest number of linearly independent row vectors it contains. It can be shown that the column and row ranks of a matrix A are equal, so we can speak unambiguously of the *rank of A*, denoted rank(A).

Clearly, given $A \in \mathbb{R}^{m \times n}$, we have

$$\text{rank}(A) = \text{rank}(A^\mathsf{T}) \leq \min(m, n). \tag{B.10}$$

If rank$(A) = \min(m, n)$, then A is said to have *full rank*. Otherwise, we say that it has *short rank*.

Useful general properties are as follows:

(P1) rank$(AB) \leq \min(\text{rank}(A), \text{rank}(B))$, $\forall A \in \mathbb{R}^{m \times q}$, $\forall B \in \mathbb{R}^{q \times n}$.

(P2) rank$(A + B) \leq \text{rank}(A) + \text{rank}(B)$, $\forall A, B \in \mathbb{R}^{m \times n}$.

(P3) If $B \in \mathbb{R}^{n \times n}$ has full rank, then rank$(AB) = \text{rank}(A)$, $\forall A \in \mathbb{R}^{m \times n}$.

(P4) The rank of a symmetric matrix is equal to the number of its nonzero eigenvalues.

The following result relates the inverse of a matrix after a low-rank perturbation to the known inverse of the original matrix.

THEOREM B.1 Sherman-Morrison-Woodbury formulae
Let $A \in \mathbb{R}^{n \times n}$ be a matrix with known inverse A^{-1}. Given $U, V \in \mathbb{R}^{n \times m}$, if $I + V^\mathsf{T} A^{-1} U$ is invertible, then so is $A + UV^\mathsf{T}$ and

$$(A + UV^\mathsf{T})^{-1} = A^{-1} - A^{-1}U(I + V^\mathsf{T}A^{-1}U)^{-1}V^\mathsf{T}A^{-1}. \tag{B.11}$$

In the special case where U and V are column vectors u and v, respectively, we have

$$(A + uv^\mathsf{T})^{-1} = A^{-1} - \alpha A^{-1}uv^\mathsf{T}A^{-1}, \tag{B.12}$$

where $\alpha = 1/(1 + v^\mathsf{T}A^{-1}u)$.

Frequently, (B.11) is called the Woodbury formula, while (B.12) is called the Sherman-Morrison formula [106]. Observe that the matrix $I + V^\mathsf{T}A^{-1}U$ which is often termed the *capacitance matrix* is $m \times m$. Consequently, the Woodbury formula becomes extremely useful in situations where m is much smaller than n, so that the effort involved in evaluating the correction $A^{-1}U(I + V^\mathsf{T}A^{-1}U)^{-1}V^\mathsf{T}A^{-1}$ is small relative to the effort involved in inverting $A + UV^\mathsf{T}$ which is $n \times n$ [106, 150]. This gain in efficiency is especially in evidence for the Sherman-Morrison formula.

THEOREM B.2

Let $A \in \mathbb{R}^{n \times n}$ be a matrix with known inverse A^{-1} and known determinant $\det(A)$. Then we have

$$\det(A + UV^\mathsf{T}) = \det(A) \det(I + V^\mathsf{T} A^{-1} U) \tag{B.13}$$

for any $U, V \in \mathbb{R}^{n \times m}$. This evidently implies that $A + UV^\mathsf{T}$ is invertible if and only if $I + V^\mathsf{T} A^{-1} U$ is. In particular, given $u, v \in \mathbb{R}^n$, we get

$$\det(A + uv^\mathsf{T}) = (1 + v^\mathsf{T} A^{-1} u) \det(A). \tag{B.14}$$

Finally, let us recall a useful result regarding partitioned matrices.

THEOREM B.3 Frobenius formulae

Let $A \in \mathbb{R}^{n \times n}$ be partitioned as follows:

$$A = \left[\begin{array}{c:c} A_{11} & A_{12} \\ \hdashline A_{21} & A_{22} \end{array} \right], \tag{B.15}$$

where A_{11} and A_{22} are square matrices. If A_{11}^{-1} exists, then

$$\det(A) = \det(A_{11}) \det(F), \tag{B.16}$$

where $F = A_{22} - A_{21} A_{11}^{-1} A_{12}$. If additionally $\det(F) \neq 0$, then

$$A^{-1} = \left[\begin{array}{c:c} A_{11}^{-1} + A_{11}^{-1} A_{12} F^{-1} A_{21} A_{11}^{-1} & -A_{11}^{-1} A_{12} F^{-1} \\ \hdashline -F^{-1} A_{21} A_{11}^{-1} & F^{-1} \end{array} \right]. \tag{B.17}$$

Similarly, if A_{22}^{-1} exists, then

$$\det(A) = \det(A_{22}) \det(G), \tag{B.18}$$

where $G = A_{11} - A_{12} A_{22}^{-1} A_{21}$. If additionally $\det(G) \neq 0$, then

$$A^{-1} = \left[\begin{array}{c:c} G^{-1} & -G^{-1} A_{12} A_{22}^{-1} \\ \hdashline -A_{22}^{-1} A_{21} G^{-1} & A_{22}^{-1} + A_{22}^{-1} A_{21} G^{-1} A_{12} A_{22}^{-1} \end{array} \right]. \tag{B.19}$$

B.2 Symmetric, nonnegative definite and positive-definite matrices

Let us denote by $\mathrm{Sym}(n)$ the subspace of all symmetric matrices in $\mathbb{R}^{n \times n}$.

THEOREM B.4

[22, Prop. A.17, p. 659] *Let $A \in \mathrm{Sym}(n)$. Then*

1. *The eigenvalues $\lambda_1, \ldots, \lambda_n$ of A are real.*

2. *The eigenvectors $v_1, \ldots, v_n$ of A are real and mutually orthogonal.*

3. *If, additionally, $\lambda_1 \geq \cdots \geq \lambda_n$, then the outermost eigenvalues define the range of the Rayleigh quotient $(a^{\mathsf{T}} A a)/(a^{\mathsf{T}} a)$:*

$$\sup_{0 \neq a \in \mathbb{R}^n} \frac{a^{\mathsf{T}} A a}{a^{\mathsf{T}} a} = \sup_{\|a\|=1} a^{\mathsf{T}} A a = \lambda_1, \tag{B.20}$$

$$\inf_{0 \neq a \in \mathbb{R}^n} \frac{a^{\mathsf{T}} A a}{a^{\mathsf{T}} a} = \inf_{\|a\|=1} a^{\mathsf{T}} A a = \lambda_n, \tag{B.21}$$

and the extrema are attained at v_1 and v_n, respectively.

Suppose that the eigenvectors of $A \in \mathrm{Sym}(n)$ are normalized, i.e., they have unit Euclidean norms. Arranging the set of n eigensolutions (eigenvalues and eigenvectors) of A in matrix form as

$$A \underbrace{[\, v_1 \; \vdots \; \cdots \; \vdots \; v_n \,]}_{V} = \underbrace{[\, v_1 \; \vdots \; \cdots \; \vdots \; v_n \,]}_{V} \underbrace{\begin{bmatrix} \lambda_1 & 0 & \ldots & 0 \\ 0 & \lambda_2 & 0 \ldots & 0 \\ & & \ldots & \\ 0 & \ldots & 0 & \lambda_n \end{bmatrix}}_{\Lambda}, \tag{B.22}$$

we get

$$AV = V\Lambda, \tag{B.23}$$

which implies

$$A = V\Lambda V^{-1}. \tag{B.24}$$

Since the columns of V are orthonormal, we have $V^{\mathsf{T}} V = I$ and hence $V^{-1} = V^{\mathsf{T}}$, i.e., V is unitary. Thus we have proven the following result:

THEOREM B.5 Spectral decomposition

Let the eigenvectors of $A \in \mathrm{Sym}(n)$ be normalized. Then we have

$$V^{\mathsf{T}} A V = \Lambda, \quad A = V\Lambda V^{\mathsf{T}}, \tag{B.25}$$

where V and Λ are defined by (B.22). Alternatively, A can be represented as

$$A = \sum_{i=1}^{n} \lambda_i v_i v_i^{\mathsf{T}}, \tag{B.26}$$

where λ_i is the eigenvalue corresponding to v_i.

Note that, in general, if an eigenvalue λ_j has multiplicity greater than one, many choices of the eigenvector v_j become feasible and the decomposition (B.26) fails to be unique.

By using (B.25), we obtain

$$\text{tr}(A) = \text{tr}(V\Lambda V^\mathsf{T}) = \text{tr}(\Lambda V^\mathsf{T}V) = \text{tr}(\Lambda I) = \text{tr}(\Lambda). \qquad (\text{B.27})$$

Since Λ is diagonal with the eigenvalues of A on its diagonal, the trace of a symmetric matrix is the sum of its eigenvalues. Furthermore, we get

$$\begin{aligned}
\det(A) &= \det(V\Lambda V^\mathsf{T}) = \det(V)\det(\Lambda)\det(V^\mathsf{T}) \\
&= \det(V^\mathsf{T})\det(V)\det(\Lambda) = \det(V^\mathsf{T}V)\det(\Lambda) \qquad (\text{B.28}) \\
&= \det(I)\det(\Lambda) = \det(\Lambda).
\end{aligned}$$

Since $\det(\Lambda)$ is just the product of its diagonal elements, the determinant of a symmetric matrix equals the product of its eigenvalues.

The space $\text{Sym}(n)$ contains two subsets which are vital in this monograph, i.e., the subset of nonnegative definite matrices, $\text{NND}(n)$, and that of positive-definite matrices, $\text{PD}(n)$, which are, respectively, given by

$$\text{NND}(n) = \{A \in \text{Sym}(n) : x^\mathsf{T}Ax \geq 0 \text{ for all } x \in \mathbb{R}^n\}, \qquad (\text{B.29})$$

$$\text{PD}(n) = \{A \in \text{Sym}(n) : x^\mathsf{T}Ax > 0 \text{ for all } 0 \neq x \in \mathbb{R}^n\}. \qquad (\text{B.30})$$

THEOREM B.6
If $A \in \text{PD}(n)$ and $B \in \text{NND}(n)$, then $A + B \in \text{PD}(n)$.

THEOREM B.7
If $A \in \text{PD}(n)$, then $A^{-1} \in \text{PD}(n)$.

An extremely useful characterization of the nonnegative definiteness and positive definiteness is given by the following results in which $\lambda_{\min}(A)$ stands for the smallest eigenvalue of A:

THEOREM B.8
[224, p. 9] *Let $A \in \text{Sym}(n)$. The following are equivalent:*

(i) $A \in \text{NND}(n)$,

(ii) $\lambda_{\min}(A) \geq 0$,

(iii) $\text{tr}(AB) \geq 0$ for all $B \in \text{NND}(n)$,

(iv) There exists a matrix $B \in \mathbb{R}^{n \times n}$ such that $A = B^\mathsf{T}B$.

THEOREM B.9

[224, p. 9] *Let $A \in \text{Sym}(n)$. The following are equivalent:*

(i) $A \in \text{PD}(n)$,

(ii) $\lambda_{\min}(A) > 0$,

(iii) $\text{tr}(AB) > 0$ for all $0 \neq B \in \text{NND}(n)$,

(iv) *There exists a* nonsingular *matrix* $B \in \mathbb{R}^{n \times n}$ *such that* $A = B^{\mathsf{T}}B$.

THEOREM B.10

If $A = \begin{bmatrix} a_{ij} \end{bmatrix} \in \text{NND}(m)$, then

$$a_{ii} \geq 0, \quad a_{ij} \leq (a_{ii} + a_{jj})/2, \quad i, j = 1, \ldots, n, \tag{B.31}$$
$$\det(A) \geq 0, \quad \text{tr}(A) \geq 0. \tag{B.32}$$

If $A = \begin{bmatrix} a_{ij} \end{bmatrix} \in \text{PD}(m)$, then

$$a_{ii} > 0, \quad a_{ij} < (a_{ii} + a_{jj})/2, \quad i, j = 1, \ldots, n, \tag{B.33}$$
$$\det(A) > 0, \quad \text{tr}(A) > 0. \tag{B.34}$$

We often use expressions involving integer powers of matrices. The spectral decomposition leads to the following two results:

THEOREM B.11

If $A \in \text{PD}(n)$, then the eigenvalues of A^{-1} are the reciprocals of those of A, whereas the eigenvectors remain the same.

PROOF   The proof is based on the observation that

$$A^{-1} = (V \Lambda V^{\mathsf{T}})^{-1} = (V^{\mathsf{T}})^{-1} \Lambda^{-1} V^{-1} = V \Lambda^{-1} V^{\mathsf{T}} \tag{B.35}$$

and Λ^{-1} is a diagonal matrix whose diagonal elements are the reciprocals of those in Λ.   ∎

THEOREM B.12

If $k \in \{0, 1, \ldots\}$ and $A \in \text{NND}(n)$, then the eigenvalues of A^k constitute the k-th powers of those of A, and the eigenvectors are the same.

PROOF   For $k = 0$, note that by convention, $A^0 = I$ for any A. But $I = VV^{\mathsf{T}} = VIV^{\mathsf{T}} = V\Lambda^0 V^{\mathsf{T}}$, which establishes the formula.

As for other k's, consider only the case of $k = 2$, as the presented technique easily extends to any positive integer by induction. We have

$$A^2 = AA = (V\Lambda V^\mathsf{T})(V\Lambda V^\mathsf{T}) = V\Lambda(V^\mathsf{T}V)\Lambda V^\mathsf{T} = V\Lambda I\Lambda V^\mathsf{T} = V\Lambda^2 V^\mathsf{T}. \tag{B.36}$$

Since Λ^2 is a diagonal matrix whose nonzero elements are the squares of those in Λ, the desired result follows. ∎

THEOREM B.13

If $A \in \mathrm{PD}(n)$, then

$$\mathrm{tr}(A^{-1}) \geq \frac{n^2}{\mathrm{tr}(A)}. \tag{B.37}$$

PROOF A well-known mathematical fact is that the geometric mean never exceeds the arithmetic mean. Applying it to the reciprocals of the eigenvalues of A and to these eigenvalues themselves, we obtain

$$\underbrace{\left(\frac{1}{\lambda_1} \cdots \frac{1}{\lambda_n}\right)^{1/n}}_{\det(A^{-1})} \leq \underbrace{\frac{1}{n}\left(\frac{1}{\lambda_1} + \cdots + \frac{1}{\lambda_n}\right)}_{\mathrm{tr}(A^{-1})} \tag{B.38}$$

and

$$\underbrace{\left(\lambda_1 \ldots \lambda_n\right)^{1/n}}_{\det(A)} \leq \underbrace{\frac{1}{n}\left(\lambda_1 + \cdots + \lambda_n\right)}_{\mathrm{tr}(A)}. \tag{B.39}$$

Consequently,

$$\begin{aligned}
\mathrm{tr}(A^{-1}) &\geq n\left[\det(A^{-1})\right]^{1/n} \\
&= \frac{n}{\left[\det(A)\right]^{1/n}} \geq \frac{n}{\mathrm{tr}(A)/n} = \frac{n^2}{\mathrm{tr}(A)}.
\end{aligned} \tag{B.40}$$

∎

If a matrix $A \in \mathrm{Sym}(n)$ is partitioned as

$$A = \left[\begin{array}{c:c} Q & S \\ \hdashline S^\mathsf{T} & R \end{array}\right] \tag{B.41}$$

and $\det(Q) \neq 0$, $Q \in \mathrm{Sym}(m)$, the matrix $R - S^\mathsf{T}Q^{-1}S$ is called the *Schur complement* of Q in A and is denoted by (A/Q) [55].

THEOREM B.14
The positive definiteness of A is equivalent to the conditions

$$Q \succ 0, \quad (A/Q) \succ 0. \tag{B.42}$$

PROOF   The matrix A is positive definite iff

$$x^T A x > 0 \quad \text{for any } 0 \neq x \in \mathbb{R}^n. \tag{B.43}$$

Partitioning x using $u \in \mathbb{R}^m$ and $v \in \mathbb{R}^{n-m}$, this amounts to the condition

$$\begin{bmatrix} u^T & \vdots & v^T \end{bmatrix} \begin{bmatrix} Q & \vdots & S \\ \cdots & & \cdots \\ S^T & \vdots & R \end{bmatrix} \begin{bmatrix} u \\ \cdots \\ v \end{bmatrix} = u^T Q u + 2 v^T S^T u + v^T R v > 0 \tag{B.44}$$

whenever $\left[\begin{smallmatrix} u \\ v \end{smallmatrix}\right] \neq 0$.

Two cases can be considered. First, if $v = 0$, then (B.44) becomes

$$u^T Q u > 0 \quad \text{for any } 0 \neq u \in \mathbb{R}^m, \tag{B.45}$$

which is just the first condition of (B.42). In turn, for $v \neq 0$ we must have

$$f(v) = \min_{u \in \mathbb{R}^{n-m}} \{ u^T Q u + 2 v^T S^T u + v^T R v \} > 0. \tag{B.46}$$

An easy computation shows that the minimum of the expression in the braces is attained at $u^\star = -Q^{-1} S v$ and equals

$$f(v) = v^T (R - S^T Q^{-1} S) v = v^T (A/Q) v. \tag{B.47}$$

The inequality $f(v) > 0$ for any nonzero v is thus equivalent to the second condition of (B.42). ∎

In much the same way, we can prove the following result:

THEOREM B.15
If $\det(Q) \neq 0$, then the nonnegative definiteness of A is equivalent to the conditions

$$Q \succ 0, \quad (A/Q) \succeq 0. \tag{B.48}$$

Schur complements play a fundamental role in converting nonlinear (convex) inequalities into LMIs.

B.3   Vector and matrix differentiation

The following conventions are adapted throughout:

1. $\partial f/\partial x$, where f is a real-valued function and x a vector, denotes a row vector with the i-th element $\partial f/\partial x_i$, x_i being the i-th component of x.

2. $\partial g/\partial x$, where g is a vector-valued function and x a vector, denotes a matrix with the (i,j)-th element $\partial g_i/\partial x_j$, where g_i, x_j are the i-th and j-th elements of g and x, respectively.

3. $\partial f/\partial M$, where f is a real-valued function and M is a matrix, denotes a matrix with the (i,j)-th element $\partial f/\partial m_{ij}$, where m_{ij} is the (i,j)-th element of M.

At times, for a scalar-valued function f of a vector $x = (x_1,\ldots,x_n)$, in lieu of $\partial f/\partial x$ we also consider the result of differentiation as a column vector, but then we call it the gradient of the function f and use the notation ∇f or $\nabla_x f$, i.e., $\nabla f = (\partial f/\partial x)^\mathsf{T}$.

We define a second-derivatives matrix or Hessian which is computed as

$$\frac{\partial^2 f}{\partial x^2} \equiv \nabla^2 f = \begin{bmatrix} \dfrac{\partial^2 f}{\partial x_1 \partial x_1} & \cdots & \dfrac{\partial^2 f}{\partial x_1 \partial x_n} \\ \vdots & \ddots & \vdots \\ \dfrac{\partial^2 f}{\partial x_n \partial x_1} & \cdots & \dfrac{\partial^2 f}{\partial x_n \partial x_n} \end{bmatrix}. \tag{B.49}$$

A frequent use of the derivatives of f is in the Taylor series approximation. A Taylor series is a polynomial approximation to f. The choice of the number of terms is arbitrary, but the more that are used, the more accurate the approximation will be. The approximation used most frequently is the linear approximation

$$f(x) \approx f(x_0) + \left.\frac{\partial f}{\partial x}\right|_{x=x_0}(x - x_0) \tag{B.50}$$

about a given point x_0.

The second-order, or quadratic, approximation adds the second-order terms in the expansion:

$$f(x) \approx f(x_0) + \left.\frac{\partial f}{\partial x}\right|_{x=x_0}(x - x_0) + \frac{1}{2}(x - x_0)^\mathsf{T}\left.\frac{\partial^2 f}{\partial x^2}\right|_{x=x_0}(x - x_0). \tag{B.51}$$

Some useful formulae follow:

$$\frac{\partial}{\partial x}(a^\mathsf{T}x) = a^\mathsf{T}, \qquad \forall a \in \mathbb{R}^n, \tag{B.52}$$

$$\frac{\partial}{\partial x}(Ax) = A, \qquad \forall A \in \mathbb{R}^{m \times n}, \tag{B.53}$$

$$\frac{\partial}{\partial x}(x^\mathsf{T}Ax) = 2x^\mathsf{T}A, \qquad \forall A \in \mathrm{Sym}(n), \tag{B.54}$$

$$\frac{\partial^2}{\partial x^2}(x^\mathsf{T}Ax) = 2A, \qquad \forall A \in \mathrm{Sym}(n). \tag{B.55}$$

Assume that the elements of a matrix $A = [a_{ij}] \in \mathbb{R}^{m \times k}$ depend on a parameter α. Then we define

$$\frac{\partial A(\alpha)}{\partial \alpha} = \left[\frac{\partial a_{ij}(\alpha)}{\partial \alpha} \right]. \tag{B.56}$$

THEOREM B.16
For any $A \in \mathbb{R}^{n \times k}$ and $B \in \mathbb{R}^{k \times \ell}$ which depend on a parameter α, we have

$$\frac{\partial}{\partial \alpha}(AB) = \frac{\partial A}{\partial \alpha} B + A \frac{\partial B}{\partial \alpha}, \tag{B.57}$$

$$\frac{\partial}{\partial \alpha} \operatorname{tr}(AB) = \operatorname{tr}\left(\frac{\partial A}{\partial \alpha} B \right) + \operatorname{tr}\left(A \frac{\partial B}{\partial \alpha} \right). \tag{B.58}$$

Application of the chain rule for differentiation yields the following result:

THEOREM B.17 Derivative of a composition
Assume that a function $\Psi : \mathbb{R}^{n \times n} \to \mathbb{R}$ is continuously differentiable with respect to individual elements of its matrix argument. Then for any $A \in \mathbb{R}^{n \times n}$ which may depend on a parameter α we have

$$\frac{\partial \Psi[A(\alpha)]}{\partial \alpha} = \operatorname{tr}\left\{ \left(\frac{\partial \Psi}{\partial A} \right)^{\mathsf{T}} \frac{\partial A}{\partial \alpha} \right\}. \tag{B.59}$$

THEOREM B.18
For any $A \in \mathbb{R}^{n \times n}$ which may depend on a parameter α, there holds

$$\frac{\partial}{\partial \alpha} \operatorname{tr}(A) = \operatorname{tr}\left(\frac{\partial A}{\partial \alpha} \right), \tag{B.60}$$

$$\frac{\partial}{\partial \alpha} A^{-1} = -A^{-1} \frac{\partial A}{\partial \alpha} A^{-1}, \tag{B.61}$$

$$\frac{\partial}{\partial \alpha} \ln \det(A) = \operatorname{tr}\left(A^{-1} \frac{\partial A}{\partial \alpha} \right), \tag{B.62}$$

$$\frac{\partial}{\partial \alpha} \det(A) = \det(A) \operatorname{tr}\left(A^{-1} \frac{\partial A}{\partial \alpha} \right). \tag{B.63}$$

PROOF   Formula (B.60) follows directly from a trivial verification. Formula (B.61) results from the differentiation of I_n, the identity matrix of order n:

$$0 = \frac{\partial I_n}{\partial \alpha} = \frac{\partial (AA^{-1})}{\partial \alpha} = \frac{\partial A}{\partial \alpha} A^{-1} + A \frac{\partial A^{-1}}{\partial \alpha}. \tag{B.64}$$

As for (B.62), recall the Laplace rule for calculating the inverse of $A = [a_{ij}]$:

$$A^{-1} = \frac{1}{\det(A)} C^{\mathsf{T}} \quad \text{for } \det(A) \neq 0, \tag{B.65}$$

where $C = [c_{ij}]$ is the matrix composed of the cofactors of A (c_{ij} is the determinant of the reduced matrix formed by crossing out the i-th row and the j-th column of A, then multiplied by $(-1)^{i+j}$).

Therefore

$$AC^{\mathsf{T}} = \det(A)I_n. \tag{B.66}$$

Taking ln of the (i, i)-th element gives

$$\ln\det(A) = \ln\left(\sum_{k=1}^{n} a_{ik}c_{ik}\right). \tag{B.67}$$

Since c_{ik}, $k = 1, \ldots, n$ do not depend on a_{ij}, we obtain

$$\frac{\partial \ln\det(A)}{\partial a_{ij}} = \frac{c_{ij}}{\sum_{k=1}^{n} a_{ik}c_{ik}} = \frac{c_{ij}}{\det(A)}. \tag{B.68}$$

Consequently,

$$\frac{\partial \ln\det(A)}{\partial A} = \left(A^{-1}\right)^{\mathsf{T}}. \tag{B.69}$$

Theorem B.17 now leads to

$$\frac{\partial \ln\det(A)}{\partial \alpha} = \text{tr}\left\{\left(\frac{\partial \ln\det(A)}{\partial A}\right)^{\mathsf{T}}\frac{\partial A}{\partial \alpha}\right\} = \text{tr}\left(A^{-1}\frac{\partial A}{\partial \alpha}\right). \tag{B.70}$$

As for (B.63), we see at once that

$$\frac{\partial \det(A)}{\partial \alpha} = \frac{\partial \exp(\ln\det(A))}{\partial \alpha}$$
$$= \exp(\ln\det(A))\frac{\partial \ln\det(A)}{\partial \alpha} = \det(A)\,\text{tr}\left(A^{-1}\frac{\partial A}{\partial \alpha}\right). \tag{B.71}$$

∎

Based on Theorem B.18, it is a simple matter to establish the following result:

THEOREM B.19

Let $A, M \in \mathbb{R}^{n \times n}$ and $a, b \in \mathbb{R}^n$. We have

$$\frac{\partial}{\partial M} \operatorname{tr}(AM) = A^{\mathsf{T}}, \tag{B.72}$$

$$\frac{\partial}{\partial M} a^{\mathsf{T}} M b = a b^{\mathsf{T}}, \tag{B.73}$$

$$\frac{\partial}{\partial M} \operatorname{tr}(AM^{-1}) = -(M^{-1} A M^{-1})^{\mathsf{T}}, \tag{B.74}$$

$$\frac{\partial}{\partial M} a^{\mathsf{T}} M^{-1} b = -(M^{-1} b a^{\mathsf{T}} M^{-1})^{\mathsf{T}}, \tag{B.75}$$

$$\frac{\partial}{\partial M} \det(M) = \det(M) (M^{-1})^{\mathsf{T}}, \tag{B.76}$$

$$\frac{\partial}{\partial M} \ln \det(M) = (M^{-1})^{\mathsf{T}}. \tag{B.77}$$

B.4 Convex sets and convex functions

A set A in a linear vector space is said to be *convex* if the line segment between any two points in A lies in A, i.e.,

$$\alpha x + (1 - \alpha) y \in A, \quad \forall x, y \in A, \quad \forall \alpha \in [0, 1]. \tag{B.78}$$

Let A be a subset of $\mathbb{R}^n$. A *convex combination* of elements $x_1, \ldots, x_\ell \in A$ is a vector of the form $\sum_{i=1}^{\ell} \alpha_i x_i$, where $\alpha_1, \ldots, \alpha_\ell$ are nonnegative scalars such that $\sum_{i=1}^{\ell} \alpha_i = 1$. It can be thought of as a mixture or weighted average of the points with α_i the fraction of x_i in the mixture [28].

The *convex hull* of $A \subset \mathbb{R}^n$, denoted $\operatorname{conv}(A)$, is the set of all convex combinations of elements of A:

$$\operatorname{conv}(A) = \left\{ \sum_{i=1}^{\ell} \alpha_i x_i : x_i \in A, \ \alpha_i \geq 0, \ i = 1, \ldots, \ell, \ \sum_{i=1}^{\ell} \alpha_i = 1 \right\}. \tag{B.79}$$

It is easy to show that $\operatorname{conv}(A)$ is the smallest convex set that contains A.

The idea of a convex combination can be generalized to include infinite sums, integrals, and, in the most general form, probability distributions [28, p. 24]. For example, if $\alpha_1, \alpha_2, \ldots$ satisfy

$$\alpha_i \geq 0, \quad i = 1, 2, \ldots, \quad \sum_{i=1}^{\infty} \alpha_i = 1, \tag{B.80}$$

and $x_1, x_2, \cdots \in A$, where $A \subset \mathbb{R}^n$ is convex, then

$$\sum_{i=1}^{\infty} \alpha_i x_i \in A \tag{B.81}$$

whenever the series converges.

In this monograph, we are concerned with the convex hull of a set A in $\mathbb{R}^n$, where

$$A = \Big\{ h(x) : x \in X \Big\}, \tag{B.82}$$

X being a compact subset of $\mathbb{R}^k$ and $h : \mathbb{R}^k \to \mathbb{R}^n$ a continuous function. Let ξ be a probability distribution on the Borel sets of X, $\Xi(X)$ the set of all ξ. Then

$$\text{conv}(A) = \Big\{ y = \int_X h(x)\,\xi(\mathrm{d}x) : \xi \in \Xi(X) \Big\}. \tag{B.83}$$

THEOREM B.20 Carathéodory's Theorem
[269, p. 72] *Let A be a subset of $\mathbb{R}^n$. Every element c of $\text{conv}(A)$ can be expressed as a convex combination of no more than $n+1$ elements of A. If c is on the boundary of $\text{conv}(A)$, $n+1$ can be replaced by n.*

A real-valued function f defined on a convex subset A of a linear vector space is called *convex* if

$$f(\alpha x + (1-\alpha)y) \leq \alpha f(x) + (1-\alpha)f(y), \quad \forall x, y \in A, \quad \forall \alpha \in [0,1]. \tag{B.84}$$

The function f is called *strictly convex* if the above inequality is strict for all $x, y \in A$ with $x \neq y$, and all $\alpha \in (0,1)$.

THEOREM B.21
[22, Prop. B.2, p. 674]

1. *A linear function is convex.*

2. *The weighted sum of convex functions, with positive weights, is' convex.*

3. *If Y is an index set, $A \subset \mathbb{R}^n$ is a convex set, and $\varphi_y : A \to \mathbb{R}$ is convex for each $y \in Y$, then the function f defined by*

$$f(x) = \sup_{y \in Y} \varphi_y(x), \quad \forall x \in A \tag{B.85}$$

is also convex.

THEOREM B.22
[215, Th. 5.2.11, p. 668] *Suppose that $f : \mathbb{R}^n \to \mathbb{R}$ is convex. Then $f(\cdot)$ is continuous.*

THEOREM B.23
[12, Th. 2.6.1] *Let $\mathcal{H}$ be a real Hilbert space. Suppose that $f : \mathcal{H} \to \mathbb{R}$ is a continuous, convex function and that A is a closed and bounded, convex subset*

of $\mathcal{H}$*. Then there exists an* $x^\star \in A$ *such that*

$$f(x^\star) = \inf_{x \in A} f(x). \tag{B.86}$$

THEOREM B.24
[22, Prop. B.10, p. 685] *Let* $\mathcal{H}$ *be a real Hilbert space. If* $A \subset \mathcal{H}$ *is a convex set and* $f : A \to \mathbb{R}$ *is a convex function, then any local minimum of* f *is also a global minimum. If, additionally,* f *is strictly convex, then there exists at most one global minimum of* f.

THEOREM B.25
Let $\mathcal{H}$ *be a Hilbert space. Suppose that* A *is a convex subset of* $\mathcal{H}$ *and* $f : A \to \mathbb{R}$ *is a convex function whose one-sided directional differential* $\delta_+ f(x^\star; x - x^\star)$ *at a point* $x^\star \in A$ *exists for all* $x \in A$*. Then*

$$\delta_+ f(x^\star; x - x^\star) \geq 0, \quad \forall x \in A \tag{B.87}$$

constitutes a necessary and sufficient condition for $x^\star$ *to minimize* f *over* A.

PROOF   *(Neccessity)* Suppose that $x^\star$ is a minimum of f and $x \in A$. Therefore for all $\alpha \in (0, 1)$ we must have

$$f(x^\star) \leq f((1 - \alpha)x^\star + \alpha x)). \tag{B.88}$$

This implies that

$$\frac{f(x^\star + \alpha(x - x^\star)) - f(x^\star)}{\alpha} \geq 0. \tag{B.89}$$

Letting $\alpha \downarrow 0$, we see that

$$\delta_+ f(x^\star; x - x^\star) \geq 0, \tag{B.90}$$

which is the desired conclusion.

   (Sufficiency) Let (B.87) be satisfied for a point $x^\star \in A$ and $x \in A$ be arbitrary. From the convexity of f it follows that

$$f((1 - \alpha)x^\star + \alpha x)) \leq (1 - \alpha)f(x^\star) + \alpha f(x), \quad \forall \alpha \in [0, 1]. \tag{B.91}$$

Rearranging the terms, we obtain

$$f(x) - f(x^\star) \geq \frac{f(x^\star + \alpha(x - x^\star)) - f(x^\star)}{\alpha}. \tag{B.92}$$

A passage to the limit implies that

$$f(x) - f(x^\star) \geq \delta_+ f(x^\star; x - x^\star) \geq 0, \tag{B.93}$$

which completes the proof.   ∎

B.5 Convexity and differentiability of common optimality criteria

THEOREM B.26 D-optimality criterion
[208, Prop. IV.2, p. 81] *The function*

$$\mathrm{NND}(x) \ni M \mapsto -\ln\det(M) \in \mathbb{R} \tag{B.94}$$

is

1. *continuous on* $\mathrm{NND}(n)$,

2. *convex on* $\mathrm{NND}(n)$ *and strictly convex on* $\mathrm{PD}(n)$, *and*

3. *differentiable whenever it is finite. Its gradient is*

$$\frac{\partial(-\ln\det(M))}{\partial M} = -M^{-1}. \tag{B.95}$$

THEOREM B.27 A-optimality criterion
[208, Prop. IV.3, p. 82] *The function*

$$\mathrm{NND}(x) \ni M \mapsto \begin{cases} \mathrm{tr}(M^{-1}) & \text{if } \det(M) \neq 0, \\ \infty & \text{otherwise,} \end{cases} \tag{B.96}$$

is

1. *continuous on* $\mathrm{NND}(n)$,

2. *convex on* $\mathrm{NND}(n)$ *and strictly convex on* $\mathrm{PD}(n)$, *and*

3. *differentiable whenever it is finite. Its gradient is*

$$\frac{\partial\,\mathrm{tr}(M^{-1})}{\partial M} = -M^{-2}. \tag{B.97}$$

THEOREM B.28 L-optimality criterion
[208, Prop. IV.10, p. 93] *The function*

$$\mathrm{NND}(x) \ni M \mapsto \begin{cases} \mathrm{tr}(WM^{-1}) & \text{if } \det(M) \neq 0, \\ \infty & \text{otherwise,} \end{cases} \tag{B.98}$$

where $W \in \mathrm{PD}(n)$, *is*

1. *continuous on* $\mathrm{NND}(n)$,

2. *convex on* $\mathrm{NND}(n)$ *and strictly convex on* $\mathrm{PD}(n)$, *and*

3. *differentiable whenever it is finite. Its gradient is*

$$\frac{\partial \operatorname{tr}(WM^{-1})}{\partial M} = -M^{-1}WM^{-1}. \tag{B.99}$$

THEOREM B.29 E-optimality criterion
[208, Prop. IV.10, p. 91] *The function*

$$\mathrm{NND}(x) \ni M \mapsto \begin{cases} \lambda_{\max}(M^{-1}) & \text{if } \det(M) \neq 0, \\ \infty & \text{otherwise,} \end{cases} \tag{B.100}$$

is

1. *continuous on* $\mathrm{NND}(n)$,

2. *convex on* $\mathrm{NND}(n)$.

Unfortunately, the strict convexity of the E-optimality criterion cannot be proven. As a counterexample, consider the definition of convexity (B.84) for $x = \mathrm{Diag}[1, 2]$ and $y = \mathrm{Diag}[1, 4]$, where $\mathrm{Diag}(c)$ is a square matrix with the elements of the vector c put on its main diagonal.

B.6 Differentiability of spectral functions

A real-valued function Ψ defined on $\mathcal{A} \subset \mathrm{Sym}(m)$ is termed *unitarily invariant* if

$$\Psi(UAU^{\mathsf{T}}) = \Psi(A) \tag{B.101}$$

for any $A \in \mathcal{A}$ and all unitary U. It is a simple matter to check that all the optimum experimental design criteria considered in this monograph (e.g., D-, E-, A-optimum design criteria, as well as the sensitivity criterion) share this property.

The unitarily invariant functions are rightly called *spectral functions*, since $\Psi(A)$ depends only on the eigenvalues of A. In fact, defining $\lambda : \mathrm{Sym}(m) \to \mathbb{R}^m$ as the eigenvalue function, i.e.,

$$\lambda(A) = \operatorname{col}[\lambda_1(A), \dots, \lambda_m(A)], \quad \forall A \in \mathrm{Sym}(m) \tag{B.102}$$

such that the eigenvalues are ordered in accordance with

$$\lambda_1(A) \geq \lambda_2(A) \geq \cdots \geq \lambda_m(A), \tag{B.103}$$

from (B.101) we obtain

$$\Psi(A) = \Psi(\lambda(A)) \tag{B.104}$$

after setting U as the unitary matrix formed out of the normalized eigenvectors corresponding to $\lambda_1(A), \ldots, \lambda_m(A)$, cf. Theorem B.5. For example, for common experimental design criteria we get

- The D-optimality criterion:

$$\Psi(A) = -\ln \det(A) = -\sum_{i=1}^{m} \ln \lambda_i, \tag{B.105}$$

- The A-optimality criterion:

$$\Psi(A) = \operatorname{tr}(A^{-1}) = \sum_{i=1}^{m} \frac{1}{\lambda_i}, \tag{B.106}$$

- The E-optimality criterion:

$$\Psi(A) = \lambda_{\max}(A^{-1}) = \max\left\{\frac{1}{\lambda_1}, \ldots, \frac{1}{\lambda_m}\right\}, \tag{B.107}$$

- The sensitivity criterion:

$$\Psi(A) = -\operatorname{tr}(A) = -\sum_{i=1}^{m} \lambda_i. \tag{B.108}$$

Associated with any spectral function is a function $f : \mathbb{R}^m \to \mathbb{R}$ called *symmetric* due to the property that it is invariant under coordinate permutations:

$$f(Px) = f(x) \tag{B.109}$$

for any $x \in \mathbb{R}^m$ and $P \in \mathcal{P}$, the set of all permutation matrices. (Note that all functions on the right-hand sides of (B.105)–(B.108) are symmetric.) Specifically, $\Psi : \mathcal{A} \to \mathbb{R}$ may be treated as the composite mapping $f \circ \lambda$,

$$\Psi(A) = (f \circ \lambda)(A) = f(\lambda(A)), \quad \forall A \in \mathcal{A}. \tag{B.110}$$

Lewis [155] examined the problem of when $(f \circ \lambda)(A)$ is differentiable at A for an arbitrary symmetric function f. He proved that this differentiability is surprisingly conditioned only by the smoothness of f, and not by that of λ.

THEOREM B.30
[155, Th. 1.1] *Let the set $\Omega \subset \mathbb{R}^m$ be open and symmetric, and suppose that the function $f : \Omega \to \mathbb{R}$ is symmetric. Then the spectral function $\Psi = f \circ \lambda$*

is differeniable at a matrix A is and only if f is differentible at $\lambda(A)$. In addition,

$$\frac{\partial \Psi(A)}{\partial A} = U \operatorname{Diag}\big[\nabla f(\lambda(A))\big]U^\mathsf{T} \tag{B.111}$$

for any unitary matrix U satisfying $A = U \operatorname{Diag}[\lambda(A)]U^\mathsf{T}$, where $\operatorname{Diag}(c)$ is a square matrix with the elements of the vector c put on its main diagonal.

Based on this quite general result, Chen *et al.* [50] proved additionally that $\Psi = f \circ \lambda$ is continuously differentiable at A iff f is so at $\lambda(A)$.

In reference to the citeria (B.105)–(B.108), choosing U as the matrix with columns being the normalized eigenvectors corresponding to $\lambda_1(A), \ldots, \lambda_m(A)$, we get

- The D-optimality criterion:

$$f(\lambda) = -\sum_{i=1}^{m} \ln \lambda_i, \quad \nabla f(\lambda) = \operatorname{col}[-\lambda_1^{-1}, \ldots, -\lambda_m^{-1}], \tag{B.112}$$

$$\begin{aligned} \frac{\partial \Psi(A)}{\partial A} &= -V \operatorname{Diag}[\lambda_1^{-1}, \ldots, \lambda_m^{-1}]V^\mathsf{T} \\ &= -\big\{V \operatorname{Diag}[\lambda_1, \ldots, \lambda_m]V^\mathsf{T}\big\}^{-1} = -A^{-1}. \end{aligned} \tag{B.113}$$

- The A-optimality criterion:

$$f(\lambda) = \sum_{i=1}^{m} \lambda_i^{-1}, \quad \nabla f(\lambda) = \operatorname{col}[-\lambda_1^{-2}, \ldots, -\lambda_m^{-2}], \tag{B.114}$$

$$\begin{aligned} \frac{\partial \Psi(A)}{\partial A} &= -V \operatorname{Diag}[\lambda_1^{-2}, \ldots, \lambda_m^{-2}]V^\mathsf{T} \\ &= -\big\{V \operatorname{Diag}[\lambda_1, \ldots, \lambda_m]V^\mathsf{T}\big\}^{-2} = -A^{-2}. \end{aligned} \tag{B.115}$$

- The sensitivity criterion:

$$f(\lambda) = -\sum_{i=1}^{m} \lambda_i, \quad \nabla f(\lambda) = \operatorname{col}[-1, \ldots, -1], \tag{B.116}$$

$$\frac{\partial \Psi(A)}{\partial A} = -V \operatorname{Diag}[1, \ldots, 1]V^\mathsf{T} = -VV^\mathsf{T} = -I. \tag{B.117}$$

In addition, we conclude that the E-optimality criterion is not differentiable, as the function $f(\lambda) = \max\{\lambda_1^{-1}, \ldots, \lambda_m^{-1}\}$ is not so.

B.7 Monotonicity of common design criteria

As it happens, many times in this monograph we have to deal with the values of a convex function $\Psi : \mathrm{NND}(m) \to \mathbb{R}$ at a matrix argument M after a non-negative perturbation A. Below we review the properties of such an operation for two commonplace functions in the following theorems:

THEOREM B.31 *Monotonicity of the D-optimality criterion*
Let $M \in \mathrm{NND}(m)$. Then we have

$$\det(M + A) \geq \det(M) \tag{B.118}$$

for all $A \in \mathrm{NND}(m)$. Additionally, if M is nonsingular, then

$$\det(M + A) > \det(M) \tag{B.119}$$

for all $0 \neq A \in \mathrm{NND}(m)$.

PROOF Assume first that $M \succ 0$ and $0 \neq A \succeq 0$. Theorem B.8 shows that there is a matrix $B \in \mathbb{R}^{n \times n}$ such that $A = B^\mathsf{T} B$. Hence from Theorem B.2 we deduce that

$$\det(M + A) = \det(M + B^\mathsf{T} B) = \det(M)\det(I + BM^{-1}B^\mathsf{T}). \tag{B.120}$$

Obviously, we have $M^{-1} \succ 0$, which implies the nonnegativeness of the matrix $C = BM^{-1}B^\mathsf{T}$. Denoting by $\mu_1, \ldots, \mu_n$ the eigenvalues of C, we get $\mu_i \geq 0$, $i = 1, \ldots, n$. What is more, at least one eigenvalue must be positive, since otherwise we would have $C = 0$ and thereby $B = 0$, cf. Theorem B.9. Consequently, (B.28) now leads to

$$\det(I + BM^{-1}B^\mathsf{T}) = (1 + \mu_1) \ldots (1 + \mu_n) > 1, \tag{B.121}$$

owing to the fact that the eigenvalues of $I + BM^{-1}B^\mathsf{T}$ are just $1 + \mu_1, \ldots, 1 + \mu_n$. This, taken in conjunction with $\det(M) > 0$, implies (B.119).
  If $M \in \mathrm{NND}(m)$, then $M + \epsilon I \in \mathrm{PD}(m)$ for any $\epsilon > 0$. But then

$$\det(M + \epsilon I + A) > \det(M + \epsilon I). \tag{B.122}$$

Letting $\epsilon \downarrow 0$, we obtain (B.118). ∎

THEOREM B.32 *Monotonicity of the A-optimality criterion*
Let $M \in \mathrm{PD}(m)$. Then we have

$$\mathrm{tr}\big[(M + A)^{-1}\big] < \mathrm{tr}\big[M^{-1}\big] \tag{B.123}$$

for all $0 \neq A \in \text{NND}(m)$.

PROOF   Theorem B.9 guarantees that there exists a nonzero matrix $B \in \mathbb{R}^{n \times n}$ such that $A = B^\mathsf{T}B$. From the Woodbury formula (B.11) it follows that

$$(M + A)^{-1} = (M + B^\mathsf{T}B)^{-1}$$
$$= M^{-1} - M^{-1}B^\mathsf{T}(I + BM^{-1}B^\mathsf{T})^{-1}BM^{-1}. \tag{B.124}$$

Therefore

$$\text{tr}\big[(M + A)^{-1}\big]$$
$$= \text{tr}\big[M^{-1}\big] - \text{tr}\big[(I + BM^{-1}B^\mathsf{T})^{-1}BM^{-2}B^\mathsf{T}\big]. \tag{B.125}$$

The matrices $BM^{-1}B^\mathsf{T}$ and $BM^{-2}B^\mathsf{T}$ are easily seen to be nonnegative definite. Consequently, $I + BM^{-1}B^\mathsf{T}$ is positive definite, and hence so is its inverse. On account of the characterization given in Theorem B.9 we thus have

$$\text{tr}\big[(I + BM^{-1}B^\mathsf{T})^{-1}BM^{-2}B^\mathsf{T}\big] > 0, \tag{B.126}$$

which yields the desired result.   ∎

B.8   Integration with respect to probability measures

Let $(\Omega, \mathfrak{F}, \xi)$ be a probability space, where $\Omega \subset \mathbb{R}^n$ is a sample space, $\mathfrak{F}$ is a σ-algebra of subsets of Ω and $\xi : \mathfrak{F} \to [0, 1]$ stands for a probability measure. Note that this implies that $\mathfrak{F}$ contains all subsets of sets of measure zero. In what follows, we shall assume that all the considered real-valued functions defined on Ω are $\mathfrak{F}$-measurable, i.e., they are random variables. Construction of integrals for such functions is usually done in three stages, cf. [58, p. 28] or [287].

DEFINITION B.1   **(Integral of a simple function)** *Assume that $v : \Omega \to \mathbb{R} \cup \{\infty\}$ is a simple function, i.e., there are nonzero constants c_i and disjoint measurable sets A_i such that it can be expressed as*

$$v(\omega) = \sum_{i=1}^{m} c_i \chi_{A_i}(\omega), \tag{B.127}$$

where χ_{A_i} is the characteristic or indicator function of the set A_i, which equals 1 on A_i and 0 on $\Omega \setminus A_i$. Then we define

$$\int_A v(\omega)\, \xi(\mathrm{d}\omega) = \sum_{i=1}^{m} c_i \xi(A_i \cap A) \tag{B.128}$$

for any $A \in \mathfrak{F}$.

DEFINITION B.2 (**Integral of a nonnegative function**) *If $v : \Omega \to [0, \infty]$, then there exists a sequence $\{v_n\}$ of monotonic increasing simple functions, i.e., $v_n \leq v_{n+1}$ on Ω for all n, such that $v(\omega) = \lim_{n \to \infty} v_n(\omega)$ for all Ω. Then we define*

$$\int_A v(\omega)\,\xi(\mathrm{d}\omega) = \lim_{n \to \infty} \int_A v_n(\omega)\,\xi(\mathrm{d}\omega) \tag{B.129}$$

for any $A \in \mathfrak{F}$. It can be shown that the value of the integral is independent of the sequence $\{v_n\}$.

DEFINITION B.3 (**Integral of any function**) *Let v be any measurable function on $(\Omega, \mathfrak{F}, \xi)$. If $\int_A |v(\omega)|\,\xi(\mathrm{d}\omega) < \infty$, then we call v integrable and define its integral over A with respect to ξ using the following decomposition:*

$$\int_A v(\omega)\,\xi(\mathrm{d}\omega) = \int_A v^+(\omega)\,\xi(\mathrm{d}\omega) - \int_A v^-(\omega)\,\xi(\mathrm{d}\omega), \tag{B.130}$$

where

$$v^+(\omega) = \max(0, v(\omega)), \quad v^-(\omega) = -\min(0, v(\omega)). \tag{B.131}$$

Note that several conventions regarding the notation exist. For example, the following equivalent forms can be encountered:

$$\int_A v\,\mathrm{d}\xi \quad \text{or} \quad \int_A v(\omega)\,\mathrm{d}\xi(\omega).$$

The notation adopted in the monograph prevails in optimum experimental design theory.

This integral has a number of useful properties which are listed below (we assume that u and v are integrable):

(P1) $\int_A (u(\omega) + v(\omega))\,\xi(\mathrm{d}\omega) = \int_A u(\omega)\,\xi(\mathrm{d}\omega) + \int_A v(\omega)\,\xi(\mathrm{d}\omega).$

(P2) $\int_A cv(\omega)\,\xi(\mathrm{d}\omega) = c \int_A v(\omega)\,\xi(\mathrm{d}\omega), \quad \forall c \in \mathbb{R},$

(P3) If $u(\omega) = v(\omega)$ *almost everywhere* on A (i.e., except possibly on some set $B \subset A$ such that $\xi(B) = 0$), then

$$\int_A u(\omega)\,\xi(\mathrm{d}\omega) = \int_A v(\omega)\,\xi(\mathrm{d}\omega).$$

(P4) If $u(\omega) \leq v(\omega)$ *almost everywhere* on A, then

$$\int_A u(\omega)\,\xi(\mathrm{d}\omega) \leq \int_A v(\omega)\,\xi(\mathrm{d}\omega).$$

(P5) $\left|\int_A v(\omega)\,\xi(\mathrm{d}\omega)\right| \leq \int_A |v(\omega)|\,\xi(\mathrm{d}\omega).$

(P6) If $A = A_1 \cup A_2$ and $A_1 \cap A_2 = \emptyset$, then

$$\int_A v(\omega)\,\xi(\mathrm{d}\omega) = \int_{A_1} v(\omega)\,\xi(\mathrm{d}\omega) + \int_{A_2} v(\omega)\,\xi(\mathrm{d}\omega).$$

(P7) If $\displaystyle\int_A v(\omega)\,\xi(\mathrm{d}\omega) = 0$ for any $A \in \mathfrak{F}$, then $v(\omega) = 0$ almost everywhere on Ω.

(P8) (*Lebesgue dominated convergence*) If $\{v_n\}$ is a sequence of measurable functions on A such that $v_n(\omega) \to v(\omega)$ as $n \to \infty$ almost everywhere on A and $|v_n(\omega)| \leq u(\omega)$ almost everywhere on A, where u is an integrable function on A, then

$$\int_A v(\omega)\,\xi(\mathrm{d}\omega) = \lim_{n\to\infty} \int_A v_n(\omega)\,\xi(\mathrm{d}\omega).$$

B.9 Projection onto the canonical simplex

The canonical simplex in $\mathbb{R}^n$ is defined as the polyhedron

$$\mathbb{S} = \left\{ p = (p_1, \ldots, p_n) : \sum_{i=1}^n p_i = 1, \ p_i \geq 0, \ i = 1, \ldots, n \right\}, \qquad \text{(B.132)}$$

which is the convex hull of its extreme points being the unit basis vectors along the consecutive rectangular coordinates (the Cartesian base vectors).

Let q be a fixed vector in $\mathbb{R}^n$ and consider the problem of finding a vector $p^\star$ in $\mathbb{S}$ which is at a minimum distance from q, i.e.,

$$p^\star = \arg\min_{p \in \mathbb{S}} \|q - p\|^2. \qquad \text{(B.133)}$$

This vector is called the *projection of q onto* $\mathbb{S}$. Since $\mathbb{S}$ is closed and convex, the existence and unicity of the projection are guaranteed, cf. [163, Th. 1, p. 69] or [22, Prop. 2.1.3, p. 201].

Although the problem does not possess a solution given explicitly, an algorithm can be developed to find $p^\star$, which is almost as simple as a closed-form solution. Its justification can be found in [297]. Here we limit ourselves only to the presentation of the details which are necessary for a functioning implementation.

Without loss of generality, we assume that $q_1 \geq q_2 \geq \cdots \geq q_n$, since this is just a matter of reordering the components of q.

Step 1. Set $s = 0$ and $m = 1$.

Step 2. Set $m^* = m$ and $s^* = s$. Increment m by one and update

$$s \leftarrow s + m^*(q_{m^*} - q_m).$$

Step 3. If $s > 1$, then go to Step 5.

Step 4. If $m = n$ then set $m^* = m$, $s^* = s$ and go to Step 5. Otherwise, go to Step 2.

Step 5. Set

$$\lambda = \frac{1 - s^*}{m^*} - q_{m^*}$$

and the components of the sought projection p^* as follows:

$$p_i^* = \begin{cases} q_i + \lambda & \text{for } 1 \leq i \leq m^*, \\ 0 & \text{otherwise,} \end{cases}$$

STOP.

B.10 Conditional probability and conditional expectation

We shall limit our attention here to discrete random variables that can take finite numbers of values. Let X and Y be two random variables taking on the values (outcomes) $x_1, \ldots, x_m \in \mathbb{R}$ and $y_1, \ldots, y_n \in \mathbb{R}$, respectively. The most important way to characterize X and Y is through the probabilities of the values that they can take, which is captured by the probability mass functions (PMF's) p_X and p_Y:

$$p_X(x_i) = \Pr(\{X = x_i\}), \quad i = 1, \ldots, m, \tag{B.134}$$
$$p_Y(y_j) = \Pr(\{Y = y_j\}), \quad j = 1, \ldots, n. \tag{B.135}$$

The expectations of X and Y are, respectively, defined by

$$E[X] = \sum_{i=1}^{m} x_i p_X(x_i), \quad E[Y] = \sum_{i=1}^{n} y_j p_Y(y_j). \tag{B.136}$$

Assume that X and Y are associated with the same experiment. We can introduce the joint PMF of X and Y as

$$p_{X,Y}(x_i, y_j) = \Pr(\{X = x_i\} \cap \{Y = y_j\}), \quad i = 1, \dots, m, \quad j = 1, \dots, n,$$
(B.137)

and then obtain the PMFs (called the marginal PMFs) using the formulae

$$p_X(x_i) = \sum_{j=1}^{n} p_{X,Y}(x_i, y_j), \quad i = 1, \dots, m,$$
(B.138)

$$p_Y(y_j) = \sum_{i=1}^{n} p_{X,Y}(x_i, y_j), \quad j = 1, \dots, n.$$
(B.139)

If we know that the experimental value of Y is some particular y_{j_0} (clearly, we must have $p_Y(y_{j_0}) > 0$, this provides partial knowledge about the value of X. Account of this fact is taken by introducing the conditional PMF of X given Y:

$$p_{X|Y}(x_i \mid y_j) = \Pr(\{X = x_i\} \mid \{Y = y_j\}) = \frac{p_{X,Y}(x_i, y_j)}{p_Y(y_j)}.$$
(B.140)

Since this is a legitimate PMF, we can define the conditional expectation of X, which is the same as an ordinary expectation, except that all probabilities are replaced by their conditional counterparts:

$$\mathrm{E}[X \mid Y = y_j] = \sum_{i=1}^{m} x_i p_{X|Y}(x_i \mid y_j).$$
(B.141)

Note, however, that the value of the conditional expectation so defined depends on the realized experimental value y_j of Y. Prior to any experiment, it is a function of Y, and therefore a random variable itself. More precisely, we define [23] $\mathrm{E}[X \mid Y]$ to be the random variable whose value is $\mathrm{E}[X \mid Y = y_j]$ when the outcome of Y is y_j.

The conditional expectation possesses a number of useful properties, e.g., it obeys the law of iterated expectations:

$$\mathrm{E}[\mathrm{E}[X \mid Y]] = \mathrm{E}(X).$$
(B.142)

Generalizations of the above to more than two random variables are rather obvious. Similarly, the concept can be extended to continuous random variables. This topic is thoroughly treated, e.g., in [23, 351]. In particular, the following result turns out to be valuable in this monograph:

LEMMA B.1
[208, Prop. V.9, p. 141] *Let $V_1, \dots, V_m$ be independent discrete random vectors, each with the same set of outcomes $\mathcal{X} = \{x^1, \dots, x^\ell\} \subset \mathbb{R}^n$ and the same PMF*

$$p_{V_1}(x^i) = \dots = p_{V_m}(x^i), \quad i = 1, \dots, \ell.$$
(B.143)

Let $h : \mathcal{X}^m \to \mathbb{R}$ be a *nonnegative function. Then*

$$\left\{ \mathrm{E}[h] \right\}^{m+1} \leq \mathrm{E}\big[h \, \mathrm{E}[h \mid V_1] \ldots \mathrm{E}[h \mid V_m] \big] \tag{B.144}$$

and the inequality is strict unless

$$\mathrm{E}[h \mid V_1] = \cdots = \mathrm{E}[h \mid V_m] = \mathrm{E}[h] \tag{B.145}$$

almost surely.

B.11 Some accessory inequalities from statistical learning theory

B.11.1 Hoeffding's inequality

Let $X_1, \ldots, X_n$ be independent random variables such that

$$\mathrm{E}[X_i] = 0, \qquad a_i \leq X_i \leq b_i, \quad i = 1, \ldots, n, \tag{B.146}$$

where the a_i's and b_i's are finite. Then the inequality

$$\Pr\left\{ \sum_{i=1}^{n} X_i \geq \alpha \right\} \leq \exp\!\left(-2\alpha^2 \Big/ \sum_{i=1}^{n} (b_i - a_i)^2 \right), \tag{B.147}$$

called Hoeffding's inequality, is satisfied for any $\alpha \in \mathbb{R}$, cf. [340, p. 24].

In particular, given a probability space $(\Theta, \mathfrak{F}, \varsigma)$, where $\Theta \subset \mathbb{R}^m$, is a sample space, $\mathfrak{F}$ is a σ-algebra of subsets of Θ and $\varsigma : \mathfrak{F} \to [0,1]$ stands for a probability measure, if $f : \Theta \to [0,1]$ is measurable with respect to $\mathfrak{F}$, then

$$\mathrm{E}[f] = \int_{\Theta} f(\theta)\, \varsigma(\mathrm{d}\theta) \tag{B.148}$$

is the expectation of f. Drawing i.i.d. samples $\theta_1, \ldots, \theta_r$ from ς, we define the sample mean

$$\widehat{\mathrm{E}}[f; \widetilde{\theta}] = \frac{1}{r} \sum_{j=1}^{r} f(\theta_j), \tag{B.149}$$

where $\widetilde{\theta} = (\theta_1, \ldots, \theta_r) \in \Theta^r$. Clearly, $f(\theta_1) - \mathrm{E}[f], \ldots, f(\theta_r) - \mathrm{E}[f]$ are all zero-mean random variables with ranges between $-\mathrm{E}[f]$ and $1 - \mathrm{E}[f]$. Applying Hoeffding's inequality, we get the following bounds:

$$\varsigma^r \left\{ \widetilde{\theta} \in \Theta^r : \widehat{\mathrm{E}}[f; \widetilde{\theta}] - \mathrm{E}[f] \geq \epsilon \right\} \leq \exp(-2r\epsilon^2), \tag{B.150}$$

$$\varsigma^r \left\{ \widetilde{\theta} \in \Theta^r : \widehat{\mathrm{E}}[f; \widetilde{\theta}] - \mathrm{E}[f] \leq -\epsilon \right\} \leq \exp(-2r\epsilon^2), \tag{B.151}$$

where ς^r stands for the product probability measure. Moreover, we have

$$\varsigma^r\left\{\tilde{\theta} \in \Theta^r : |\hat{E}[f;\tilde{\theta}] - E[f]| \geq \epsilon\right\}$$
$$\leq \varsigma^r\left\{\tilde{\theta} \in \Theta^r : \hat{E}[f;\tilde{\theta}] - E[f] \leq -\epsilon\right\} + \varsigma^r\left\{\tilde{\theta} \in \Theta^r : \hat{E}[f;\tilde{\theta}] - E[f] \geq \epsilon\right\}$$
$$\leq 2\exp(-2r\epsilon^2).$$

$$(\text{B.152})$$

B.11.2   Estimating the minima of functions

LEMMA B.2

(Cf. [340, Lem. 11.1, p. 357]) *Given a probability space* $(X, \mathfrak{F}, \nu)$, *where* X, *is a sample space,* $\mathfrak{F}$ *is a σ-algebra of subsets of X and $\nu : \mathfrak{F} \to [0,1]$ stands for a probability measure, suppose that $f : X \to \mathbb{R}$ is measurable with respect to $\mathfrak{F}$, i.e., it is a random variable. If $x_1, \ldots, x_n \in X$ is an i.i.d. sample taken from ν and*

$$\bar{f}(\tilde{x}) = \min_{1 \leq i \leq n} f(x_i),$$

$$(\text{B.153})$$

where $\tilde{x} = (x_1, \ldots, x_n) \in X^n$, *then*

$$\nu^n\left\{\tilde{x} \in X^n : \nu\left\{x \in X : f(x) < \bar{f}(\tilde{x})\right\} > \alpha\right\} \leq (1 - \alpha)^n$$

$$(\text{B.154})$$

for any $\alpha \in [0,1]$. *Here ν^n denotes the product probability defined for the product of n copies of the probability space* $(X, \mathfrak{F}, \nu)$.

C

Statistical properties of estimators

C.1 Best linear unbiased estimators in a stochastic-process setting

Suppose that over the interval $[0, t_f]$ we may observe a realization of a continuous-time stochastic process having the form

$$z(t) = f^\mathsf{T}(t)\theta + \varepsilon(t), \quad t \in [0, t_f], \tag{C.1}$$

where the constant vector $\theta \in \mathbb{R}^m$ is unknown, $f \in L^2(0, t_f; \mathbb{R}^m)$ is a known function and ε is a stochastic process with zero mean and a known covariance kernel κ,

$$\mathrm{E}\big[\varepsilon(t)\big] = 0, \qquad \forall t \in [0, t_f], \tag{C.2}$$

$$\mathrm{E}\big[\varepsilon(t)\varepsilon(s)\big] = \kappa(t, s), \quad \forall t, s \in [0, t_f]. \tag{C.3}$$

It is required that the components of f, i.e., functions $f_1, \ldots, f_m$, be linearly independent in $L^2(0, t_f)$, which constitutes a rather natural condition.

Our basic assumption is that the realizations of the process ε satisfy

$$\mathrm{E}\left[\left(\int_0^{t_f} a(t)\varepsilon(t)\,\mathrm{d}t\right)^2\right] < \infty, \quad \forall a \in L^2(0, t_f) \tag{C.4}$$

with probability 1, which guarantees the existence of the mean and covariance operators being basic stochastic properties needed in what follows.

Given an observed realization $z(\cdot)$ and the model (C.1), we are interested in estimating the 'true' (though unknown) value of θ. To do so, we wish to construct a best linear unbiased estimator (BLUE) $\hat{\theta}$ of θ defined as the estimator that produces the minimum mean-square (MS) error among the class of all linear unbiased estimators. For our current problem, we thus restrict attention to the $\hat{\theta}$'s being linear functions of z, i.e.,

$$\hat{\theta} = \int_0^{t_f} h(t)z(t)\,\mathrm{d}t, \tag{C.5}$$

where $h \in L^2(0, t_f; \mathbb{R}^m)$, which additionally satisfy the unbiasedness condition

$$\mathrm{E}\big[\hat{\theta}\big] = \theta, \quad \forall \theta \in \mathbb{R}^m. \tag{C.6}$$

We wish to find a function h such that, if θ is estimated using (C.5), the resulting MS error

$$e = \mathrm{E}\big[\|\hat{\theta} - \theta\|^2\big] \tag{C.7}$$

is minimum.

We start constructing the BLUE by the following useful characterization of unbiasedness:

LEMMA C.1
The estimator (C.5) is unbiased if and only if

$$\int_0^{t_f} h(t) f^{\mathsf{T}}(t)\,\mathrm{d}t = I, \tag{C.8}$$

where I denotes the identity matrix.

PROOF We have

$$
\begin{aligned}
\mathrm{E}[\hat{\theta}] &= \mathrm{E}\left[\int_0^{t_f} h(t) z(t)\,\mathrm{d}t\right] \\
&= \int_0^{t_f} h(t) f^{\mathsf{T}}(t)\theta\,\mathrm{d}t + \int_0^{t_f} h(t) E\left[\varepsilon(t)\right]\,\mathrm{d}t \\
&= \left\{\int_0^{t_f} h(t) f^{\mathsf{T}}(t)\,\mathrm{d}t\right\}\theta.
\end{aligned}
\tag{C.9}
$$

Hence, from (C.6) we see that there must be

$$\left\{\int_0^{t_f} h(t) f^{\mathsf{T}}(t)\,\mathrm{d}t\right\}\theta = \theta, \quad \forall \theta \in \mathbb{R}^m, \tag{C.10}$$

which implies the desired conclusion. ∎

Define the operator

$$\mathcal{A}: \begin{cases} L^2(0,t_f) \longrightarrow L^2(0,t_f), \\[2mm] v \longmapsto \displaystyle\int_0^{t_f} \kappa(\cdot,s)v(s)\,\mathrm{d}s. \end{cases} \tag{C.11}$$

Note that the evident symmetry $\kappa(t,s) = k(s,t)$ implies that $\mathcal{A}$ is self-adjoint, i.e.,

$$(\mathcal{A}v, w)_{L^2(0,t_f)} = (v, \mathcal{A}w)_{L^2(0,t_f)}, \quad \forall v, w \in L^2(0,t_f). \tag{C.12}$$

Furthermore, we define the self-adjoint operator

$$\mathcal{T}: \begin{cases} L^2(0,t_f;\mathbb{R}^m) \longrightarrow L^2(0,t_f;\mathbb{R}^m), \\[2mm] g \longmapsto (\mathcal{A}g_1,\ldots,\mathcal{A}g_m). \end{cases} \tag{C.13}$$

For any $g \in L^2(0, t_f; \mathbb{R}^m)$ we have

$(g, \mathcal{T}g)_{L^2(0,t_f;\mathbb{R}^m)}$

$$
\begin{aligned}
&= \int_0^{t_f} g^\mathsf{T}(t) [\mathcal{T}g](t) \, dt = \sum_{i=1}^m \int_0^{t_f} g_i(t) [\mathcal{A}g_i](t) \, dt \\
&= \sum_{i=1}^m \int_0^{t_f} \int_0^{t_f} g_i(t) g_i(s) \kappa(t, s) \, dt \, ds \\
&= \sum_{i=1}^m \int_0^{t_f} \int_0^{t_f} g_i(t) g_i(s) \, \mathrm{E}[\varepsilon(t)\varepsilon(s)] \, dt \, ds &&\text{(C.14)} \\
&= \sum_{i=1}^m \mathrm{E} \left[\int_0^{t_f} \int_0^{t_f} g_i(t) g_i(s) \varepsilon(t) \varepsilon(s) \, dt \, ds \right] \\
&= \sum_{i=1}^m \mathrm{E} \left[\left\{ \int_0^{t_f} g_i(t) \varepsilon(t) \, dt \right\}^2 \right] = \mathrm{E} \left[\left\| \int_0^{t_f} g(t) \varepsilon(t) \, dt \right\|^2 \right] \geq 0,
\end{aligned}
$$

i.e., the operator $\mathcal{T}$ is nonnegative. Thus, using (C.5) and Lemma C.1, we get

$$
\begin{aligned}
\mathrm{E}[\|\hat{\theta} - \theta\|^2] &= \mathrm{E} \left[\left\| \int_0^{t_f} h(t) z(t) \, dt - \int_0^{t_f} h(t) f^\mathsf{T}(t) \theta \, dt \right\|^2 \right] \\
&= \mathrm{E} \left[\left\| \int_0^{t_f} h(t) \varepsilon(t) \, dt \right\|^2 \right] = \int_0^{t_f} h^\mathsf{T}(t) [\mathcal{T}h](t) \, dt \geq 0.
\end{aligned}
$$

(C.15)

Also observe that $\varepsilon \in (\ker \mathcal{A})^\perp$, as for any $v \in \ker \mathcal{A}$ we have

$$
\begin{aligned}
&\mathrm{E} \left[(v, \varepsilon)^2_{L^2(0,t_f)} \right] \\
&= \int_0^{t_f} \int_0^{t_f} v(t) v(s) \, \mathrm{E}[\varepsilon(t)\varepsilon(s)] \, dt \, ds &&\text{(C.16)} \\
&= \int_0^{t_f} \int_0^{t_f} v(t) v(s) \kappa(t, s) \, dt \, ds = \int_0^{t_f} v(t) \underbrace{[\mathcal{A}v](t)}_{=0} \, dt = 0.
\end{aligned}
$$

This justifies the following assumption:

$$
\{f_i\}_{i=1}^m \subset \operatorname{ran} \mathcal{A} \tag{C.17}
$$

since it amounts to the impossibility of getting precise information about θ based on the adopted form of the observation equation. Note that this is not particularly restrictive, as $\operatorname{ran} \mathcal{A} \subset \overline{\operatorname{ran} \mathcal{A}} = (\ker \mathcal{A})^\perp$, i.e., $\operatorname{ran} \mathcal{A}$ is dense in $(\ker \mathcal{A})^\perp$.

In general, $\mathcal{A}$ is not invertible. However, as $L^2(0, t_f)$ can be decomposed as the direct sum $\overline{\ker \mathcal{A}} \oplus (\ker \mathcal{A})^\perp$, for any $w \in \operatorname{ran} \mathcal{A}$ we can define $\mathcal{A}^{-1}w$ as

v being the orthogonal projection of w onto $(\ker \mathcal{A})^{\perp} = \overline{\operatorname{ran} \mathcal{A}}$. Clearly, v is unique, which results from the properties of direct sums, cf. [163, p. 53], and $\mathcal{A}v = w$.

The solution to the problem of finding the BLUE is given by the following result:

THEOREM C.1
Let $\varphi = (\mathcal{A}^{-1}f_1, \ldots, \mathcal{A}^{-1}f_m)$, i.e.,

$$\int_0^{t_f} \kappa(t, s)\varphi(s) \, ds = f(t), \tag{C.18}$$

and

$$M = \int_0^{t_f} f(t)\varphi^{\mathsf{T}}(t) \, dt. \tag{C.19}$$

Then

$$\hat{\theta} = M^{-1} \int_0^{t_f} \varphi(t)z(t) \, dt \tag{C.20}$$

defines the BLUE. In other words, in (C.5) we have to set $h(t) = M^{-1}\varphi(t)$.

PROOF   Before proceeding with the proof, note that the matrix M is symmetric. What is more, since $\mathcal{T}$ is self-adjoint, it follows easily that

$$\int_0^{t_f} \{g(t) - M^{-1}\varphi(t)\}^{\mathsf{T}}[\mathcal{T}(g - M^{-1}\varphi)](t) \, dt$$

$$= \int_0^{t_f} g^{\mathsf{T}}(t)[\mathcal{T}g](t) \, dt - \operatorname{tr}(M^{-1}), \quad \forall g \in L^2(0, t_f; \mathbb{R}^m). \tag{C.21}$$

This implies that for any unbiased estimator (C.5) we must have

$$\mathrm{E}\left[\|\hat{\theta} - \theta\|^2\right]$$

$$= \int_0^{t_f} h^{\mathsf{T}}(t)[\mathcal{T}h](t) \, dt$$

$$= \operatorname{tr}(M^{-1}) + \underbrace{\int_0^{t_f} \{h(t) - M^{-1}\varphi(t)\}^{\mathsf{T}}[\mathcal{T}(h - M^{-1}\varphi)](t) \, dt}_{\geq 0, \text{ cf. (C.14)}} \tag{C.22}$$

$$\geq \operatorname{tr}(M^{-1}),$$

so we must set $h_*(t) = M^{-1}\varphi(t)$.

Now, let us show the unicity of this solution. To this end, suppose that we have another estimator

$$\tilde{\theta} = \int_0^{t_f} h_\diamond(t) z(t) \, dt \tag{C.23}$$

such that

$$\int_0^{t_f} h_\star^{\mathsf{T}}(t) [\mathcal{T} h_\star](t) \, dt = \int_0^{t_f} h_\diamond^{\mathsf{T}}(t) [\mathcal{T} h_\diamond](t) \, dt = \operatorname{tr} M^{-1}. \tag{C.24}$$

Then

$$
\begin{aligned}
\mathrm{E}\big[\|\tilde{\theta} - \hat{\theta}\|^2\big] &= \mathrm{E}\left[\left\| \int_0^{t_f} \big[h_\diamond(t) - h_\star(t)\big] z(t) \, dt \right\|^2\right] \\
&= \left\| \int_0^{t_f} \big[h_\diamond(t) - h_\star(t)\big] f^{\mathsf{T}}(t) \theta \, dt \right\|^2 \\
&\quad + \mathrm{E}\left[\left\| \int_0^{t_f} \big[h_\diamond(t) - h_\star(t)\big] \varepsilon(t) \, dt \right\|^2\right] \\
&= \mathrm{E}\left[\left\| \int_0^{t_f} \big[h_\diamond(t) - h_\star(t)\big] \varepsilon(t) \, dt \right\|^2\right],
\end{aligned}
\tag{C.25}
$$

the last equality resulting from (C.1).
From (C.14) it may be further concluded that

$$
\begin{aligned}
\mathrm{E}&\left[\left\| \int_0^{t_f} \big[h_\diamond(t) - h_\star(t)\big] \varepsilon(t) \, dt \right\|^2\right] \\
&= \int_0^{t_f} h_\diamond^{\mathsf{T}}(t) [\mathcal{T} h_\diamond](t) \, dt + \int_0^{t_f} h_\star^{\mathsf{T}}(t) [\mathcal{T} h_\star](t) \, dt \\
&\quad - 2 \int_0^{t_f} h_\diamond^{\mathsf{T}}(t) [\mathcal{T} h_\star](t) \, dt \\
&= 2 \operatorname{tr}(M^{-1}) - 2 \int_0^{t_f} h_\diamond^{\mathsf{T}}(t) M^{-1} f(t) \, dt \\
&= 2 \operatorname{tr}(M^{-1}) - 2 \operatorname{tr}\left\{ M^{-1} \int_0^{t_f} f(t) h_\diamond^{\mathsf{T}}(t) \, dt \right\} \\
&= 2 \operatorname{tr}(M^{-1}) - 2 \operatorname{tr}(M^{-1}) = 0.
\end{aligned}
\tag{C.26}
$$

Consequently, $\mathrm{E}\big[\|\tilde{\theta} - \hat{\theta}\|^2\big] = 0$, which means that $\tilde{\theta} = \hat{\theta}$ almost surely. This completes the proof. ∎

In the analogy of Section 3.1, the symmetric matrix M in Theorem C.1 is called the information matrix as it serves to quantify statistical accuracy in the estimates. We can now consider various ways in which we might wish to make M large and thereby, appropriately altering its elements by modifying the functions $f_1, \ldots, f_m$ we can influence the precision of the solutions to our problem.

Unfortunately, a major complication is caused by the necessity of solving the system of Fredholm equations of the first kind (C.18), which constitutes a problem with many possible pitfalls. For example, such equations are often extremely ill-conditioned. Applying the kernel $\kappa(\cdot, \cdot)$ to a function φ is generally a smoothing operation, so the solution of (C.18), which requires inverting the operator, will be extremely sensitive to small changes in f. Smoothing often actually loses information, and there is no way to get it back in an inverse operation. Specialized methods have been developed for such equations, which make use of some prior knowledge of the nature of the solution to restore lost information [4, 118, 218, 291]. Owing to these difficulties, the framework of correlated continuous-time observations has been considered in the optimum experimental design literature rather sporadically [31, 44, 181–183, 189, 204, 209, 258–260, 277, 344, 345].

C.2 Best linear unbiased estimators in a partially uncorrelated framework

Consider the continuous-time system modelled as follows:

$$z(t) = F^{\mathsf{T}}(t; \zeta)\theta + \varepsilon(t; \zeta), \quad t \in [0, t_f], \tag{C.27}$$

where t denotes time and t_f is a fixed finite time horizon, $\zeta \in \Delta$ is a fixed parameter, Δ being a given compact subset of $\mathbb{R}^n$, $F(\cdot; \zeta) : [0, t_f] \to \mathbb{R}^{m \times N}$ denotes a continuous function and $\theta \in \mathbb{R}^m$ stands for an unknown parameter to be recovered based on the observations $z(t) \in \mathbb{R}^N$, $t \in [0, t_f]$.

The measurement process z has associated with it the zero-mean disturbance process ε being white in time [6, 25, 203, 205]:

$$\mathrm{E}\big[\varepsilon(t; \zeta)\big] = 0, \tag{C.28}$$
$$\mathrm{E}\big[\varepsilon(t; \zeta)\varepsilon^{\mathsf{T}}(s; \zeta)\big] = C(t; \zeta)\delta(t - s), \tag{C.29}$$

where $C(t; \zeta) \in \mathrm{PD}(N)$ is a covariance matrix and δ signifies the Dirac delta function.

In what follows, we shall construct a BLUE for θ, as its form is frequently used in this monograph. Note, however, that no attempt at rigour is made here. We omit some technicalities (which are of considerable importance, but not in this context) and proceed formally. A theoretically oriented reader interested in the assumptions required for a proper mathematical formulation is referred to specialized literature on stochastic processes [6, 32, 205, 339]. Finally, for notational simplicity, we ignore the dependence of F, ε and C on the parameter ζ, i.e., we simply write $F(t)$, $\varepsilon(t)$ and $C(t)$, respectively.

As usual, we require the estimator $\hat{\theta}$ of θ to be a linear function of the observations:

$$\hat{\theta} = \int_0^{t_f} H(t)z(t)\,dt, \tag{C.30}$$

where $H : [0, t_f] \to \mathbb{R}^{m \times N}$. Just as in Appendix C.1, the BLUE defined here as the estimator that guarantees the minimum mean-square (MS) error $\mathrm{E}\left[\|\hat{\theta} - \theta\|^2\right]$ among the class of all linear unbiased estimators.

Note that for any linear estimator of the form (C.30) we have

$$
\begin{aligned}
\mathrm{E}&\left[\left\|\int_0^{t_f} H(t)\varepsilon(t)\,dt\right\|^2\right] \\
&= \mathrm{E}\left[\int_0^{t_f}\int_0^{t_f} \varepsilon^{\mathsf{T}}(t)H^{\mathsf{T}}(t)H(s)\varepsilon(s)\,dt\,ds\right] \\
&= \mathrm{E}\left[\int_0^{t_f}\int_0^{t_f} \mathrm{tr}\left\{H(s)\varepsilon(s)\varepsilon^{\mathsf{T}}(t)H^{\mathsf{T}}(t)\right\}\,dt\,ds\right] \\
&= \int_0^{t_f}\int_0^{t_f} \mathrm{tr}\left\{H(s)\,\mathrm{E}\left[\varepsilon(s)\varepsilon^{\mathsf{T}}(t)\right]H^{\mathsf{T}}(t)\right\}\,dt\,ds \\
&= \int_0^{t_f} \mathrm{tr}\left[H(t)C(t)H^{\mathsf{T}}(t)\right]dt \\
&= \mathrm{tr}\left[\int_0^{t_f} H(t)C(t)H^{\mathsf{T}}(t)\,dt\right].
\end{aligned}
\tag{C.31}
$$

The unbiasedness means that

$$\mathrm{E}\left[\hat{\theta}\right] = \theta, \quad \forall \theta \in \mathbb{R}^m. \tag{C.32}$$

Expanding the left-hand side, we obtain

$$
\begin{aligned}
\mathrm{E}\left[\hat{\theta}\right] &= \mathrm{E}\left[\int_0^{t_f} H(t)z(t)\,dt\right] \\
&= \int_0^{t_f} H(t)F^{\mathsf{T}}(t)\theta\,dt + \mathrm{E}\left[\int_0^{t_f} H(t)\varepsilon(t)\,dt\right] \\
&= \left\{\int_0^{t_f} H(t)F^{\mathsf{T}}(t)\,dt\right\}\theta,
\end{aligned}
\tag{C.33}
$$

which, taken in conjunction with (C.32), gives the following necessary and sufficient condition for H to assure the unbiasedness:

$$\int_0^{t_f} H(t)F^{\mathsf{T}}(t)\,dt = I. \tag{C.34}$$

Our objective is to minimize the MS error which simplifies to the following form:

$$E[\|\hat{\theta} - \theta\|^2]$$

$$= E\left[\left\|\underbrace{\int_0^{t_f} H(t)z(t)\,dt}_{\text{c.f. (C.30)}} - \underbrace{\int_0^{t_f} H(t)F^{\mathsf{T}}(t)\,dt\,\theta}_{=I}\right\|^2\right] \tag{C.35}$$

$$= E\left[\left\|\int_0^{t_f} H(t)\varepsilon(t)\,dt\right\|^2\right]$$

$$= \operatorname{tr}\left\{\int_0^{t_f} H(t)C(t)H^{\mathsf{T}}(t)\,dt\right\}.$$

Consequently, to find a BLUE means to discover a matrix-valued function H for which the criterion

$$J(H) = \operatorname{tr}\left\{\int_0^{t_f} H(t)C(t)H^{\mathsf{T}}(t)\,dt\right\} \tag{C.36}$$

is minimum subject to the condition (C.34).

Define the symmetric matrix

$$M = \int_0^{t_f} F(t)C^{-1}(t)F^{\mathsf{T}}(t)\,dt \tag{C.37}$$

and assume that it is nonsingular. It is easy to check that

$$\operatorname{tr}\left\{\int_0^{t_f} \left[H(t) - M^{-1}F(t)C^{-1}(t)\right]C(t)\left[H(t) - M^{-1}F(t)C^{-1}(t)\right]^{\mathsf{T}}\,dt\right\}$$

$$= \operatorname{tr}\left\{\int_0^{t_f} H(t)C(t)H^{\mathsf{T}}(t)\,dt\right\} - \operatorname{tr}\left[M^{-1}\right] \tag{C.38}$$

(it suffices to expand the left-hand side). Thus we can rewrite (C.36) as follows:

$$J(H) = \operatorname{tr}\left[M^{-1}\right] + \operatorname{tr}\left\{\int_0^{t_f} G(t)C(t)G^{\mathsf{T}}(t)\,dt\right\}$$

$$= \operatorname{tr}\left[M^{-1}\right] + \int_0^{t_f} \operatorname{tr}\left\{C(t)G^{\mathsf{T}}(t)G(t)\right\}\,dt, \tag{C.39}$$

where

$$G(t) = H(t) - M^{-1}F(t)C^{-1}(t). \tag{C.40}$$

Owing to the positive definiteness of $C(t)$, from Theorem B.9 we have

$$\operatorname{tr}\left\{C(t)G^{\mathsf{T}}(t)G(t)\right\} > 0 \tag{C.41}$$

whenever $G^\mathsf{T}(t)G(t) \neq 0$ since $G^\mathsf{T}(t)G(t) \succeq 0$.

The only way of making the left-hand side of (C.41) zero, thereby letting $J(H)$ attain its lowest possible value equal to $\mathrm{tr}\big[M^{-1}\big]$, is to select H so as to yield $G^\mathsf{T}(t)G(t) = 0$. Observe, however, that this can take place only if $G(t) = 0$, as $\|G(t)\| = \big(\mathrm{tr}\big[G^\mathsf{T}(t)G(t)\big]\big)^{\frac{1}{2}} = 0$. This implies that the optimal choice of H in (C.30) is

$$H_\star(t) = M^{-1}F(t)C^{-1}(t). \tag{C.42}$$

Therefore, the BLUE has the form

$$\hat{\theta} = M^{-1}\int_0^{t_f} F(t)C^{-1}(t)z(t)\,\mathrm{d}t. \tag{C.43}$$

Since

$$\hat{\theta} - \theta = M^{-1}\int_0^{t_f} F(t)C^{-1}(t)z(t)\,\mathrm{d}t - M^{-1}M\theta$$
$$= M^{-1}\int_0^{t_f} F(t)C^{-1}(t)\varepsilon(t)\,\mathrm{d}t, \tag{C.44}$$

we see that the covariance matrix of $\hat{\theta}$ is given by

$$\mathrm{cov}\{\hat{\theta}\} = \mathrm{E}\Big[(\hat{\theta}-\theta)(\hat{\theta}-\theta)^\mathsf{T}\Big]$$
$$= M^{-1}\int_0^{t_f}\int_0^{t_f} F(t)C^{-1}(t)\,\mathrm{E}\{\varepsilon(t)\varepsilon^\mathsf{T}(s)\}C^{-1}(s)F^\mathsf{T}(s)\,\mathrm{d}t\,\mathrm{d}s\,M^{-1}$$
$$= M^{-1}MM^{-1} = M^{-1}. \tag{C.45}$$

In order to prove the unicity of the estimator, suppose that there exists another estimator

$$\hat{\theta} = \int_0^{t_f} H_\circ(t)z(t)\,\mathrm{d}t, \tag{C.46}$$

for which

$$J(H_\circ) = \mathrm{tr}\left\{\int_0^{t_f} H_\circ(t)C(t)H_\circ^\mathsf{T}(t)\,\mathrm{d}t\right\} = \mathrm{tr}\big[M^{-1}\big]. \tag{C.47}$$

Then we have

$$\mathrm{E}\big[\|\tilde{\theta}-\hat{\theta}\|^2\big]$$
$$= \mathrm{E}\Big[\big\|\int_0^{t_f}[H_\circ(t)-H_\star(t)]z(t)\,\mathrm{d}t\big\|^2\Big]$$
$$= \big\|\int_0^{t_f}[H_\circ(t)-H_\star(t)]F^\mathsf{T}(t)\theta\,\mathrm{d}t\big\|^2 + \mathrm{E}\Big[\big\|\int_0^{t_f}[H_\circ(t)-H_\star(t)]\varepsilon(t)\,\mathrm{d}t\big\|^2\Big]$$
$$= \mathrm{E}\Big[\big\|\int_0^{t_f}[H_\circ(t)-H_\star(t)]\varepsilon(t)\,\mathrm{d}t\big\|^2\Big], \tag{C.48}$$

the last inequality resulting from the fact that both $\hat{\theta}$ and $\tilde{\theta}$ obey (C.34). Furthermore, from (C.31) we have

$$
\mathrm{E}\left[\left\|\int_0^{t_f} \left[H_\diamond(t) - H_\star(t)\right]\varepsilon(t)\,\mathrm{d}t\right\|^2\right]
$$

$$
= \mathrm{tr}\left\{\int_0^{t_f} \left[H_\diamond(t) - H_\star(t)\right]C(t)\left[H_\diamond(t) - H_\star(t)\right]^\mathsf{T}\mathrm{d}t\right\}
$$

$$
= \mathrm{tr}\left[\int_0^{t_f} H_\diamond(t)C(t)H_\diamond^\mathsf{T}(t)\,\mathrm{d}t\right] + \mathrm{tr}\left[\int_0^{t_f} H_\star(t)C(t)H_\star^\mathsf{T}(t)\,\mathrm{d}t\right]
$$

$$
- 2\,\mathrm{tr}\left[\int_0^{t_f} H_\diamond(t)C(t)H_\star^\mathsf{T}(t)\,\mathrm{d}t\right]
$$

$$
= 2\,\mathrm{tr}\left[M^{-1}\right] - 2\,\mathrm{tr}\left[\int_0^{t_f} H_\diamond(t)F^\mathsf{T}(t)\,\mathrm{d}t\,M^{-1}\right]
$$

$$
= 2\,\mathrm{tr}\left[M^{-1}\right] - 2\,\mathrm{tr}\left[M^{-1}\right] = 0.
$$

$$
(\mathrm{C}.49)
$$

Consequently, we get $\mathrm{E}\left[\|\tilde{\theta} - \hat{\theta}\|^2\right] = 0$, which implies $\tilde{\theta} = \hat{\theta}$ almost surely.

D

Analysis of the largest eigenvalue

D.1 Directional differentiability

Assume that a matrix $A(q) \in \text{Sym}(m)$ is continuously differentiable with respect to $q \in \mathbb{R}^n$. The maximum eigenvalue function is often defined as the first component of the eigenvalue function $\lambda : \text{Sym}(m) \to \mathbb{R}^m$, where for any $B \in \text{Sym}(m)$, $\lambda(B)$ is the vector of the eigenvalues of B in nonincreasing order, i.e., $\lambda_1(B) \geq \lambda_2(B) \geq \cdots \geq \lambda_m(B)$. We are interested here in the question of a differentiability of the composite mapping $q \mapsto J(q) = \lambda_1(A(q))$ as it arises in the context of E-optimum experimental designs.

The problem is not as easy as it may seem at first sight and there is no close analogy, e.g., to differentiation of the determinant of A. As an example, consider the matrix

$$A(q) = \begin{bmatrix} 1 + q_1 & q_2 \\ q_2 & 1 - q_1 \end{bmatrix}. \tag{D.1}$$

Its eigenvalues are

$$\lambda_1 = 1 + \sqrt{q_1^2 + q_2^2}, \quad \lambda_2 = 1 - \sqrt{q_1^2 + q_2^2}, \tag{D.2}$$

so that the largest eigenvalue λ_1 is minimized by $q = 0$ where $\lambda_1 = \lambda_2 = 1$, i.e., the eigenvalues coalesce. $J(\cdot)$ is not a smooth function at $q = 0$, cf. Fig. D.1, and only directional derivatives exist, as we have

$$\delta_+ J(0; h) = \sqrt{h_1^2 + h_2^2}, \quad \forall h \in \mathbb{R}^2. \tag{D.3}$$

Note, however, that these directional derivatives are not linear in h.

More generally, the largest eigenvalue cannot be written as the pointwise maximum of two smooth functions at $q = 0$ and standard minimax optimization techniques cannot be applied. This makes the eigenvalue optimization problem both particularly interesting and particularly difficult. In what follows, we analyse this topic more thoroughly.

D.1.1 Case of the single largest eigenvalue

Consider first the special case where the largest eigenvalue λ_1 is simple (not repeated) for some q^*. The aforementioned difficulties with differentiation of

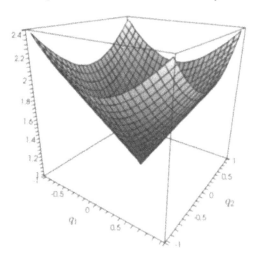

FIGURE D.1
3–D plot of the largest eigenvalue of the matrix (D.1). Note the spike at the origin, which indicates the nondifferentiability for this point.

λ_1 are not experienced then, as it can be shown [110, Th. 1.3.1, p. 50] that both the eigenvalue λ_1 and the associated eigenvector v_1 are continuously differentiable with respect to q.

As for concrete formulae, we can argue as follows: By definition, λ_1 and v_1 treated as functions of q satisfy

$$A(q^\star)v_1(q^\star) = \lambda_1(q^\star)v_1(q^\star), \tag{D.4}$$

where the normalization via

$$v_1^{\mathsf{T}}(q^\star)v_1(q^\star) = 1 \tag{D.5}$$

is imposed. Note that by abuse of notation, we briefly write $\lambda_1(q)$ and $v_1(q)$ instead of $\lambda_1[A(q)]$ and $v_1[A(q)]$. This convenient convention will not lead to ambiguity, as the distinction in the interpretation should be clear from the context.

Let $w \in \mathbb{R}^m$ be an arbitrary vector. From (D.4) it follows that

$$w^{\mathsf{T}}A(q^\star)v_1(q^\star) = \lambda_1(q^\star)w^{\mathsf{T}}v_1(q^\star). \tag{D.6}$$

Now consider the perturbations in $\lambda_1(q)$ and $v_1(q)$ due to the perturbation in $q^\star$ of the form $q^\star + \alpha\delta q$ for $\alpha \in \mathbb{R}$ and a fixed vector $\delta q \in \mathbb{R}^n$. Clearly, we must have

$$w^{\mathsf{T}}A(q^\star + \alpha\delta q)v_1(q^\star + \alpha\delta q) = \lambda_1(q^\star + \alpha\delta q)w^{\mathsf{T}}v_1(q^\star + \alpha\delta q). \tag{D.7}$$

Differentiation of both the sides with respect to α at $\alpha = 0$ gives

$$
\frac{\partial}{\partial q}\left\{w^{\mathsf{T}}A(q)v_1(q^\star)\right\}\bigg|_{q=q^\star} \delta q + w^{\mathsf{T}}A(q^\star)\frac{\partial v_1(q)}{\partial q}\bigg|_{q=q^\star} \delta q
$$
$$
= \frac{\partial \lambda_1(q)}{\partial q}\bigg|_{q=q^\star} \delta q\, w^{\mathsf{T}}v_1(q^\star) + \lambda_1(q^\star)w^{\mathsf{T}}\frac{\partial v_1(q)}{\partial q}\bigg|_{q=q^\star} \delta q. \quad \text{(D.8)}
$$

Substituting now $w = v_1(q^\star)$ and making use of (D.4) and (D.5), we conclude that

$$
\frac{\partial \lambda_1(q)}{\partial q}\bigg|_{q=q^\star} = \frac{\partial}{\partial q}\left\{v_1^{\mathsf{T}}(q^\star)A(q)v_1(q^\star)\right\}\bigg|_{q=q^\star}. \quad \text{(D.9)}
$$

In particular, it is easily seen that

$$
\frac{\partial \lambda_1(A)}{\partial A} = \frac{\partial(v_1^{\mathsf{T}}Av_1)}{\partial A} = v_1 v_1^{\mathsf{T}}. \quad \text{(D.10)}
$$

Note that the same properties and characterizations can be derived for any eigenvalue, not only the largest one. For example, if the smallest eigenvalue λ_m is simple, then

$$
\frac{\partial \lambda_m(q)}{\partial q}\bigg|_{q=q^\star} = \frac{\partial}{\partial q}\left\{v_m^{\mathsf{T}}(q^\star)A(q)v_m(q^\star)\right\}\bigg|_{q=q^\star}. \quad \text{(D.11)}
$$

Consequently, differentiation of simple eigenvalues presents no difficult problems. What is more, a straightforward iterative method of computing one selected eigenvalue of interest along with the corresponding eigenvector can be employed. It is called the *power method* [4, 150, 230], and its version for determining approximations to λ_1 and v_1 starts from any $0 \neq w^{(0)} \in \mathbb{R}^m$. Then a sequence of vectors $w^{(k)}$ and a sequence of associated numbers $\lambda_1^{(k)}$ are defined by

$$
w^{(k+1)} = Aw^{(k)}, \quad \lambda_1^{(k)} = \frac{\left[w^{(k)}\right]^{\mathsf{T}}w^{(k+1)}}{\left[w^{(k)}\right]^{\mathsf{T}}w^{(k)}}. \quad \text{(D.12)}
$$

Provided that $v_1^{\mathsf{T}}w^{(0)} \neq 0$, we obtain [4, 150]

$$
\lim_{k\to\infty} \lambda_1^{(k)} = \lambda_1 \quad \text{(D.13)}
$$

and

$$
\lim_{k\to\infty} \frac{w^{(k)}}{\|w^{(k)}\|} = \pm v_1 \quad \text{or} \quad \lim_{k\to\infty} (-1)^k \frac{w^{(k)}}{\|w^{(k)}\|} = \pm v_1. \quad \text{(D.14)}
$$

D.1.2 Case of the repeated largest eigenvalue

For a long time, the occurence of repeated eigenvalues has been neglected on practical grounds since many practitioners claim that a precisely repeated eigenvalue is an extremely unlikely accident. Although multiple eigenvalues may indeed be unlikely for randomly specified matrices, they become by far more likely when eigenvalue optimization problems are considered. As inicated by many authors [156–158, 201], minimizing the largest eigenvalue will potentially lead to the coalescence of eigenvalues, which means that λ_1 will have mutiplicity greater than one. The same applies to the E-optimum design problems considered in this monograph.

As indicated in the example at the beginning of this appendix, the gradient of λ_1 does not exist in general. Nevertheless, even a repeated eigenvalue is directionally differentiable in this case. Indeed, suppose that the maximum eigenvalue $\lambda(q^\star)$ has multiplicity $s \geq 1$, i.e.,

$$\lambda_1(q^\star) = \cdots = \lambda_s(q^\star) > \lambda_{s+1}(q^\star) \geq \cdots \geq \lambda_m(q^\star). \tag{D.15}$$

If we denote by $v_1(q^\star), \ldots, v_s(q^\star)$ the corresponding orthonormal basis of eigenvectors, then the one-sided directional differential $\delta_1(q^\star; \delta q)$ coincides with the largest eigenvalue of the matrix $\mathfrak{T} \in \mathrm{Sym}(s)$ with elements (see, e.g., [110, Th. 1.32, p. 61] or [201, Th. 4]

$$\begin{aligned}
\mathfrak{T}_{ij} &= \frac{\partial}{\partial q}\left\{ v_i^{\mathsf{T}}(q^\star) A(q) v_j(q^\star)\right\}\bigg|_{q=q^\star} \delta q \\
&= \sum_{r=1}^{n} v_i^{\mathsf{T}}(q^\star)\frac{\partial A(q^\star)}{\partial q_r} v_j(q^\star)\delta q_r, \quad i,j = 1,\ldots,s.
\end{aligned} \tag{D.16}$$

Note that $\delta_+\lambda_1(q^\star; \delta q)$ is thus nonlinear in δq. For example, in case $s = 2$ (i.e., the eigenvalue λ_1 is double), its determination requires solving the characteristic equation

$$\det\left(\begin{bmatrix} \mathfrak{T}_{11} - \zeta & \mathfrak{T}_{12} \\ \mathfrak{T}_{21} & \mathfrak{T}_{22} - \zeta \end{bmatrix}\right) = \zeta^2 - (\mathfrak{T}_{11} + \mathfrak{T}_{22})\zeta + \mathfrak{T}_{11}\mathfrak{T}_{22} - \mathfrak{T}_{12}^2 = 0 \tag{D.17}$$

for ζ. We thus obtain a pair of roots and the larger of them provides the one-sided directional differential

$$\begin{aligned}
&\delta_+\lambda_1(q; \delta q) \\
&= \frac{1}{2}\left\{ (\mathfrak{T}_{11} + \mathfrak{T}_{22}) + \left[(\mathfrak{T}_{11} + \mathfrak{T}_{22})^2 - 4\left(\mathfrak{T}_{11}\mathfrak{T}_{22} - \mathfrak{T}_{12}^2\right)\right]^{1/2}\right\}. \tag{D.18}
\end{aligned}$$

A detailed treatment of the discussed topics is contained in specialized literature [156, 157].

D.1.3 Smooth approximation to the largest eigenvalue

It goes without saying that the nondifferentiability of the largest eigenvalue inevitably creates severe difficulties while trying to minimize it, as common nonlinear programming algorithms set strong requirements regarding the smoothness of the cost functions (e.g., they must be continuously differentiable functions). This gives rise to numerous specialized algorithmic developments [158, 201]. Other attempts consist in replacing the original nonsmooth problem by minimization of a smooth function being an approximation to the largest eigenvalue. In the sequel, we shall characterize the latter approach which is much less cumbersome from the conceptual point of view.

At first, observe that the function $\lambda_1 : \mathrm{Sym}(m) \to \mathbb{R}$ can be treated as the composite mapping $f \circ \lambda$, where $\lambda : \mathrm{Sym}(m) \to \mathbb{R}^m$ is the eigenvalue function, i.e.,

$$\lambda(A) = \mathrm{col}\big[\lambda_1(A), \ldots, \lambda_m(A)\big], \quad \forall A \in \mathrm{Sym}(m) \tag{D.19}$$

such that

$$\lambda_1(A) \geq \lambda_2(A) \geq \cdots \geq \lambda_m(A), \tag{D.20}$$

and $f : \mathbb{R}^m \to \mathbb{R}$ is defined as follows:

$$f(x) = \max\{x_1, \ldots, x_m\}, \quad \forall x \in \mathbb{R}^m. \tag{D.21}$$

Consequently, $\lambda_1(A) = (f \circ \lambda)(A) = f(\lambda(A))$ for all $A \in \mathrm{Sym}(m)$. Note that the function f is symmetric, i.e., invariant with respect to coordinate permutations, cf. Appendix B.6.

Theorem B.30 asserts that the differentiability of the composite mapping $f \circ \lambda$ is conditioned only by the smoothness of f and not by that of λ. Difficulties in differentiating $\lambda_1(\cdot)$ are thus experienced due to the nondifferentiability of the function f. Based on this quite general result, Chen *et al.* [50] proposed to approximate f by a smooth symmetric exponential penalty function [215, p. 248]

$$f_\varepsilon(x) = \varepsilon \ln\left(\sum_{i=1}^{m} e^{x_i/\varepsilon}\right), \tag{D.22}$$

where $\varepsilon > 0$ is a parameter.

It is a C^∞ convex function and it is easy to check that it possesses the following uniform approximation property to f [215, p.249]:

$$0 \leq f_\varepsilon(x) - f(x) \leq \varepsilon \ln(m). \tag{D.23}$$

Indeed, we have

$$f_\varepsilon(x) = \varepsilon \ln\left(\sum_{i=1}^{m} e^{x_i/\varepsilon}\right)$$

$$= f(x) - \varepsilon \ln e^{f(x)/\varepsilon} + \varepsilon \ln\left(\sum_{i=1}^{m} e^{x_i/\varepsilon}\right) \tag{D.24}$$

$$= f(x) + \varepsilon \ln\left(\sum_{i=1}^{m} e^{(x_i - f(x))/\varepsilon}\right).$$

But

$$1 \leq \sum_{i=1}^{m} e^{(x_i - f(x))/\varepsilon} \leq m, \tag{D.25}$$

which proves (D.23).

It is easy to show that

$$\nabla f_\varepsilon(x) = \left(\sum_{j=1}^{m} e^{x_j/\varepsilon}\right)^{-1} \operatorname{col}\left[e^{x_1/\varepsilon}, \ldots, e^{x_m/\varepsilon}\right]. \tag{D.26}$$

Clearly, it follows that

$$0 \leq (f_\varepsilon \circ \lambda)(A) - \lambda_1(A) \leq \varepsilon \ln(m), \quad \forall \varepsilon > 0, \quad \forall A \in \operatorname{Sym}(m). \tag{D.27}$$

Defining

$$\Phi_\varepsilon(A) = (f_\varepsilon \circ \lambda)(A), \quad \forall A \in \operatorname{Sym}(m), \tag{D.28}$$

we get

$$\lim_{\varepsilon \downarrow 0} \Phi_\varepsilon(A) = \lambda_1(A), \quad \forall A \in \operatorname{Sym}(m), \tag{D.29}$$

i.e., we have a family of smooth functions which are uniform approximations to the largest eigenvalue function with the accuracy controlled by the smoothing parameter ε.

Consequently, instead of operating on the criterion

$$\Phi(A) = \lambda_1(A), \tag{D.30}$$

we may work with its smoothed version

$$\Phi_\varepsilon(A) = \varepsilon \ln\left(\sum_{i=1}^{m} e^{\lambda_i(A)/\varepsilon}\right) \tag{D.31}$$

for a sufficiently small ε. It follows that

$$\frac{\partial \Psi_\varepsilon(A)}{\partial A} = \sum_{i=1}^{m} \pi_i(A, \varepsilon) v_i(A) v_i^\mathsf{T}(A), \tag{D.32}$$

where

$$\pi_i(A, \varepsilon) = \frac{e^{\lambda_i(A)/\varepsilon}}{\sum\limits_{j=1}^{m} e^{\lambda_j(A)/\varepsilon}} \tag{D.33}$$

and $v_1, \ldots, v_m$ are the eigenvectors corresponding to $\lambda_1, \ldots, \lambda_m$. Observe that for the weighting coefficients we have

$$\pi_i(A, \varepsilon) > 0, \quad i = 1, \ldots, m, \quad \sum_{i=1}^{m} \pi_i(A, \varepsilon) = 1. \tag{D.34}$$

As indicated in [215, p. 249], it turns out that when ε is set in the range of 10^{-4}–10^{-6}, Φ_ε yields and excellent approximation to Φ. However, care must be taken when trying to implement a computer code, since the exponentials may sometimes lead to numerical complications such as, e.g., problems with a highly ill-conditioned Hessian.

E

Differentiation of nonlinear operators

E.1 Gâteaux and Fréchet derivatives

The idea of derivative or differential of a scalar function of a scalar variable can be profitably extended to general mappings. The value of these abstract differentials and derivatives is both practical and theoretical. Practically, the theory allows for first-order approximation or 'linearization' of nonlinear functionals. From a theoretical point of view, differentials and derivatives are frequently used to prove existence results and properties of dependence of state variables on system parameters (see Appendix F.2) [110].

Let X and Y be two Banach spaces and F be an operator which maps an open subset V of X into Y. If given $x_0 \in V$ and $\delta x \in X$ the limit

$$\delta F(x_0; \delta x) = \lim_{\lambda \to 0} \frac{F(x_0 + \lambda \delta x) - F(x_0)}{\lambda} \qquad \text{(E.1)}$$

exists, then F is said to be Gâteaux differentiable and $\delta F(x_0; \delta x)$ is called the *Gâteaux differential* of F at x_0 in the direction δx.

We say that F is Gâteaux differentiable at x_0 if it is Gâteaux differentiable in every direction. If, additionally, the operator $U : X \ni \delta x \mapsto \delta F(x_0; \delta x) \in Y$ is linear and bounded, i.e., $U \in \mathcal{L}(X, Y)$, then U is called the *Gâteaux derivative* of F at x_0 and we write $U = F'(x_0)$. Accordingly,

$$\delta F(x_0; \delta x) = F'(x_0)\delta x. \qquad \text{(E.2)}$$

Next, if the extra requirement that the convergence in (E.1) is uniform for all $\delta x \in X$ from the unit sphere (i.e., $\|\delta x\|_X = 1$), then we call $F'(x_0)$ the *Fréchet derivative* of F at x_0 (in turn, F is said to be Fréchet differentiable at x_0). It is a simple matter to show that the existence of the Fréchet derivative is equivalent to the fulfilment of the conditions

$$F'(x_0) \in \mathcal{L}(X, Y) \qquad \text{(E.3)}$$

and

$$\lim_{\delta x \to 0} \frac{\|F(x_0 + \delta x) - F(x_0) - F'(x_0)\delta x\|_Y}{\|\delta x\|_X} = 0. \qquad \text{(E.4)}$$

297

E.2 Chain rule of differentiation

Let X, Y and Z be Banach spaces. Suppose that F is an operator which maps an open set $V \subset X$ into an open set $W \subset Y$ and G is an operator which maps W into Z. If G is Fréchet differentiable at $y_0 = F(x_0)$, $x_0 \in V$, and F is Gâteaux (resp. Fréchet) differentiable at x_0, then the composite $G \circ F$ is Gâteaux (resp. Fréchet) differentiable at x_0 and we have [126, 173]

$$\left[G \circ F\right]'(x_0) = G'(F(x_0))F'(x_0). \tag{E.5}$$

E.3 Partial derivatives

Let X, Y and Z be Banach spaces. If F is an operator which maps an open subset $V \subset X \times Y$ into Z, then for a fixed $y_0 \in Y$ we may introduce the operator

$$F^{(y_0)} : x \mapsto F(x, y_0) \tag{E.6}$$

which maps $V^{(y_0)} = \{x \in X : (x, y_0) \in V\}$ into Z. Similarly, for a given $x_0 \in X$ we may consider

$$F^{(x_0)} : y \mapsto F(x_0, y) \tag{E.7}$$

which maps $V^{(x_0)} = \{y \in Y : (x_0, y) \in V\}$ into Z.

Obviously, if $x_0 \in X$ is an interior point of $V^{(y_0)}$, then we may speak of the (Gâteaux or Fréchet) derivative $F^{(y_0)\prime}(x_0, y_0) \in \mathcal{L}(X, Z)$ which we call the *partial derivative* of F at (x_0, y_0) and write $F'_x(x_0, y_0)$. Similarly, we introduce the partial derivative with respect to y: $F^{(x_0)\prime}(x_0, y_0) = F'_y(x_0, y_0) \in \mathcal{L}(Y, Z)$.

If at least one of the partial Gâteaux (resp. Fréchet) derivatives is continuous in a neighbourhood of (x_0, y_0), then we have a very useful formula [126, 173]

$$F'(x_0, y_0)(\delta x, \delta y) = F'_x(x_0, y_0)\delta x + F'_y(x_0, y_0)\delta y, \tag{E.8}$$

where F' stands for the Gâteaux (resp. Fréchet) derivative of F. This permits calculations with individual variables and yields the differential of a mapping as the sum of its partial differentials.

E.4 Differentiability of mappings with one-dimensional domains

Let us assume now that Y is a Banach space and consider $U \in \mathcal{L}(\mathbb{R}, Y)$. We see at once that U may be rewritten as

$$U(t) = t y_0, \quad t \in \mathbb{R}, \tag{E.9}$$

where $y_0 = U(1) \in Y$. Moreover, from $\|U(t)\| = |t| \|y_0\|$ we obtain

$$\|U\| = \|y_0\|. \tag{E.10}$$

Conversely, for any $y_0 \in Y$, (E.9) defines an operator $U \in \mathcal{L}(\mathbb{R}, Y)$. Clearly, the correspondence between the elements of Y and $\mathcal{L}(\mathbb{R}, Y)$ is one-to-one, linear and, based on (E.10), it preserves the norm. Hence $\mathcal{L}(\mathbb{R}, Y)$ and Y are isometrically isomorphic and we may thus identify $\mathcal{L}(\mathbb{R}, Y)$ with Y.

Consider now an operator $F : \mathbb{R} \supset \Omega \to Y$. The existence of the Gâteaux derivative $F'(t_0) = y_0 \in Y$ means that for each $t \in \mathbb{R}$

$$\lim_{\lambda \to 0} \frac{F(t_0 + \lambda t) - F(t_0)}{\lambda} = t y_0. \tag{E.11}$$

If we set $\Delta t = \lambda t$, then we may rewrite (E.11) in the form

$$\lim_{\Delta t \to 0} \frac{F(t_0 + \Delta t) - F(t_0)}{\Delta t} = y_0, \tag{E.12}$$

so in the case considered the definition of $F'(t_0)$ is the same as for the derivative of a real function of one real variable. Of course, (E.12) implies that the Gâteaux derivative is also the Fréchet one, i.e., both the notions are equivalent (in general, this is not the case on higher-dimensional spaces, i.e., the Gâteaux derivative may exist, whereas the Fréchet one may not).

E.5 Second derivatives

When $F : X \to Y$ and both X, Y are Banach spaces, we have seen that the Gâteaux (or Fréchet) derivative of F is a bounded linear operator, i.e., $F'(x) \in \mathcal{L}(X, Y)$. But $\mathcal{L}(X, Y)$ itself is a Banach space and hence we may consider the derivative of $F'(\cdot) : X \to \mathcal{L}(X, Y)$ at a point $x_0 \in X$. If it exists, we call it the second derivative of F at x_0 and write $F''(x_0)$. It is clear that

$$F''(x_0) \in \mathcal{L}(X, \mathcal{L}(X, Y)). \tag{E.13}$$

However, it can be shown that $\mathcal{L}(X, \mathcal{L}(X, Y))$ is isometrically isomorphic to $\mathcal{L}(X \times X, Y)$ [58, 126], so that we may think of $F''(x_0)$ as an element of $\mathcal{L}(X \times X, Y)$ and

$$F''(\cdot) : X \to \mathcal{L}(X \times X, Y). \tag{E.14}$$

Interpreting $F''(x_0)$ as a bilinear operator, we obtain the following modified definition of the second Gâteaux derivative:

$$\lim_{\lambda \to 0} \frac{F'(x_0 + \lambda \delta x') - F'(x_0)}{\lambda} = F''(x_0)(\cdot, \delta x'). \tag{E.15}$$

Consequently, for any $\delta x \in X$ we get

$$F''(x_0)(\delta x, \delta x') = \lim_{\lambda \to 0} \frac{F'(x_0 + \lambda \delta x')\delta x - F'(x_0)\delta x}{\lambda}. \tag{E.16}$$

Note that (E.15) and (E.16) are not equivalent, i.e., it may happen that we have a bilinear operator $B \in \mathcal{L}(X \times X, Y)$ such that

$$\lim_{\lambda \to 0} \frac{F'(x_0 + \lambda \delta x')\delta x - F'(x_0)\delta x}{\lambda} = B(\delta x, \delta x'), \quad \forall \, \delta x, \delta x' \in X \tag{E.17}$$

and yet $F''(x_0)$ does not exist. Nevertheless, it is straightforward to show that the necessary and sufficient condition for F to have the second Gâteaux derivative is that the convergence in (E.16) is uniform on the unit sphere (i.e., for $\|\delta x\| = 1$). Moreover, the uniform convergence for both $\|\delta x\| = 1$ and $\|\delta x'\| = 1$ guarantees that B is also the second Fréchet derivative of F.

E.6 Functionals on Hilbert spaces

Given a real Hilbert space V with inner product $\langle \cdot, \cdot \rangle$, for any bounded linear functional $f : V \to \mathbb{R}$, there is a vector $y \in V$ such that

$$f(x) = \langle y, x \rangle, \quad \forall x \in V. \tag{E.18}$$

This result is known as the Riesz representation theorem.

Therefore, it follows that if $J : V \to \mathbb{R}$ possesses either a Fréchet or a Gâteaux derivative at $x_0 \in V$, then there exists a vector $\nabla J(x_0) \in V$ called the *gradient* of J at x_0, such that

$$J'(x_0)\delta x = \langle \nabla J(x_0), \delta x \rangle, \quad \forall \delta x \in V. \tag{E.19}$$

Next, for any continuous bilinear function $g : V \times V \to \mathbb{R}$, there exists a bounded linear operator $G : V \to V$, such that

$$g(x, y) = \langle x, Gy \rangle, \quad \forall x, y \in V. \tag{E.20}$$

Therefore, when $J : V \to \mathbb{R}$ has the second derivative, its second differential at $x_0 \in V$ is given by

$$J'(x_0)(\delta x, \delta x') = \langle \delta x, H(x_0)\delta x' \rangle, \tag{E.21}$$

where $H(x_0) \in \mathcal{L}(V)$ is called the *Hessian* of J at x_0.

E.7 Directional derivatives

At last, we introduce the notion of a directional derivative which may exist even when an operator fails to have a Gâteaux differential. Let X and Y be Banach spaces and suppose that $F : V \to Y$, where V is an open subset of X. We define the *one-sided directional differential* of F at a point $x_0 \in X$ in the direction $\delta x \in X$ by

$$\delta_+ F(x_0; \delta x) = \lim_{\lambda \downarrow 0} \frac{F(x_0 + \lambda \delta x) - F(x_0)}{\lambda} \tag{E.22}$$

if this limit exists. We say that F is directionally differentiable at x_0 if the directional differential exists for all $\delta x \in X$. Clearly, when F is Gâteaux differentiable at x_0, then it is directionally differentiable at x_0 and $\delta_+ F(x_0; \delta x) = \delta F(x_0; \delta x)$ for all $\delta x \in X$.

E.8 Differentiability of max functions

Max functions functions of the form $f(x) = \max_{y \in Y} \varphi(x, y)$ play a central role in optimization-based engineering where the needs for reliable, high-performance, worst-case designs have to be met. In general, such functions are not differentiable everywhere, and a good example of such a situation is given in Appendix D.1 where the largest eigenvalue is considered. However, they do possess directional differentiability properties, which is characterized by the following theorem, cf. [215, Th. 5.4.7, p. 688] or [22, Prop. B.25, p. 717]:

THEOREM E.1
Let $Y \subset \mathbb{R}^m$ be a compact set and let $\varphi : \mathbb{R}^n \times Y \to \mathbb{R}$ be continuous and such that $\partial \varphi / \partial x$ exists and is continuous. Then the one-sided directional derivative of f exists for all $x_0, \delta x \in \mathbb{R}^n$ and is given by

$$\delta_+ f(x_0; \delta x) = \max_{y \in \widehat{Y}(x_0)} \frac{\partial \varphi}{\partial x}(x_0, y)\delta x, \tag{E.23}$$

where

$$\widehat{Y}(x) = \{y \in Y : f(x) = \varphi(x, y)\}. \tag{E.24}$$

Consequently, the directional derivative $\delta_+ f(x_0; \delta x)$ is equal to the largest of the directional derivatives of the functions $x \mapsto \varphi(x, y)$ that are 'active' at x_0, i.e., for which there exists a $y_0 \in Y$ such that $\varphi(x_0, y_0) = f(x_0)$. The set $\widehat{Y}(x_0)$ is sometimes called the *answering set*: the elements y in $\widehat{Y}(x_0)$ 'answer x_0.'

F

Some accessory results for partial-differential equations

F.1   Green formulae

Let Ω be a bounded open domain in $\mathbb{R}^n$ with a Lipschitz boundary Γ. If $u, v \in H^1(\Omega)$, then we have the following integration-by-parts formula [171,200,249]:

$$\int_\Omega \frac{\partial u}{\partial x_i} v \, \mathrm{d}x = -\int_\Omega u \frac{\partial v}{\partial x_i} \, \mathrm{d}x + \int_\Gamma u v \nu_i \, \mathrm{d}\sigma, \quad 1 \le i \le n, \qquad \text{(F.1)}$$

where ν_i's signify the direction cosines of the unit outward normal to Γ which exist a.e. due to the Lipschitz assumption.

Moreover, if $u \in H^2(\Omega)$, then

$$-\int_\Omega (\Delta u) v \, \mathrm{d}x = \sum_{i=1}^n \int_\Omega \frac{\partial u}{\partial x_i} \frac{\partial v}{\partial x_i} \, \mathrm{d}x - \int_\Gamma \frac{\partial u}{\partial \nu} v \, \mathrm{d}\sigma, \qquad \text{(F.2)}$$

where the normal derivative

$$\left. \frac{\partial u}{\partial \nu} \right|_\Gamma = \sum_{i=1}^n \nu_i \frac{\partial u}{\partial x_i} \bigg|_\Gamma$$

is well defined as a function of $L^2(\Gamma)$ since $\nu_i \in L^\infty(\Gamma)$, $1 \le i \le n$.

Now, let us introduce the second-order elliptic differential operator with spatially varying coefficients

$$Au = -\sum_{i,j=1}^n \frac{\partial}{\partial x_i} \left\{ a_{ij} \frac{\partial u}{\partial x_j} \right\} + a_0 u, \qquad \text{(F.3)}$$

where $u \in H^2(\Omega)$, $a_0 \in L^\infty(\Omega)$, $a_{ij} \in C^1(\bar{\Omega})$, $1 \le i, j \le n$. The regularity of a_{ij}'s implies $\sum_{j=1}^n a_{ij} \partial u / \partial x_j \in H^1(\Omega)$, $1 \le i \le n$. Consequently, from (F.1), we deduce that

$$\int_\Omega Au \, v \, \mathrm{d}x + \int_\Gamma \frac{\partial u}{\partial \nu_A} v \, \mathrm{d}\sigma = \int_\Omega \left\{ \sum_{i,j=1}^n a_{ij} \frac{\partial u}{\partial x_j} \frac{\partial v}{\partial x_i} + a_0 uv \right\} \mathrm{d}x, \qquad \text{(F.4)}$$

where

$$\frac{\partial u}{\partial \nu_A} = \sum_{i,j=1}^{n} a_{ij} \frac{\partial u}{\partial x_j} \nu_i$$

stands for the co-normal derivative associated with operator A.

F.2 Differentiability of the solution of a linear parabolic equation with respect to parameters

Let Ω be a bounded domain in $\mathbb{R}^n$ with a Lipschitz boundary Γ and T be the time horizon $]0, t_f[$. In this section we wish to investigate in detail the differentiability with respect to a functional parameter ϑ of the solution to the linear parabolic equation

$$\frac{\partial y}{\partial t}(x,t) - \sum_{i=1}^{n} \frac{\partial}{\partial x_i}\left\{\vartheta(x)\frac{\partial y}{\partial x_i}(x,t)\right\} = f(x,t) \quad \text{in } Q = \Omega \times T \qquad \text{(F.5)}$$

subject to the boundary conditions

$$y(x,t) = 0 \quad \text{on } \Sigma = \Gamma \times T \qquad \text{(F.6)}$$

and the initial condition

$$y(x,0) = y_0(x) \quad \text{in } \Omega. \qquad \text{(F.7)}$$

At first, however, we have to clarify how the notion 'solution' is to be interpreted in our case. Indeed, we might look for the so-called classical solution to (F.5)–(F.7) which is defined to be a function $y = y(x,t)$ such that all the derivatives which appear in (F.5) exist, are continuous and boundary and initial conditions are all satisfied [58,177,251,252]. It turns out, however, that the story is not quite so simple, as in many problems arising naturally in differential equations such requirements regarding regularity are too stringent. This prompts an introduction of generalized definitions of functions, derivatives, convergence, integrals, etc., and leads to the so-called variational formulation of problems described by PDEs with corresponding *weak* solutions.

In the case considered here, let us assume that $\vartheta \in C^1(\bar{\Omega})$, $f \in C(\bar{\Omega})$, and take a classical solution $y \in C^{2,1}(\bar{Q})$. We choose some $\psi \in H_0^1(\Omega)$ (called a *test* or *trial* function) arbitrarily, multiply (F.5) by it and integrate the result on Ω. We thus get

$$\int_\Omega \frac{\partial y}{\partial t}(x,t)\psi(x)\,dx - \int_\Omega \sum_{i=1}^{n} \frac{\partial}{\partial x_i}\left\{\vartheta(x)\frac{\partial y}{\partial x_i}(x,t)\right\}\psi(x)\,dx$$

$$= \int_\Omega f(x,t)\psi(x)\,dx. \qquad \text{(F.8)}$$

Making use of the Green formula (F.4) and noticing that

$$\int_\Omega \frac{\partial y}{\partial t}(x,t)\psi(x)\,dx = \frac{d}{dt}\int_\Omega y(x,t)\psi(x)\,dx,$$

we conclude that

$$\frac{d}{dt}\int_\Omega y(x,t)\psi(x)\,dx + \sum_{i=1}^n \int_\Omega \vartheta(x)\frac{\partial y}{\partial x_i}(x,t)\frac{\partial\psi}{\partial x_i}(x)\,dx$$
$$= \int_\Omega f(x,t)\psi(x)\,dx, \quad \forall\psi \in H_0^1(\Omega), \quad \text{(F.9)}$$

which is called the *weak* or *variational* formulation of the problem.

Obviously, (F.9) makes sense whenever weaker assumptions on data are made, i.e., $\vartheta \in L^\infty(\Omega)$ and $f \in L^2(Q)$, and the derivatives are interpreted in the distributional sense. The classical solution need not exist under these more general hypotheses, but when it exists, it coincides with the variational one. A proper mathematical statement of the weak formulation of (F.5)–(F.7) requires, however, rather profound knowledge of vector-valued distributions, generalized derivatives and Sobolev spaces [63,160,200,252], so the reader who is not familiar with those notions or is not interested in such technicalities may skip what follows and proceed directly with Theorem F.1 which is the main result of this section.

From now on, we make the folowing assumption regarding the set of admissible parameters Θ_{ad}:

$$\Theta_{ad} = \left\{\vartheta \in L^\infty(\Omega) : \exists\alpha > 0,\ \vartheta(x) \geq \alpha > 0 \text{ a.e. in } \Omega\right\}. \quad \text{(F.10)}$$

Furthermore, we introduce the notation $V = H_0^1(\Omega)$, $H = L^2(\Omega)$,

$$\begin{cases} ((\cdot,\cdot)) \text{ the scalar product, } \|\cdot\| \text{ the norm in } V, \\ (\cdot,\cdot) \text{ the scalar product, } |\cdot| \text{ the norm in } H. \end{cases}$$

We select H as the pivot space and therefore identify H with its dual H'. Denoting by V' the dual of V with norm $\|\cdot\|_*$ (in our case $V' = H^{-1}(\Omega)$), we get

$$V \hookrightarrow H \hookrightarrow V' \quad \text{(F.11)}$$

where $\hookrightarrow$ stands for a continuous and dense injection. The pairing between V' and V is denoted by $\langle w,v\rangle$ for $w \in V'$ and $v \in V$, and it coincides with the scalar product (w,v) if $w,v \in H$.

Additionally, we define a_ϑ to be the bounded bilinear form

$$a_\vartheta(u,v) = \sum_{i=1}^n \int_\Omega \vartheta\frac{\partial u}{\partial x_i}\frac{\partial v}{\partial x_i}\,dx \quad \text{on } V \times V. \quad \text{(F.12)}$$

Let us note that a_ϑ defines a bounded linear operator $A_\vartheta : V \to V'$ according to the formula

$$\langle A_\vartheta u, v \rangle = a_\vartheta(u, v), \quad \forall u, v \in V. \tag{F.13}$$

We have

$$
\begin{aligned}
\|A_\vartheta\|_{\mathcal{L}(V,V')} &= \sup_{\|u\|=1} \|A_\vartheta u\|_\star = \sup_{\|u\|=1} \sup_{\|v\|=1} |\langle A_\vartheta u, v \rangle| \\
&= \sup_{\|u\|=1} \sup_{\|v\|=1} |a_\vartheta(u, v)| \le \|\vartheta\|_{L^\infty(\Omega)}.
\end{aligned}
\tag{F.14}
$$

Finally, we introduce the space

$$W(T) = \left\{ u \in L^2(T; V) : \frac{du}{dt} \in L^2(T; V') \right\}, \tag{F.15}$$

where the derivative is understood in the vector-valued distributional sense. It is a Hilbert space when equipped with the norm

$$
\begin{aligned}
\|u\|_{W(T)} &= \left(\|u\|_{L^2(T;V)}^2 + \|\frac{du}{dt}\|_{L^2(T;V')}^2 \right)^{1/2} \\
&= \left(\int_0^{t_f} \left[\|u(t)\|^2 + \|\frac{du}{dt}(t)\|_\star^2 \right] dt \right)^{1/2}.
\end{aligned}
\tag{F.16}
$$

If we assume that $y_0 \in H$ and $f \in L^2(T; V')$, then the following problem may be formulated:

Problem (P) Find y satisfying

$$y \in W(T), \tag{F.17}$$

$$
\begin{cases}
\dfrac{d}{dt}(y(\,\cdot\,), \psi) + a_\vartheta(y(\,\cdot\,), \psi) = \langle f(\,\cdot\,), \psi \rangle \\
\text{in the sense of } \mathcal{D}'(T) \text{ for all } \psi \in V,
\end{cases}
\tag{F.18}
$$

$$y(0) = y_0. \tag{F.19}$$

It can be shown that it always possesses a unique solution [63, 160, 191, 200]. We remark that it is interesting to rewrite (F.18) in the vector form

$$\frac{d}{dt}y(\,\cdot\,) + A_\vartheta y(\,\cdot\,) = f(\,\cdot\,) \quad \text{in the sense of } L^2(T; V'). \tag{F.20}$$

THEOREM F.1

The mapping $\Theta_{\mathrm{ad}} \ni \vartheta \longmapsto y \in W(T)$ defined by (F.17)–(F.19) is Fréchet differentiable and the variation $\delta y \in W(T)$ in y due to a variation $\delta\vartheta$ at a point $\vartheta^0 \in \Theta_{\mathrm{ad}}$ is the solution to the problem

$$\frac{d}{dt}(\delta y(\,\cdot\,), \psi) + a_{\vartheta^0}(\delta y(\,\cdot\,), \psi) = -a_{\delta\vartheta}(y^0(\,\cdot\,), \psi), \quad \delta y(0) = 0 \tag{F.21}$$

in the sense of $\mathcal{D}'(T)$ for all $\psi \in V$, where y^0 is the solution to (F.17)–(F.19) for $\vartheta = \vartheta^0$.

PROOF The main idea of the proof was suggested by Chavent [45]. We begin by introducing the mapping

$$P : \begin{cases} \Theta_{\mathrm{ad}} \times W(T) \longrightarrow & L^2(T;V') \times H, \\ (\vartheta,y) & \longmapsto \left(\dfrac{dy}{dt} + A_\vartheta y - f, y(0) - y_0\right). \end{cases} \quad \text{(F.22)}$$

Let us take an arbitrary $\vartheta^0 \in \Theta_{\mathrm{ad}}$ and denote by y^0 the corresponding solution of (F.17)–(F.19). In this setting, the following conditions are satisfied:

(i) *P is continuous at* (ϑ^0, y^0).

Indeed, if $y \to y^0$ in $W(T)$, then $y(0) \to y^0(0)$ from the continuity of the trace operator (or, in other words, since $W(T)$ is continuously embedded in $C([0, t_f]; H)$) and $dy/dt \to dy^0/dt$ in $L^2(T;V')$ from the definition of $W(T)$.

If we assume additionally that $\vartheta \to \vartheta^0$ in $L^\infty(\Omega)$, then from the estimate

$$\|A_\vartheta y - A_{\vartheta^0} y^0\|_{L^2(T;V')}$$
$$\leq \|A_\vartheta y - A_{\vartheta^0} y\|_{L^2(T;V')} + \|A_{\vartheta^0} y - A_{\vartheta^0} y^0\|_{L^2(T;V')}$$
$$\leq \|A_\vartheta - A_{\vartheta^0}\|_{\mathcal{L}(V,V')} \|y\|_{L^2(T;V)} + \|A_{\vartheta^0}\|_{\mathcal{L}(V,V')} \|y - y^0\|_{L^2(T;V)}$$
$$\leq \|\vartheta - \vartheta^0\|_{L^\infty(\Omega)} \|y\|_{L^2(T;V)} + \|\vartheta^0\|_{L^\infty(\Omega)} \|y - y^0\|_{L^2(T;V)}$$

we get $A_\vartheta y \to A_{\vartheta^0} y^0$ in $L^2(T;V')$, which completes the proof of the desired property.

(ii) $P(\vartheta^0, y^0) = 0$.

This is an immediate consequence of the definitions of y^0 and P.

(iii) *The partial Gâteaux derivative P'_y exists in $\Theta_{\mathrm{ad}} \times W(T)$ and is continuous at* (ϑ^0, y^0).

Given any $\bar{\vartheta} \in \Theta_{\mathrm{ad}}$ and any $\bar{y} \in W(T)$, consider an increment $\delta y \in W(T)$. As for the limit in formula (E.1), we get

$$P'_y(\bar{\vartheta}, \bar{y}) \delta y = \lim_{\lambda \to 0} \frac{P(\bar{\vartheta}, \bar{y} + \lambda \delta y) - P(\bar{\vartheta}, \bar{y})}{\lambda} = \left(\frac{d\delta y}{dt} + A_{\bar{\vartheta}} \delta y, \delta y(0)\right).$$

It follows immediately that $P'_y(\bar{\vartheta}, \bar{y}) \in \mathcal{L}(W(T), L^2(T;V') \times H)$, therefore it does define a partial Gâteaux derivative. From the estimate

$$\|P_y'(\bar\vartheta, \bar y) - P_y'(\vartheta^0, y^0)\|_{\mathcal{L}(W(T), L^2(T;V') \times H)}$$

$$= \sup_{\|\delta y\|_{W(T)}=1} \|A_{\bar\vartheta}\delta y - A_{\vartheta^0}\delta y\|_{L^2(T;V')}$$

$$\leq \|A_{\bar\vartheta} - A_{\vartheta^0}\|_{\mathcal{L}(V;V')} \sup_{\|\delta y\|_{W(T)}=1} \|\delta y\|_{L^2(T;V)}$$

$$\leq \|A_{\bar\vartheta} - A_{\vartheta^0}\|_{\mathcal{L}(V;V')} \leq \|\bar\vartheta - \vartheta^0\|_{L^\infty(\Omega)}$$

we conclude that this Gâteaux derivative is continuous at (ϑ^0, y^0).

(iv) $P_y'(\vartheta^0, y^0)$ *is invertible and*

$$\left[P_y'(\vartheta^0, y^0)\right]^{-1} \in \mathcal{L}(L^2(T;V') \times H, W(T)).$$

The existence of the inverse results from the fact that for each $r \in L^2(T;V')$ and $\eta \in H$ the problem

$$\frac{d\delta y}{dt} + A_{\vartheta^0}\delta y = r, \quad \delta y(0) = \eta$$

has a unique solution in $W(T)$ (cf. (F.17)–(F.19)).

Clearly, $\left[P_y'(\vartheta^0, y^0)\right]^{-1}$ is linear and its continuity follows from that of the solution to (F.17)–(F.19) with respect to data, as there exists $C > 0$ depending only on ϑ^0 such that [63, Th. 3, p. 520]

$$\|\delta y\|_{L^2(T;V)} \leq C \left\{|\eta|^2 + \|r\|^2_{L^2(T;V')}\right\}^{1/2}$$

and we have

$$\|\frac{d\delta y}{dt}\|_{L^2(T;V')} = \|r - A_{\vartheta^0}\delta y\|_{L^2(T;V')}$$

$$\leq \|r\|_{L^2(T;V')} + \|A_{\vartheta^0}\|_{\mathcal{L}(V;V')}\|\delta y\|_{L^2(T;V)}.$$

An alternative proof of Property (iv) is based on the observation that the bounded linear operator $P_y'(\vartheta^0, y^0)$ is one-to-one and onto its range, so from the Banach Theorem [30, 256, p. 19] it follows that its inverse is also a bounded linear operator.

(v) *The partial Gâteaux derivative P_ϑ' exists in $\Theta_{ad} \times W(T)$ and is continuous at (ϑ^0, y^0).*

For arbitrary $\bar\vartheta \in \Theta_{ad}$ and $\bar y \in W(T)$ consider an increment $\delta\vartheta \in L^\infty(\Omega)$. As regards formula (E.1), we have

$$P_\vartheta'(\bar\vartheta, \bar y)\delta\vartheta = \lim_{\lambda \to 0} \frac{P(\bar\vartheta + \lambda\delta\vartheta, \bar y) - P(\bar\vartheta, \bar y)}{\lambda}$$

$$= \lim_{\lambda \to 0} \frac{1}{\lambda}\left(A_{\bar\vartheta + \lambda\delta\vartheta}\bar y - A_{\bar\vartheta}\bar y, 0\right) = \left(A_{\delta\vartheta}\bar y, 0\right)$$

since the mapping $L : \vartheta \mapsto A_\vartheta y$ is linear and bounded. Moreover,

$$\|L\|_{\mathcal{L}(L^\infty(\Omega), L^2(T;V'))} \leq \|y\|_{L^2(T;V)}.$$

Consequently, $P'_\vartheta(\bar{\vartheta}, \bar{y}) \in \mathcal{L}(L^\infty(\Omega), L^2(T; V') \times H)$, i.e., indeed, we deal with a partial Gâteaux derivative.

From the estimate

$$\|P'_\vartheta(\bar{\vartheta}, \bar{y}) - P'_\vartheta(\vartheta^0, y^0)\|_{\mathcal{L}(L^\infty(\Omega), L^2(T;V') \times H)}$$
$$= \sup_{\|\delta\vartheta\|_{L^\infty(\Omega)}=1} \|A_{\delta\vartheta}(\bar{y} - y^0)\|_{L^2(T;V')}$$
$$\leq \|\bar{y} - y^0\|_{L^2(T;V)} \sup_{\|\delta\vartheta\|_{L^\infty(\Omega)}=1} \|A_{\delta\vartheta}\|_{\mathcal{L}(V;V')}$$
$$\leq \|\bar{y} - y^0\|_{L^2(T;V)}$$

it follows that P'_ϑ is continuous at (ϑ^0, y^0).

Properties (i)–(v) constitute the assumptions of the Implicit Function Theorem [126, Th. 3, p. 673] from which we thus conclude that there exists an operator F defined in a neighbourhood $G \subset \Theta_{\mathrm{ad}}$ of the point (ϑ^0, y^0), $F : G \to W(T)$, such that

(a) $P(\vartheta, F(\vartheta)) = 0 \quad (\vartheta \in G)$,

(b) $F(\vartheta^0) = y^0$,

(c) F is continuous at ϑ^0,

(d) F is Fréchet differentiable at ϑ^0 (markedly, this implies (c) per se) and

$$F'(\vartheta^0) = -[P'_y(\vartheta^0, y^0)]^{-1} P'_\vartheta(\vartheta^0, y^0).$$

The last conclusion signifies that calculation of the variation $\delta y = F'_\vartheta(\vartheta^0)\delta\vartheta$ amounts to solving the problem

$$\frac{d\delta y}{dt} + A_{\vartheta^0}\delta y = -A_{\delta\vartheta}y^0, \quad \delta y(0) = 0$$

in the sense of $L^2(T; V')$. This completes the proof. ∎

Let us now consider the case when $\vartheta \in L^\infty(\Omega)$ is parameterized by a parameter vector $\theta \in \mathbb{R}^m$, which is defined by a mapping $R : \mathbb{R}^m \to L^\infty(\Omega)$. Such a situation arises naturally in practice or is a consequence of applied parameterization which aims at simplifying the structure of an infinite- or a high-dimensional parameter space. Clearly, θ has to be chosen in such a way as to satisfy the imposed condition $\vartheta(\theta) \in \Theta_{\mathrm{ad}}$, so we introduce the set $\bar{\Theta}_{\mathrm{ad}} = R^{-1}(\Theta_{\mathrm{ad}})$. We wish to investigate the differentiability of the solution to (F.17)–(F.19) with respect to individual components θ_q, $1 \le q \le m$ of θ at θ^0. For that purpose, let us orient the mapping $Q = R|\bar{\Theta}_{ad}$ by the requirement that it possess all the partial Fréchet derivatives at θ^0. Based on the remarks of Appendix E.4, $\mathcal{L}(\mathbb{R}, L^\infty(\Omega))$ is isometrically isomorphic to $L^\infty(\Omega)$ and therefore $Q'_{\theta_q}(\theta^0) \in \mathcal{L}(\mathbb{R}, L^\infty(\Omega))$ may be identified with an element $\vartheta_q(\theta^0) = Q'_{\theta_q}(\theta^0)(1) \in L^\infty(\Omega)$.

The variation s_q in y at y^0 produced by $\vartheta_q(\theta^0)$ is then given as the solution to the problem

$$\frac{d}{dt}(s_q(\cdot), \psi) + a_{Q(\theta^0)}(s_q(\cdot), \psi) = -a_{\vartheta_q(\theta^0)}(y^0(\cdot), \psi), \quad s_q(0) = 0 \quad (F.23)$$

in the sense of $\mathcal{D}'(T)$ for all $\psi \in V$, which results from the chain rule of differentiation (E.5). Obviously, s_q is at the same time the element of $W(T)$ which is identified with the partial Fréchet derivative of the mapping assigning to each $\theta \in \bar{\Theta}_{ad}$ the corresponding solution to (F.17)–(F.19).

Equations (F.23), $1 \le q \le m$ are said to be the *sensitivity equations* and their solutions s_q's are called the *sensitivity coefficients*. A key step in finding an optimal sensor location is to solve both the state and sensitivity equations, so attention is now directed toward a technique to obtain their approximate solutions. To this end, we shall apply first the finite-element method in discretization of space variables (i.e., we apply the so-called semidiscrete Galerkin scheme).

Let us denote by V_h a finite-element subspace of V obtained, e.g., after introducing a triangular mesh on Ω (for a detailed description of possible choices of V_h, see [122, 191, 249]); a reader who has had no exposure to the finite-element method can interpret what follows in terms of a finite-dimensional approximation of V. To obtain the finite-element approximation of (F.17)–(F.19) simply amounts to finding $y_h = y_h(x, t)$ which belongs to V_h for every $t \in T$ and satisfies

$$\frac{d}{dt}(y_h(\cdot), v_h) + a_\vartheta(y_h(\cdot), v_h) = \langle f(\cdot), v_h \rangle, \quad \forall v_h \in V_h, \quad \text{a.e. in } T. \quad (F.24)$$

Seeking y_h in the form

$$y_h(x, t) = \sum_{i=1}^{I} Y_i(t)\varphi_i(x), \quad (F.25)$$

where $\{\varphi_i\}_{i=1}^{I}$ is a basis in V_h, we plug this expansion into (F.24) and get the system of first-order ODEs:

$$\sum_{j=1}^{I}(\varphi_i, \varphi_j)\frac{dY_j}{dt} + \sum_{j=1}^{I} a_\vartheta(\varphi_i, \varphi_j)Y_j = \langle f, \varphi_i \rangle, \quad 1 \le i \le I, \quad \text{in } T \quad (F.26)$$

for unknown functions $Y_1(\cdot), \ldots, Y_I(\cdot)$ ($I = \dim V_h$). Let us note that the boundary conditions are already included into this system via the space V.

We have only to deal with the initial condition (F.19). Writing $y_{0h}(x) = y_h(x, 0)$, we set $y_{0h} = y_0$ when $y_0 \in V_h$. If $y_0 \notin V_h$, then y_{0h} may be the standard interpolant of y_0 [191, 249] or a projection of y_0 onto V_h which can be determined, e.g., by the set of linear equations

$$(y_{0h}, v_h) = (y_0, v_h), \quad \forall v_h \in V_h. \tag{F.27}$$

The initial values for Y_i are uniquely determined by the relation

$$\sum_{i=1}^{I} Y_i(0)\varphi_i(x) = y_{0h} \tag{F.28}$$

since $\{\varphi_i\}_{i=1}^{I}$ is a basis of V_h. Introducing the vector of unknowns $Y = (Y_1, \ldots, Y_I)$, we rewrite (F.26) in matrix form

$$M\frac{dY}{dt} + KY = F, \tag{F.29}$$

where $F = (\langle f, \varphi_1 \rangle, \ldots, \langle f, \varphi_I \rangle)$ is the *load vector*, $M = [(\varphi_i, \varphi_j)]_{i,j=1}^{I}$ is said to be the *mass matrix*, and $K = [a_\vartheta(\varphi_i, \varphi_j)]_{i,j=1}^{I}$ is termed the *stiffness matrix*. This method is traditionally called the *method-of-lines* semidiscretization.

As for the sensitivity equations, the procedure is exactly the same, i.e., each s_q is approximated by the quantity

$$s_{qh}(x, t) = \sum_{i=1}^{I} S_{qi}(t)\varphi_i(x). \tag{F.30}$$

Setting $S_q = (S_{q1}, \ldots, S_{qI})$ as the vector of additional unknowns, we get the sets of ODE's in matrix form

$$M\frac{dS_q}{dt} + KS_q = -K_q Y, \quad S_q(0) = 0, \quad 1 \leq q \leq m, \tag{F.31}$$

where $K_q = [a_{\vartheta_q(\theta^0)}(\varphi_i, \varphi_j)]_{i,j=1}^{I}$.

Equations (F.29) and (F.31) can be further discretized by forming a partition

$$0 = t_0 < t_1 < \cdots < t_N = t_f \tag{F.32}$$

of the time interval $[0, t_f]$ with equidistant nodes $t_k = k\tau$ and time step $\tau = t_f/N$. Denoting by $f^{(k)}$ the value of F at the k-th time step, we can approximate (F.29), e.g., with the use of the classical Crank-Nicolson scheme [122, 191]:

$$M\frac{Y^{(k+1)} - Y^{(k)}}{\tau} + K\frac{Y^{(k+1)} + Y^{(k)}}{2} = \frac{f^{(k+1)} + f^{(k)}}{2}, \tag{F.33}$$

where $Y^{(k)}$ signifies the vector of approximate values of Y at the k-th time step. Thus $Y^{(k+1)}$ can be calculated from the equation

$$\left(M + \frac{\tau K}{2}\right) Y^{(k+1)} = \left(M - \frac{\tau K}{2}\right) Y^{(k)} + \tau \frac{f^{(k+1)} + f^{(k)}}{2},$$

$$k = 0, 1, \ldots, N - 1, \quad \text{(F.34)}$$

with the initial condition

$$Y^{(0)} = (Y_1(0), \ldots, Y_I(0)). \quad \text{(F.35)}$$

Of course, equations (F.31) can be treated in much the same way, which leads to solving the equations

$$\left(M + \frac{\tau K}{2}\right) S_q^{(k+1)} = \left(M - \frac{\tau K}{2}\right) S_q^{(k)} - \tau K_q \frac{Y^{(k+1)} + Y^{(k)}}{2},$$

$$k = 0, 1, \ldots, N - 1, \quad q = 1, \ldots, m, \quad \text{(F.36)}$$

subject to the initial condition

$$S_q^{(0)} = 0. \quad \text{(F.37)}$$

A thorough error analysis of the method presented here can be made following standard textbooks on numerical methods for PDEs [122, 191, 249] and is therefore omitted.

Finally, let us note that the form of the sensitivity problems (F.36) is exactly the same as that of the simulation problem (F.34), so we can use the same computer code to solve all of them.

G

Interpolation of tabulated sensitivity coefficients

The numerical treatment of the sensor-location problem requires an efficient procedure to evaluate the sensitivity coefficents at arbitrarily picked points of the space-time domain. It is extremely costly to resolve numerically the sensitivity equations whenever the aforementioned values are demanded (after all, this happens continually while planning). A more judicious approach is to solve those equations once and to store the solution in the form of a sequence of values on a finite grid resulting from the appropriate space-time discretization. But doing so, we also have to settle the problem of estimating, if necessary, the missing values at points out of the grid. Of course, the issue pertains to interpolation of functions in multidimensions and we can employ numerous well-developed techniques to deal with this task, see, e.g., [219]. As partial derivatives of the sensitivities with respect to spatial variables will also be required while using gradient techniques to find optimal sensor locations, cubic spline interpolation seems especially efficient in achieving our aim. Since the topic of interpolation in multidimensions is usually limited in the literature to the case of two independent variables, below we delineate the procedure for trivariate functions, which is particularly suited for our purposes (two spatial variables and time). But first, some elementary notions regarding the one-dimensional situation are briefly recalled, because interpolation in three dimensions boils down to a sequence of univariate interpolations.

G.1 Cubic spline interpolation for functions of one variable

Given a tabulated function $f_i = f(x_i)$, $i = 0, \ldots, n$ on a set of points, called *nodes*, $a = x_0 \leq x_1 \leq \cdots \leq x_n = b$, a *natural cubic spline* interpolant g for f on the interval $[a, b]$ is a function which satisfies the following conditions [35, 170]:

(a) $g \in C^2([a, b])$,

(b) g is a (generally different) cubic polynomial on each of the subintervals $[x_{i-1}, x_i]$, $i = 1, \ldots, n$,

(c) $g(x_i) = f_i$, $i = 0, 1, \ldots, n$, and

(d) $g''(a) = g''(b) = 0$ (the so-called *free* boundary conditions).

Its graph approximates the shape which a long flexible rod would assume if forced to go through each of the data points $\{(x_i, f_i)\}$.

To construct the cubic-spline interpolant, let us observe that the second derivative g'' is to be continuous and piecewise linear. This clearly forces the conditions

$$g''(x) = m_{i-1}\frac{x_i - x}{\Delta x_i} + m_i\frac{x - x_{i-1}}{\Delta x_i} \quad \text{on } [x_{i-1}, x_i], \qquad (\text{G.1})$$

where $\Delta x_i = x_i - x_{i-1}$, $m_i = g''(x_i)$, $1 \le i \le n$.

Integrating (G.1) twice while bearing in mind Condition (c), we deduce that

$$
\begin{aligned}
g(x) &= m_{i-1}\frac{(x_i - x)^3}{6\Delta x_i} + m_i\frac{(x - x_{i-1})^3}{6\Delta x_i} \\
&\quad + \left(f_{i-1} - \frac{m_{i-1}\Delta x_i^2}{6}\right)\frac{x_i - x}{\Delta x_i} + \left(f_i - \frac{m_i\Delta x_i^2}{6}\right)\frac{x - x_{i-1}}{\Delta x_i}
\end{aligned}
\qquad (\text{G.2})
$$

and

$$
\begin{aligned}
g'(x) &= -m_{i-1}\frac{(x_i - x)^2}{2\Delta x_i} + m_i\frac{(x - x_{i-1})^2}{2\Delta x_i} + \frac{f_i - f_{i-1}}{\Delta x_i} \\
&\quad - \frac{m_i - m_{i-1}}{6}\Delta x_i.
\end{aligned}
\qquad (\text{G.3})
$$

From (G.3) and the postulated continuity of the first derivatives at points $x_1, \ldots, x_{n-1}$, we obtain the set of $n - 1$ linear equations

$$
\begin{aligned}
\frac{\Delta x_i}{6}m_{i-1} &+ \frac{\Delta x_i + \Delta x_{i+1}}{3}m_i + \frac{\Delta x_{i+1}}{6}m_{i+1} \\
&= \frac{f_{i+1} - f_i}{\Delta x_{i+1}} - \frac{f_i - f_{i-1}}{\Delta x_i}, \quad 1 \le i \le n-1 \quad (\text{G.4})
\end{aligned}
$$

in the $n-1$ unknowns m_i, $i = 1, \ldots, n-1$ (Condition (d) yields $m_0 = m_n = 0$). Its matrix of coefficients is tridiagonal and strictly diagonally dominant, so the linear system (G.4) has a unique solution.

G.2 Tricubic spline interpolation

In the three-dimensional interpolation, we seek an estimate of a function $f = f(x, y, z)$ from a three-dimensional grid of tabulated values of f. For simplicity, we focus our attention only on the problem of interpolating on the so-called *Cartesian mesh*, i.e., the one which has tabulated function values at the vertices of a rectangular array.

Let $D = [a_l, a_u] \times [b_l, b_u] \times [c_l, c_u] \subset \mathbb{R}^3$ be a bounded cuboid. Given I, J, K and partitions

$$
\begin{cases}
a_l = x_0 < x_1 < \ldots < x_I = a_u, \\
b_l = y_0 < y_1 < \ldots < y_J = b_u, \\
c_l = z_0 < z_1 < \ldots < x_K = c_u,
\end{cases} \tag{G.5}
$$

we introduce the grid

$$
D_h = \{(x_i, y_j, z_k) : 0 \leq i \leq I,\ 0 \leq j \leq J,\ 0 \leq k \leq K\}. \tag{G.6}
$$

On such assumptions, tricubic spline interpolation of a real-valued function $f = f(x, y, z)$ represented by the values $f_{i,j,k}$ at the points of D_h consists in constructing a function $g : D \to \mathbb{R}$ which satisfies the following conditions [170]:

(a) $g \in C^2(D)$,

(b) in each cell $[x_{i-1}, x_i] \times [y_{j-1}, y_j] \times [z_{k-1}, z_k]$, $(1 \leq i \leq I,\ 1 \leq j \leq J,\ 1 \leq k \leq K)$ g is a tricubic polynomial of the form

$$
g(x, y, z) = g_{i,j,k}(x, y, z) = \sum_{p,q,r=0}^{3} a_{p,q,r}^{i,j,k} (x_i - x)^p (y_j - y)^q (z_k - z)^r,
$$

(c) $g(x_i, y_j, z_k) = f_{i,j,k}$, $0 \leq i \leq I, 0 \leq j \leq J, 0 \leq k \leq K$,

(d) if we denote by ν the vector outward normal to the boundary ∂D of D, then

$$
\frac{\partial^2 g}{\partial \nu^2}\bigg|_{\partial D} = 0.
$$

It can be shown that such an interpolatory function exists and is unique, but here we are interested above all in the practical problem of its determination. First of all, let us observe that it is a simple matter to calculate the values of the second derivatives g_{xx}, g_{yy} and g_{zz} on D_h. In fact, in Appendix G.1 (cf. (G.1) and (G.2)), we have shown how to obtain the values m_i's of the second derivative of the spline interpolant (let us recall that the affair boils down to solving a tridiagonal system of linear equations). The derivative g_{xx} on D_h here can be handled in much the same way, the only difference being the necessity of solving $(J + 1)(K + 1)$ linear systems of type (G.4) while performing one-dimensional interpolations along the lines $\{y = y_j, z = z_k\}$, $0 \leq j \leq J, 0 \leq k \leq K$ of the grid. Similarly, by solving $(I + 1)(K + 1)$ one-dimensional problems along the lines $\{x = x_i, z = z_k\}$, $0 \leq i \leq I, 0 \leq k \leq K$ and $(I+1)(J+1)$ similar problems along the lines $\{x = x_i, y = y_j\}$, $0 \leq i \leq I$, $0 \leq j \leq J$, we obtain g_{yy} and g_{zz} on D_h, respectively.

Continuing in the same fashion, we calculate the arrays of g_{xxyy}, g_{yyzz} and g_{xxzz} on D_h based on the newly tabulated values of g_{xx} and g_{yy}. Finally, the delinated procedure permits determination of the values of the derivative g_{xxyyzz} on the grid, as may be readily guessed, from the tabulated values of g_{xxyy}. Consequently, seven three-dimensional arrays are formed. They can be precomputed and stored in computer memory.

Suppose now that it is necessary to determine the value of the interpolant g at a point (x, y, z) such that $x_{i-1} \leq x \leq x_i$, $y_{j-1} \leq y \leq y_j$ and $z_{k-1} \leq z \leq z_k$ for some i, j and k. First, let us take notice of the fact that this can be accomplished as a result of one-dimensional spline interpolation

$$g(x, y, z) = g_{xx}(x_{i-1}, y, z)\frac{(x_i - x)^3}{6\Delta x_i} + g_{xx}(x_i, y, z)\frac{(x - x_{i-1})^3}{6\Delta x_i}$$
$$+ \left(g(x_{i-1}, y, z) - \frac{g_{xx}(x_{i-1}, y, z)\Delta x_i^2}{6} \right) \frac{x_i - x}{\Delta x_i} \qquad \text{(G.7)}$$
$$+ \left(g(x_i, y, z) - \frac{g_{xx}(x_i, y, z)\Delta x_i^2}{6} \right) \frac{x - x_{i-1}}{\Delta x_i}.$$

But to make this formula useful, we have to indicate how to compute the missing quantities $g(x_i, y, z)$ and $g_{xx}(x_i, y, z)$ (more precisely, $g(x_{i-1}, y, z)$ and $g_{xx}(x_{i-1}, y, z)$ as well, but the corresponding alterations in the formulae for $g(x_i, y, z)$ and $g_{xx}(x_i, y, z)$ are obvious, so they are omitted for brevity; the same simplification is applied everywhere in what follows). For instance, $g(x_i, y, z)$ can be produced based on the following chain of dependences:

$$g(x_i, y, z) = g_{yy}(x_i, y_{j-1}, z)\frac{(y_j - y)^3}{6\Delta y_j}$$
$$+ g_{yy}(x_i, y_j, z)\frac{(y - y_{j-1})^3}{6\Delta y_j}$$
$$+ \left(g(x_i, y_{j-1}, z) - \frac{g_{yy}(x_i, y_{j-1}, z)\Delta y_j^2}{6} \right) \frac{y_j - y}{\Delta y_j} \qquad \text{(G.8)}$$
$$+ \left(g(x_i, y_j, z) - \frac{g_{yy}(x_i, y_j, z)\Delta y_j^2}{6} \right) \frac{y - y_{j-1}}{\Delta y_j}$$

in conjunction with

$$g(x_i, y_j, z) = g_{zz}(x_i, y_j, z_{k-1})\frac{(z_k - z)^3}{6\Delta z_k}$$
$$+ g_{zz}(x_i, y_j, z_k)\frac{(z - z_{k-1})^3}{6\Delta z_k}$$
$$+ \left(g(x_i, y_j, z_{k-1}) - \frac{g_{zz}(x_i, y_j, z_{k-1})\Delta z_k^2}{6} \right) \frac{z_k - z}{\Delta z_k} \qquad \text{(G.9)}$$
$$+ \left(g(x_i, y_j, z_k) - \frac{g_{zz}(x_i, y_j, z_k)\Delta z_k^2}{6} \right) \frac{z - z_{k-1}}{\Delta z_k}$$

and

$$g_{yy}(x_i, y_j, z) = g_{yyzz}(x_i, y_j, z_{k-1}) \frac{(z_k - z)^3}{6\Delta z_k}$$

$$+ g_{yyzz}(x_i, y_j, z_k) \frac{(z - z_{k-1})^3}{6\Delta z_k}$$

$$+ \left(g_{yy}(x_i, y_j, z_{k-1}) - \frac{g_{yyzz}(x_i, y_j, z_{k-1})\Delta z_k^2}{6} \right) \frac{z_k - z}{\Delta z_k}$$

$$+ \left(g_{yy}(x_i, y_j, z_k) - \frac{g_{yyzz}(x_i, y_j, z_k)\Delta z_k^2}{6} \right) \frac{z - z_{k-1}}{\Delta z_k}. \qquad (G.10)$$

Similarly, we can establish the formulae to calculate $g_{xx}(x_i, y, z)$:

$$g_{xx}(x_i, y, z) = g_{xxyy}(x_i, y_{j-1}, z) \frac{(y_j - y)^3}{6\Delta y_j}$$

$$+ g_{xxyy}(x_i, y_j, z) \frac{(y - y_{j-1})^3}{6\Delta y_j}$$

$$+ \left(g_{xx}(x_i, y_{j-1}, z) - \frac{g_{xxyy}(x_i, y_{j-1}, z)\Delta y_j^2}{6} \right) \frac{y_j - y}{\Delta y_j}$$

$$+ \left(g_{xx}(x_i, y_j, z) - \frac{g_{xxyy}(x_i, y_j, z)\Delta y_j^2}{6} \right) \frac{y - y_{j-1}}{\Delta y_j}, \qquad (G.11)$$

$$g_{xx}(x_i, y_j, z) = g_{xxzz}(x_i, y_j, z_{k-1}) \frac{(z_k - z)^3}{6\Delta z_k}$$

$$+ g_{xxzz}(x_i, y_j, z_k) \frac{(z - z_{k-1})^3}{6\Delta z_k}$$

$$+ \left(g_{xx}(x_i, y_j, z_{k-1}) - \frac{g_{xxzz}(x_i, y_j, z_{k-1})\Delta z_k^2}{6} \right) \frac{z_k - z}{\Delta z_k}$$

$$+ \left(g_{xx}(x_i, y_j, z_k) - \frac{g_{xxzz}(x_i, y_j, z_k)\Delta z_k^2}{6} \right) \frac{z - z_{k-1}}{\Delta z_k}, \qquad (G.12)$$

$$g_{xxyy}(x_i, y_j, z) = g_{xxyyzz}(x_i, y_j, z_{k-1}) \frac{(z_k - z)^3}{6\Delta z_k}$$

$$+ g_{xxyyzz}(x_i, y_j, z_k) \frac{(z - z_{k-1})^3}{6\Delta z_k}$$

$$+ \left(g_{xxyy}(x_i, y_j, z_{k-1}) - \frac{g_{xxyyzz}(x_i, y_j, z_{k-1})\Delta z_k^2}{6} \right) \frac{z_k - z}{\Delta z_k}$$

$$+ \left(g_{xxyy}(x_i, y_j, z_k) - \frac{g_{xxyyzz}(x_i, y_j, z_k)\Delta z_k^2}{6} \right) \frac{z - z_{k-1}}{\Delta z_k}. \qquad (G.13)$$

As regards the calculation of the derivative $g_x(x, y, z)$ (this is indispensable while using gradient techniques of sensor location outlined in this monograph), from (G.3) it follows that

$$
\begin{aligned}
g_x(x, y, z) = &-g_{xx}(x_{i-1}, y, z)\frac{(x_i - x)^2}{2\Delta x_i} + g_{xx}(x_i, y, z)\frac{(x - x_{i-1})^2}{2\Delta x_i} \\
&+ \frac{g(x_i, y, z) - g(x_{i-1}, y, z)}{\Delta x_i} \\
&- \frac{g_{xx}(x_i, y, z) - g_{xx}(x_{i-1}, y, z)}{6}\Delta x_i.
\end{aligned}
\tag{G.14}
$$

In order to evaluate the derivative $g_y(x, y, z)$, the most convenient approach is to employ the alternative form of (G.7), i.e.,

$$
\begin{aligned}
g(x, y, z) = &\, g_{yy}(x, y_{j-1}, z)\frac{(y_j - y)^3}{6\Delta y_j} + g_{yy}(x, y_j, z)\frac{(y - y_{j-1})^3}{6\Delta y_j} \\
&+ \left(g(x, y_{j-1}, z) - \frac{g_{yy}(x, y_{j-1}, z)\Delta y_j^2}{6} \right)\frac{y_j - y}{\Delta y_j} \\
&+ \left(g(x, y_j, z) - \frac{g_{xx}(x, y_j, z)\Delta y_j^2}{6} \right)\frac{y - y_{j-1}}{\Delta y_j},
\end{aligned}
\tag{G.15}
$$

which yields

$$
\begin{aligned}
g_y(x, y, z) = &-g_{yy}(x, y_{j-1}, z)\frac{(y_j - y)^2}{2\Delta y_j} + g_{yy}(x, y_j, z)\frac{(y - y_{j-1})^2}{2\Delta y_j} \\
&+ \frac{g(x, y_j, z) - g(x, y_{j-1}, z)}{\Delta y_j} \\
&- \frac{g_{yy}(x, y_j, z) - g_{yy}(x, y_{j-1}, z)}{6}\Delta y_j.
\end{aligned}
\tag{G.16}
$$

This is related to the additional formulae

$$
\begin{aligned}
g(x, y_j, z) = &\, g_{xx}(x_{i-1}, y_j, z)\frac{(x_i - x)^3}{6\Delta x_i} + g_{xx}(x_i, y_j, z)\frac{(x - x_{i-1})^3}{6\Delta x_i} \\
&+ \left(g(x_{i-1}, y_j, z) - \frac{g_{xx}(x_{i-1}, y_j, z)\Delta x_i^2}{6} \right)\frac{x_i - x}{\Delta x_i} \\
&+ \left(g(x_i, y_j, z) - \frac{g_{xx}(x_i, y_j, z)\Delta x_i^2}{6} \right)\frac{x - x_{i-1}}{\Delta x_i}
\end{aligned}
\tag{G.17}
$$

and

$$
\begin{aligned}
g_{yy}(x, y_j, z) = {} & g_{yyxx}(x_{i-1}, y_j, z) \frac{(x_i - x)^3}{6\Delta x_i} \\
& + g_{yyxx}(x_i, y_j, z) \frac{(x - x_{i-1})^3}{6\Delta x_i} \\
& + \left(g_{yy}(x_{i-1}, y_j, z) - \frac{g_{yyxx}(x_{i-1}, y_j, z)\Delta x_i^2}{6} \right) \frac{x_i - x}{\Delta x_i} \\
& + \left(g_{yy}(x_i, y_j, z) - \frac{g_{yyxx}(x_i, y_j, z)\Delta x_i^2}{6} \right) \frac{x - x_{i-1}}{\Delta x_i}.
\end{aligned}
\tag{G.18}
$$

Let us note that $g_{yyxx} = g_{xxyy}$, so no additional calculations and storage are necessary.

H

Calculation of the differentials introduced in Section 4.3.3

H.1   Derivation of formula (4.126)

Setting $M(s) = \{\mu_{ij}\}_{m \times m}$ and $\overset{\circ}{\Psi}(s) = \{c_{ij}\}_{m \times m} = \partial\Psi(M)/\partial M\big|_{M=M(s)}$, we get

$$\delta J = \sum_{i=1}^{m}\sum_{j=1}^{m} c_{ij}\delta\mu_{ij} = \int_{0}^{t_f}\sum_{i=1}^{m}\sum_{j=1}^{m} c_{ij}\frac{\partial\chi_{ij}}{\partial s}\,\delta s\,\mathrm{d}t, \qquad \text{(H.1)}$$

the last equality being a consequence of the dependence

$$\delta\chi_{ij} = \int_{0}^{t_f}\frac{\partial\chi_{ij}}{\partial s}\,\delta s\,\mathrm{d}t$$

with

$$\chi_{ij}(s(t),t) = \frac{1}{Nt_f}\sum_{\ell=1}^{N} g_i(x^\ell(t),t)g_j(x^\ell(t),t)$$

From (4.125) and the Lagrange identity (i.e., the integration-by-parts formula) of the form

$$\int_{0}^{t_f}\langle\zeta(t),\delta\dot{s}(t) - f_s(t)\delta s(t)\rangle\,\mathrm{d}t + \int_{0}^{t_f}\langle\dot{\zeta}(t) + f_s^\mathsf{T}(t)\zeta(t),\delta s(t)\rangle\,\mathrm{d}t$$
$$= \langle\zeta(t_f),\delta s(t_f)\rangle - \langle\zeta(0),\delta s(0)\rangle, \qquad \text{(H.2)}$$

we deduce that

$$\delta J = \langle\zeta(0),\delta s_0\rangle + \int_{0}^{t_f}\langle f_u^\mathsf{T}(t)\zeta(t),\delta u(t)\rangle\,\mathrm{d}t \qquad \text{(H.3)}$$

after setting ζ as the solution to the Cauchy problem

$$\dot{\zeta}(t) + f_s^\mathsf{T}(t)\zeta(t) = -\sum_{i=1}^{m}\sum_{j=1}^{m} c_{ij}\left(\frac{\partial\chi_{ij}}{\partial s}\right)^\mathsf{T}_{s=s(t)}, \qquad \zeta(t_f) = 0. \qquad \text{(H.4)}$$

H.2   Derivation of formula (4.129)

For brevity, we omit all higher-order terms. The perturbed quantities yield

$$h(s_0 + \delta s_0, u + \delta u) = \max_{(\ell,t)\in\bar{\nu}\times T} \{\gamma_\ell(s(t) + \delta s(t))\}$$

$$= \max_{(\ell,t)\in\bar{\nu}\times T} \left\{ \gamma_\ell(s(t)) + \left\langle \left(\frac{\partial\gamma_\ell}{\partial s}\right)^{\mathsf{T}}_{s=s(t)}, \delta s(t) \right\rangle \right\}. \qquad (\text{H.5})$$

Write $S = \{(\ell,t) \in \bar{\nu} \times T : \gamma_\ell(s(t)) = h(s_0, u)\}$. Consequently,

$$h(s_0 + \delta s_0, u + \delta u)$$

$$= h(s_0, u) + \max_{(\ell,t)\in S} \left\{ \left\langle \left(\frac{\partial\gamma_\ell}{\partial s}\right)^{\mathsf{T}}_{s=s(t)}, \delta s(t) \right\rangle \right\}$$

$$= h(s_0, u) + \max_{(\ell,t)\in S} \left\{ \int_0^{t_f} \left\langle \left(\frac{\partial\gamma_\ell}{\partial s}\right)^{\mathsf{T}}_{s=s(\tau)}, \delta s(\tau) \right\rangle \delta(t-\tau)\,\mathrm{d}\tau \right\}$$

$$= h(s_0, u) + \max_{(\ell,t)\in S} \left\{ \langle \zeta_h^\ell(0;t), \delta s_0 \rangle + \int_0^{t_f} \langle f_u^{\mathsf{T}}(\tau)\zeta_h^\ell(\tau;t), \delta u(\tau) \rangle \,\mathrm{d}\tau \right\}, \qquad (\text{H.6})$$

where δ is the Dirac delta function and $\zeta_h^\ell(\,\cdot\,;t)$ is the solution to the Cauchy problem

$$\frac{\mathrm{d}\zeta_h^\ell(\tau;t)}{\mathrm{d}\tau} + f_s^{\mathsf{T}}(\tau)\zeta_h^\ell(\tau;t) = -\left(\frac{\partial\gamma_\ell}{\partial s}\right)^{\mathsf{T}}_{s=s(\tau)} \delta(\tau-t), \quad \zeta_h^\ell(t_f;t) = 0. \quad (\text{H.7})$$

I

Solving sensor-location problems using MAPLE and MATLAB

The expanding accessibility of modern computing tools to the wide community of engineers and scientists made it possible to easily solve realistic nontrivial problems in scientific computing [53, 97, 111, 165, 172, 293]. The way we think about doing mathematical computations has dramatically changed, as high-performance numeric and symbolic computation programs allow their users to compute not only with numbers, but also with symbols, formulae, equations, etc. Many mathematical computations such as differentiation, integration, minimization of functions and matrix algebra with both numeric and symbolic entries can be carried out quickly, with emphasis on the exactness of results and without much human effort.

Addressing the optimum experimental design problems discussed in this monograph involves coping with time-consuming tasks such as solving PDEs and intricate nonlinear programming problems. This appendix is intended as an attempt to demonstrate how high-performance computing tools such as MAPLE (a computer algebra system) and MATLAB (a flexible computer environment for numeric computation and visualization) can be used in a straightforward manner to obtain solutions to nontrivial sensor-location problems. Emphasis is put on using both the programs to obtain solutions, and not on the technical subject matter *per se*, since the latter can be found in the main body of this work. The material in this section has been verified to work correctly through MAPLE 9 and Version 6.5 (Release 13) of MATLAB.

I.1 Optimum experimental effort for a 1D problem

In this example we consider a nonlinear system described by the PDE

$$\frac{\partial y}{\partial t} = a\frac{\partial^2 y}{\partial x^2} - by^3, \quad x \in (0,1), \quad t \in (0, t_f), \tag{I.1}$$

subject to the conditions

$$y(x,0) = c\psi(x), \qquad x \in (0,1), \tag{I.2}$$

$$y(0,t) = y(1,t) = 0, \quad t \in (0, t_f), \tag{I.3}$$

where $t_f = 0.8$,

$$\psi(x) = \sin(\pi x) + \sin(3\pi x). \tag{I.4}$$

The vector of unknown parameters θ is composed of a, b and c. Assuming that the state y can be observed directly, we wish to determine a D-optimum design of the form

$$\xi^\star = \begin{Bmatrix} x^1, \ \dots, \ x^\ell \\ p_1^\star, \ \dots, \ p_\ell^\star \end{Bmatrix} \tag{I.5}$$

for recovering θ, where the support points are fixed *a priori* and given by $x_i = 0.05(i-1)$, $i = 1, \dots, \ell = 21$. To this end, the multiplicative algorithm described in detail in Section 3.1.3.2 (Algorithm 3.2) will be used.

I.1.1 Forming the system of sensitivity equations

The FIM associated with θ is

$$M(\xi) = \sum_{i=1}^{\ell} p_i \Upsilon(x^i), \tag{I.6}$$

where

$$\Upsilon(x) = \frac{1}{t_f} \int_0^{t_f} \left(\frac{\partial y}{\partial \theta} \right)^{\mathsf{T}} \left(\frac{\partial y}{\partial \theta} \right) \Bigg|_{\theta = \theta^0} dt. \tag{I.7}$$

Consequently, a way for determining the sensitivity coefficients

$$\frac{\partial y}{\partial \theta} = \left[\frac{\partial y}{\partial a}, \frac{\partial y}{\partial b}, \frac{\partial y}{\partial c} \right] \equiv \left[y_a, y_b, y_c \right] \tag{I.8}$$

has to be found. The direct-differentiation method can be applied for that purpose, which reduces to the differentiation of both the sides of (I.1) with respect to individual components of θ:

$$\begin{cases} \dfrac{\partial y_a}{\partial t} = \dfrac{\partial}{\partial a} \left(a\dfrac{\partial^2 y}{\partial x^2} - by^3 \right), \quad \dfrac{\partial y_b}{\partial t} = \dfrac{\partial}{\partial b} \left(a\dfrac{\partial^2 y}{\partial x^2} - by^3 \right), \\[3mm] \dfrac{\partial y_c}{\partial t} = \dfrac{\partial}{\partial b} \left(a\dfrac{\partial^2 y}{\partial x^2} - by^3 \right), \end{cases} \tag{I.9}$$

subject to the initial conditions

$$y_a(x,0) = y_b(x,0) = 0, \quad y_c(x,0) = \psi(x) \tag{I.10}$$

and homogeneous boundary conditions.

The right-hand sides of (I.9) can be easily determined using MAPLE's symbolic computation facilities as follows:

```
> alias(yxx = yxx(a, b, c), y = y(a, b, c)):
> RHS := a * yxx - b * y^3;
```

$$RHS := a\,yxx - b\,y^3$$

```
> RHSa := diff(RHS, a);
```

$$RHSa := yxx + a\left(\frac{\partial}{\partial a}yxx\right) - 3\,b\,y^2\left(\frac{\partial}{\partial a}y\right)$$

```
> RHSb := diff(RHS, b);
```

$$RHSb := a\left(\frac{\partial}{\partial b}yxx\right) - y^3 - 3\,b\,y^2\left(\frac{\partial}{\partial b}y\right)$$

```
> RHSc := diff(RHS, c);
```

$$RHSc := a\left(\frac{\partial}{\partial c}yxx\right) - 3\,b\,y^2\left(\frac{\partial}{\partial c}y\right)$$

The sensitivity equations are to be solved simultaneously with the state equation (I.1) as one system of PDEs. This will be performed numerically in MATLAB, and thus defining the new state vector $w = [y, y_a, y_b, y_c]$, we effect the change of variables on the right-hand sides of (I.1) and (I.9), and translate the resulting MAPLE code to a MATLAB code:

```
> with(CodeGeneration):
> rulesFlux := [yxx = dwdx[1], diff(yxx, a) = dwdx[2],
               diff(yxx, b) = dwdx[3], diff(yxx, c) = dwdx[4]]:
> e1 := map2(select, has, [RHS, RHSa, RHSb, RHSc], yxx):
> Matlab(subs(rulesFlux, e1), resultname = "f");
```

$$f = [a*dwdx(1) \quad dwdx(1) + a*dwdx(2) \quad a*dwdx(3) \quad a*dwdx(4)];$$

```
> rulesSource := [y = w[1], diff(y, a) = w[2],
                 diff(y, b) = w[3], diff(y, c) = w[4]]:
> e2 := map2(remove, has, [RHS, RHSa, RHSb, RHSc], yxx):
> Matlab(subs(rulesSource, e2), resultname = "s");
```

$$s = [-b*w(1)\wedge 3 \quad -3*b*w(1)\wedge 2*w(2)$$
$$-w(1)\wedge 3 - 3*b*w(1)\wedge 2*w(3) \quad -3*b*w(1)\wedge 2*w(4)];$$

I.1.2 Solving the sensitivity equations

The numerical solution of the above system of sensitivity equations is based on the standard MATLAB PDE solver **pdepe**, which solves initial-boundary value problems for systems of parabolic and elliptic PDEs in one space variable and time. The solver converts the PDEs to ODEs using a second-order accurate spatial discretization based on a set of nodes specified by the user. The time integration is then performed with a multistep variable-order method based on the numerical differentiation formulae, see [172, p. 14–108] for details.

Define a mesh of 21 equidistant spatial points coinciding with the support of the sought design. The sensitivity coefficients will be computed at them and stored for 31 consecutive time moments evenly distributed in the time horizon:

```
xmesh = linspace(0, 1, 21);
tspan = linspace(0, 0.8, 31);
```

We assume that the nominal values of the identified parameters are $a^0 = 0.1$, $b^0 = 2.0$ and $c^0 = 5.0$, and call the solver:

```
theta = [0.1 2.0 5.0];
sol = pdepe(0, @pdefun, @icfun, @bcfun, ...
                        xmesh, tspan, [], theta);
```

The first input argument is zero since we deal with a standard problem (other values would correspond, e.g., to cylindrical or spherical coordinates). The second argument is the function defining our system of PDEs. Its code is listed below (note the use of the formulae produced earlier by MAPLE):

```
function [h, f, s] = pdefun(x, t, w, dwdx, theta)
a = theta(1); b = theta(2); c = theta(3);
h = ones(4, 1);
f = [a * dwdx(1); dwdx(1) + a * dwdx(2); ...
     a * dwdx(3); a * dwdx(4)];
s = [-b * w(1)^3; -3 * b * w(1)^2 * w(2); ...
     -w(1)^3 - 3 * b * w(1)^2 * w(3); ...
     -3 * b * w(1)^2 * w(4)];
```

The third and fourth input arguments specify the initial and boundary conditions in terms of the functions icfun and bcfun, respectively.

```
function w = icfun(x, theta)
c = theta(3);
r = sin(pi * x) + sin(3 * pi * x);
w = [c * r; 0; 0; r];

function [pl, ql, pr, qr] = bcfun(xl, wl, xr, wr, t, theta)
pl = wl;
ql = zeros(size(wl));
pr = wr;
qr = zeros(size(wr));
```

After passing space and time meshes, the seventh input argument is the empty matrix which constitutes only a placeholder because we do not change any default parameters steering the numerical integration. Finally, the nominal value of θ is specified as the last input argument and the solver will then pass it to pdefun, icfun and bcfun every time it calls them.

The output argument sol is a three-dimensional array such that sol(i, j, 1), sol(i, j, 2), sol(i, j, 3) and sol(i, j, 4) approximate the state y and its sensitivities y_a, y_b and y_c, respectively, at time tspan(i) and the spatial mesh point xmesh(j).

I.1.3 Determining the FIMs associated with individual spatial points

For simplicity, we assumed that the spatial mesh coincides with the support of the sought D-optimum design, but if other spatial points are to be considered, the procedure pdeval can be used to perform the required interpolation. The trapezoidal rule is then employed to compute $\Upsilon(x^i)$, $i = 1, \ldots, \ell$. The results are stored in the three-dimensional array atomfim such that atomfim(:, :, i) approximates $\Upsilon(x^i)$.

```
supp = xmesh;
nsens = length(supp);
m = 3;
nt = length(tspan);
atomfim = zeros(m, m, nsens);
for i = 1: nsens
    V = 0.5 * fimterm(sol, i, 1);
    for k = 2: nt - 1
        V = V + fimterm(sol, i, k);
    end
    V = V + 0.5 * fimterm(sol, i, nt);
    V = V / (nt - 1);
    atomfim(:, :, i) = V;
end
```

Here the function fimterm is used:

```
function S = fimterm(sol, ix, it)
g = squeeze(sol(it, ix, 2: end));
S = g * g';
```

I.1.4 Optimizing the design weights

Algorithm 3.2 can be easily and efficiently implemented using vectorization. We start from the weights which are equal to one another.

```
eta = 1e-4;
p = ones(1, nsens) / nsens;
detM = [];
while 1
    M = sum(repmat(permute(p, [3 1 2]), m, m) .* atomfim, 3);
```

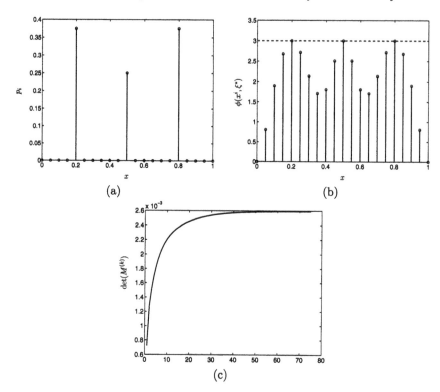

FIGURE I.1
Plots of the optimal weights (a), the function $\phi(x, \xi^\star)$ (b) and the monotonic increase in the determinant of the FIM (c) for the system (I.1).

```
    detM = [detM det(M)];
    phi = squeeze(sum(sum( ...
            repmat(inv(M), [1 1 nsens]) .* atomfim, 1), 2))';
    if max(phi) / m < 1 + eta
        break
    end
    p = p .* phi / m;
end
```

I.1.5   Plotting the results

Visualization is particularly easy in MATLAB. We plot the weight values at all support points of the ultimate design $\xi^\star$, the values of $\phi(x^i, \xi^\star)$ and the value of $\det(M^{(k)})$ in consecutive iterations, cf. Fig. I.1.

```
    stem(xmesh, p)
```

```
figure
stem(xmesh, phi)
hold on
plot([xmesh(1) xmesh(end)], [m m], 'r')
figure
plot(detM)
```

Note that the values of $\phi(x^i, \xi^*)$ do not exceed the number of the estimated parameters, i.e., three, and they are equal to this bound just where the weights are nonzero. Moreover, the values of $\det(M^{(k)})$ are increasing. This perfectly conforms to the theory developed in Section 3.1.3.2.

I.2 Clusterization-free design for a 2D problem

Consider a dynamic system described by the parabolic PDE

$$\frac{\partial y}{\partial t} = \nabla \cdot (\kappa \nabla y) \quad \text{in } \Omega \times T, \tag{I.11}$$

where $\Omega \subset \mathbb{R}^2$ is the spatial domain with boundary Γ which are shown in Fig. I.2, $T = (0, t_f)$, $y = y(x, t)$ is the state, and

$$\kappa(x) = a + b\psi(x), \quad \psi(x) = x_1 + x_2, \tag{I.12}$$

a and b being unknown parameters which have to be estimated based on the measuremements from $N = 90$ pointwise sensors whose D-optimum locations are to be determined. Throughout the design, $a^0 = 0.1$ and $b^0 = 0.3$ will be assumed as nominal values of a and b, respectively.

Equation (I.11) is supplemented with the initial and boundary conditions

$$y(x, 0) = 5 \qquad \text{in } \Omega, \tag{I.13}$$
$$y(x, t) = 5(1 - t) \quad \text{on } \Gamma \times T. \tag{I.14}$$

I.2.1 Solving sensitivity equations using the PDE Toolbox

It is a simple matter to establish the sensitivity equations associated with system (I.11) and parameters a and b:

$$\frac{\partial y_a}{\partial t} = \nabla \cdot \nabla y \quad + \nabla \cdot (\kappa \nabla y_a), \tag{I.15}$$

$$\frac{\partial y_b}{\partial t} = \nabla \cdot (\psi \nabla y) + \nabla \cdot (\kappa \nabla y_b), \tag{I.16}$$

which are supplemented with homogeneous initial and Dirichlet boundary conditions. We have to solve them simultaneously with the state equation (I.11)

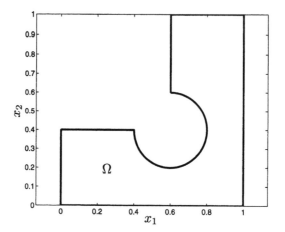

FIGURE I.2
Domain in which a DPS is described by (I.11).

and to store the solution in memory for its further use in the sensor location
algorithm. Defining $w(x,t) = \text{col}[y(x,t), y_a(x,t), y_b(x,t)]$, we reformulate this
problem as that of solving the system of simultaneous PDEs:

$$\frac{\partial w}{\partial t} - \nabla \cdot (\mathfrak{H} \otimes \nabla w) = 0 \tag{I.17}$$

subject to

$$w(x,0) = w_0(x,t) \quad \text{in } \Omega, \tag{I.18}$$
$$w(x,t) = w_\Gamma(x,t) \quad \text{on } \Gamma \times T, \tag{I.19}$$

where $\mathfrak{H}$ is a $3 \times 3 \times 2 \times 2$ tensor with components

$$\mathfrak{H}_{ijpq}(x) = \begin{cases} \kappa(x) & \text{if } i = j \text{ and } p = q, \\ 1 & \text{if } i = 2, j = 1 \text{ and } p = q, \\ \psi(x) & \text{if } i = 3, j = 1 \text{ and } p = q, \\ 0 & \text{otherwise,} \end{cases} \tag{I.20}$$

$i, j = 1, 2, 3$ and $p, q = 1, 2$, $\nabla \cdot (\mathfrak{H} \otimes \nabla w)$ denotes the three-element column
vector with the i-th component

$$\sum_{j=1}^{3} \left(\frac{\partial}{\partial x_1} \mathfrak{H}_{ij11} \frac{\partial}{\partial x_1} + \frac{\partial}{\partial x_1} \mathfrak{H}_{ij12} \frac{\partial}{\partial x_2} + \frac{\partial}{\partial x_2} \mathfrak{H}_{ij21} \frac{\partial}{\partial x_1} + \frac{\partial}{\partial x_2} \mathfrak{H}_{ij22} \frac{\partial}{\partial x_2} \right) w_j,$$

$$\tag{I.21}$$

$w_0(x) = \text{col}[5, 0, 0]$, $w_\Gamma(x,t) = \text{col}[5(1 - t), 0, 0]$. MATLAB's PDE Toolbox
turns out to be very well suited for this task, as it is designed to solve systems
of PDEs just in the form (I.17), cf. the entry on **parabolic** in [53, p. 5-43].

The PDE Toolbox extends MATLAB's technical computing environment with powerful and flexible tools for the solution of systems of PDEs in two spatial dimensions and time. It provides a set of command-line functions and a graphical-user interface for preprocessing, solving and postprocessing generic elliptic, parabolic and hyperbolic PDEs using the finite-element method. The toolbox also provides automatic and adaptive meshing capabilities and solves the resulting system of equations using MATLAB's sparse matrix solver and a state-of-the-art stiff solver from the ODE Suite.

The solution process requires the following six steps:

(S1) geometry definition (*Draw Mode*),

(S2) finite-element discretization (*Mesh Mode*),

(S3) specification of boundary conditions (*Boundary Condition Mode*),

(S4) setting the PDE coefficients (*PDE Mode*),

(S5) specification of initial conditions and soluton of the PDE (*Solve Mode*), and

(S6) postprocessing of the solution (*Plot Mode*).

The PDE Toolbox includes a complete graphical user interface (GUI) which covers all the above steps and provides quick and easy access to the functionality of efficient specialized tools. A weakness of the GUI is its limitation to the solution of only one PDE, but for a system of PDEs we can still make use of the flexibility of command-line functions. On the other hand, the GUI remains a valuable tool since part of the modelling can be done using it and it then will be made available for command-line use through the extensive data export facilities. It is even suggested [53, p. 1-40] to exploit the export facilities as much as possible, as they provide data structures with the correct syntax.

We shall now give a procedure for implementing each of the above steps required while solving the sensitivity system (I.17)–(I.19).

Geometry definition. We use the GUI as a drawing tool to make a drawing of the two-dimensional geometry of Ω. To access the GUI, we simply type `pdetool` in the Matlab command window. After the start, `pdetool` is in the *Draw* mode in which we can form a planar shape representing Ω by combining elementary solid objects (rectangles/squares, ellipses/circles and polygons) and set algebra. The solid objects can be easily created using the mouse and the click-and-drag technique. Unique labels are automatically given to each of them. They can then be used to specify the set formula definining Ω in which the set union, intersection and difference operators represented by '+', '*' and '−', respectively, can be used. The domain Ω is thus defined by the formula

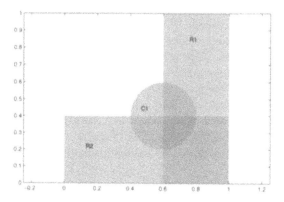

FIGURE I.3
Rectangles and a circle used to define the domain Ω.

```
(R1 + R2) - C1
```

where elementary rectangles R1, R2 and a circle C1 are shown in Fig. I.3.

Using the GUI simplifies to a great extent the geometry definition, since otherwise the user must construct the geometry model and decompose it into a minimal form using specialized functions such as `decsg`, `csgchk` or `csgdel` which require, however, much experience.

Having defined the geometry and the respective set formula, we select *Remove All Subdomain Borders* from the *Boundary* menu and then export the key matrix describing the geometry of Ω, the so-called *decomposed geometry matrix*, to the main workspace of MATLAB by selecting *Export Decomposed Geometry, Boundary Cond's...* from the same menu. The default name of this matrix is **g** (clearly, it might be altered at this juncture). We then exit the `pdetool` interface.

Finite-element discretization. The basic idea in any numerical method for a PDE is to discretize the given continuous problem with infinitely many degrees of freedom to obtain a discrete problem or a system of ODEs with only a finite number of unknowns which can be solved using a computer. The discretization process using a finite-element method (FEM) starts from expressing the problem in the equivalent variational form. Then we divide the region Ω into small subregions (elements). With two space dimensions these are usually triangles and their vertices are called the nodes of the resulting mesh. The FEM has the advantage that it is usually possible to discretize Ω in such a way that a very close approximation to the shape of the boundary is retained. In order to make this process automatic, several FE-mesh generation procedures have been proposed [122, 162, 191]. What is more, adaptive mesh refinement techniques have been developed which ensure a higher density of nodes in some subdomains chosen in accordance with specific difficulties and requirements of the problem at hand.

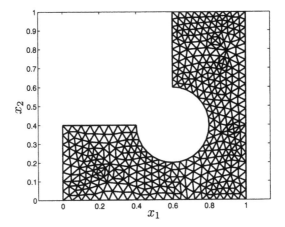

FIGURE I.4
Partition of Ω into triangular elements.

Given the shape of the geometry (in the form of the decomposed geometry matrix g), the PDE Toolbox generates a triagular mesh using a Delaunay trangulation algorithm [162]. The specific command is

```
[p, e, t] = initmesh(g);
```

The output parameters here are as follows:

p — the *point matrix* whose first and second rows constitute the x- and y-coordinates of the nodes;

e — the *edge matrix* defining the boundary of the domain built out of all triangles; clearly, this boundary approximates the true boundary Γ; the first and second rows of e constitute the indices of the nodes which start and end the consecutive edge segments;

t — the *triangle matrix* such that the first three rows of t contain the indices to the three nodes defining the consecutive triangles.

The mesh so generated contains only 126 nodes, so we refine it using refinemesh:

```
[p, e, t] = refinemesh(g, p, e, t);
```

Note that the default refinement method divides all triangles into four triangles of the same shape. The number of nodes increases now to 449. The resulting mesh can be plotted using the command pdemesh(p, e, t), cf. Fig. I.4.

Specification of boundary conditions. The Dirichlet boundary conditions (I.19) vary in time and thus the best way to describe them is to create the so-called *boundary M-file* which will accept time as one of its arguments. The relevant Matlab code is encapsulated in the following function:

```
function [q, g, h, r] = pdebound(p, e, aux, time)
m = 2;
m1 = m + 1;
ne = size(e, 2);
q = sparse(m1^2, ne);
g = sparse(m1, ne);
ident = eye(m1);
h = repmat(ident(:), 1, 2 * ne);
r = repmat([5 * (1 - time); 0; 0], 1, 2 * ne);
```

The input parameters are the point matrix p, the edge matrix e, the place-holder aux which will not be used (it is designed to be used only by the built-in solver for nonlinear problems), and the current t value time. The format of the output parameters is outlined in detail in [53, p. 5-48], cf. the description of the function pdebound. Let us only remark here that q and g characterize a generalized Neumann condition which does not appear in our formulation. Anyhow, these matrices must be defined with all elements set as zero, and hence storing them as sparse matrices constitutes the best so-lution. In turn, the parameters h and r characterize the Dirichlet condition. Note the presence of the vector $w_\Gamma(x, t)$ in the last line of the code (vector [5 * (1 - time); 0; 0]).

Setting the PDE coefficients. The PDE Toolbox allows for a slightly more general form of the system to be solved than (I.17), but the terms which are not present in the current form will be set either to zero or to unity when the solver is called. At this juncture, we only have to define the tensor $\mathfrak{H}$. Since, unlike the coefficients present in the boundary conditions, it is independent of time, the best solution is to define it through a matrix representing its values at the triangle centres of mass. A detailed description of the format of such a matrix is given in [53, p. 5-21] (the entry on assempde). The following function call fulfils this task:

```
mu = difftensor(p, t);
```

Here difftensor has the point and triangle matrices as the input parameters:

```
function mu = difftensor(p, t)
a = 0.1; b = 0.3;
m = 2;
% x & y are coordinates of triangle centres of mass
x = pdeintrp(p, t, p(1, :)');
y = pdeintrp(p, t, p(2, :)');
kappa = a + b * (x + y);
dkappa = [ones(size(kappa)); (x + y)];
```

```
m1 = m + 1;
mu = sparse(4 * m1 * m1, length(kappa));
i = 4 * (m1 + 1) * (0: m) + 1;
mu([i, i + 3], :) = repmat(kappa, 2 * m1, 1);
j = 1;
for q = 1: m
    j = j + 4;
    mu([j, j + 3], :) = repmat(dkappa(q, :), 2, 1);
end
```

Specification of initial conditions and solution of the PDEs. Owing to the triangulation, the approximation of the true solution $w : \bar{\Omega} \times \to \mathbb{R}^3$ will be obtained only at the spatial points coinciding with the mesh nodes. More precisely, if there are Q nodes, in MATLAB's representation the approximated solution at a time instant t will be a column vector whose first Q components approximate $y(\,\cdot\,, t)$ at the consecutive nodes with coordinates $(p(1, i), p(2, i))$, $i = 1, \ldots, Q$. In the same way, the following Q components of this vector will correspond to the nodal values of $y_a(\,\cdot\,, t)$, and the last Q components will do the same for $y_b(\,\cdot\,, t)$. Thus we set the initial value of w as follows:

```
w0 = [repmat(5, size(p, 2), 1); zeros(2 * size(p, 2), 1)];
```

Prior to solving the system (I.17)–(I.19), we also have to specify a list of time moments at which the solution is desired:

```
tf = 1; ntdivs = 20;
tspan = linspace(0, tf, ntdivs + 1);
```

Now the function **parabolic**, see [53, p. 5-43] for details, produces the solution to the FEM formulation via numerically solving the system of ODEs resulting from the FEM discretization:

```
w = parabolic(w0, tspan, 'pdebound', ...
              p, e, t, mu, 0, [0; 0; 0], 1);
```

Here **w** is the matrix whose columns contain the approximated solutions at the corresponding time instants stored in **tspan**.

Having obtained the sensitivity coefficients, we focus our attention on the sensor-location problem.

I.2.2   Determining potential contributions to the FIM from individual spatial points

For simplicity, we assume that the points where the sensors can potentially be located coincide with the nodes of the mesh generated by **refinemesh**. (If this is not the case, the function **griddata** can always perform the relevant interpolation.) Thus we have 449 candidate measurement points at which we wish to determine the values of

$$\Upsilon(x) = \frac{1}{t_f} \int_0^{t_f} \begin{bmatrix} y_a^2(x,t) & y_a(x,t)y_b(x,t) \\ y_a(x,t)y_b(x,t) & y_b^2(x,t) \end{bmatrix} dt. \qquad (I.22)$$

This is accomplished by the following code:

```
m = 2;
dt = tf / ntdivs;
nmpts = size(p, 2);
fim = zeros(m, m, nmpts);
ix = nmpts * (1: m);
for loop = 1: nmpts
    ix = ix + 1;
    fim(:, :, loop) = 0.5 * (w(ix, 1) * w(ix, 1)' ...
                    + w(ix, end) * w(ix, end)'));
    for k = 2: ntdivs
        fim(:, :, loop) = fim(:, :, loop) ...
                    +  w(ix, k) * w(ix, k)';
    end
    fim(:, :, loop) = dt * fim(:, :, loop);
end
```

where $\texttt{fim}(:,:,i)$ is just an approximation to $\Upsilon(x^i)$, $i = 1, \ldots, Q$, resulting from application of the trapezoidal rule for numerical integration.

I.2.3 Iterative optimization of sensor locations

A prerequisite for application of the algorithm determining a D-optimum clusterization-free configuration of $N = 90$ sensors is a starting configuration which will then be iteratively improved. The indices of the points included in the initial configuration can be generated at random:

```
nsens = 90;
rp = randperm(nmpts);
imeas = rp(1: nsens);
icand = rp(nsens + 1: nmpts);
```

The script implementing the exchange algorithm of Section 3.4 is then

```
maxiter = 200;
epsilon = 1e-6;
loop = 1;
phi = zeros(1, nmpts);
while 1
    M = sum(fim(:, :, imeas), 3);
```

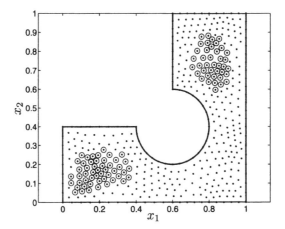

FIGURE I.5
D-optimal clusterization-free sensor configuration.

```
Minv = inv(M);
loop
disp(['det(M) = ', num2str(det(M))])
for i = 1: nmpts
    phi(i) = sum(sum(Minv .* fim(:, :, i)));
end
[phiworst, iworst] = min(phi(imeas));
[phibest,  ibest]  = max(phi(icand));
if (phibest < phiworst * (1.0 + epsilon))  ...
                        | (loop >= maxiter)
    break
end
imem = imeas(iworst);
imeas(iworst) = icand(ibest);
icand(ibest) = imem;
loop = loop + 1;
end
```

The optimal sensor configuration can now be plotted, cf. Fig. I.5:

```
pdegplot(g)
axis equal
hold on
plot(p(1, :), p(2, :), '.y')
plot(p(1, imeas), p(2, imeas), 'ob')
```

References

[1] S. I. Aihara. Consistency of extended least square parameter estimate for stochastic distributed parameter systems. In G. Bastin and M. Gevers, editors, *Proc. 4th European Control Conf.*, Brussels, Belgium, 1–4 July 1997. EUCA, 1997. Published on CD-ROM.

[2] M. Amouroux and J. P. Babary. Sensor and control location problems. In M. G. Singh, editor, *Systems & Control Encyclopedia. Theory, Technology, Applications*, volume 6, pages 4238–4245. Pergamon Press, Oxford, 1988.

[3] B. Andó, G. Cammarata, A. Fichera, S. Graziani, and N. Pitrone. A procedure for the optimization of air quality monitoring networks. *IEEE Transactions on Systems, Man, and Cybernetics — Part C: Applications and Reviews*, 29(1):157–163, Feb. 1999.

[4] H. M. Antia. *Numerical Methods for Scientists and Engineers*. Birkhäuser, Basel, 2nd edition, 2002.

[5] M. Armstrong. *Basic Linear Geostatistics*. Springer-Verlag, Berlin, 1998.

[6] K. J. Åström. *Introduction to Stochastic Control Theory*. Academic Press, New York, 1970.

[7] A. C. Atkinson. Optimum experimental designs for parameter estimation and for discrimination between models in the presence of prior information. In V. Fedorov, W. G. Müller, and I. N. Vuchkov, editors, *Model Oriented Data-Analysis. A Survey of Recent Methods*, Proc. 2nd IIASA Workshop in St. Kyrik, Bulgaria, May 28–June 1 1990, pages 3–30, Heidelberg, 1992. Physica-Verlag.

[8] A. C. Atkinson and A. N. Donev. *Optimum Experimental Designs*. Clarendon Press, Oxford, 1992.

[9] A. C. Atkinson and V. V. Fedorov. The design of experiments for discriminating between two rival models. *Biometrika*, 62(1):57–70, 1975.

[10] A. C. Atkinson and V. V. Fedorov. Optimal design: Experiments for discriminating between several models. *Biometrika*, 62(2):289–303, 1975.

[11] V. V. Azhogin, M. Z. Zgurovski, and J. Korbicz. *Filtration and Control Methods for Stochastic Distributed-Parameter Processes.* Vysha Shkola, Kiev, 1988. (In Russian).

[12] A. V. Balakrishnan. *Applied Functional Analysis.* Springer-Verlag, New York, 2nd edition, 1981.

[13] N. V. Banichuk. *Introduction to Optimization of Structures.* Springer-Verlag, New York, 1990.

[14] H. T. Banks. Computational issues in parameter estimation and feedback control problems for partial differential equation systems. *Physica D*, 60:226–238, 1992.

[15] H. T. Banks and B. G. Fitzpatrick. Statistical methods for model comparison in parameter estimation problems for distributed systems. *Journal of Mathematical Biology*, 28:501–527, 1990.

[16] H. T. Banks and K. Kunisch. *Estimation Techniques for Distributed Parameter Systems.* Systems & Control: Foundations & Applications. Birkhäuser, Boston, 1989.

[17] H. T. Banks, R. C. Smith, and Y. Wang. *Smart Material Structures: Modeling, Estimation and Control.* Research in Applied Mathematics. Masson, Paris, 1996.

[18] J. Baumeister, W. Scondo, M. A. Demetriou, and I. G. Rosen. On-line parameter estimation for infinite-dimensional dynamical systems. *SIAM Journal on Control and Optimization*, 35(2):679–713, Mar. 1997.

[19] M. B. Beck. Uncertainty, identifiability, and predictability in environmental models. In P. C. Young, editor, *Concise Encyclopedia of Environmental Systems*, pages 638–644. Pergamon Press, Oxford, 1993.

[20] A. F. Bennett. *Inverse Methods in Physical Oceanography.* Cambridge Monographs on Mechanics and Applied Mathematics. Cambridge University Press, Cambridge, 1992.

[21] E. Berger. Asymptotic behaviour of a class of stochastic approximation procedures. *Probability Theory and Related Fields*, 71:517–552, 1986.

[22] D. P. Bertsekas. *Nonlinear Programming.* Optimization and Computation Series. Athena Scientific, Belmont, MA, 2nd edition, 1999.

[23] D. P. Bertsekas and J. N. Tsitsiklis. *Introduction to Probability.* Athena Scientific, Belmont, MA, 2003.

[24] V. D. Blondel and J. N. Tsitsiklis. A survey of computational complexity results in systems and control. *Automatica*, 36(9):1249–1274, 2000.

[25] J. A. Borrie. *Stochastic Systems for Engineers. Modelling, Estimation and Control.* Prentice Hall, New York, 1992.

[26] G. E. P. Box and W. J. Hill. Discrimination among mechanistics models. *Technometrics*, 9(1):57–71, 1967.

[27] S. Boyd, L. El Ghaoui, E. Feron, and V. Balakrishnan. *Linear Matrix Inequalities in System and Control Theory*. Society for Industrial and Applied Mathematics, Philadelphia, 1994.

[28] S. Boyd and L. Vandenberghe. *Convex Optimization*. Cambridge University Press, Cambridge, 2004.

[29] D. W. Brewer. The differentiability with respect to a parameter of the solution of a linear abstract Cauchy problem. *SIAM Journal on Mathematical Analysis*, 13(4):607–620, July 1982.

[30] H. Brezis. *Analyse fonctionnelle. Théorie et applications*. Mathématiques appliquées pour la maîtrise. Masson, Paris, 1987.

[31] U. N. Brimkulov, G. K. Krug, and V. L. Savanov. *Design of Experiments in Investigating Random Fields and Processes*. Nauka, Moscow, 1986. (In Russian).

[32] A. E. Bryson. *Applied Linear Optimal Control. Examples and Algorithms*. Cambridge University Press, Cambridge, 2002.

[33] A. E. Bryson, Jr. *Dynamic Optimization*. Addison-Wesley, New York, 1999.

[34] A. E. Bryson, Jr. and Y.-C. Ho. *Applied Optimal Control. Optimization, Estimation and Control*. Hemisphere Publishing Corp., New York, 2nd edition, 1975.

[35] R. L. Burden and J. D. Faires. *Numerical Analysis*. Prindle, Weber & Schmidt, Boston, 1985.

[36] A. L. Burke, T. A. Duever, and A. Pendilis. Model discrimination via designed experiments: Discriminating between the terminal and penultimate models on the basis of composition data. *Macromolecules*, 27:386–399, 1994.

[37] J. B. Burl. *Linear Optimal Control. $\mathcal{H}_2$ and $\mathcal{H}_\infty$ Methods*. Addison-Wesley, Menlo Park, CA, 1999.

[38] A. G. Butkovskiy and A. M. Pustyl'nikov. *Mobile Control of Distributed Parameter Systems*. John Wiley & Sons, New York, 1987.

[39] L. Carotenuto, P. Muraca, and G. Raiconi. Optimal location of a moving sensor for the estimation of a distributed-parameter process. *International Journal of Control*, 46(5):1671–1688, 1987.

[40] M. Carter and B. van Brunt. *The Lebesgue-Stieltjes Integral. A Practical Introduction*. Springer-Verlag, New York, 2000.

[41] W. F. Caselton, L. Kan, and J. V. Zidek. Quality data networks that minimize entropy. In A. Walden and P. Guttorp, editors, *Statistics in the Environmental and Earth Sciences*, chapter 2, pages 10–38. Halsted Press, New York, 1992.

[42] W. F. Caselton and J. V. Zidek. Optimal monitoring network design. *Statistics & Probability Letters*, 2:223–227, 1984.

[43] T. K. Chandra. *A First Course in Asymptotic Theory of Statistics*. Narosa Publishing House, New Delhi, 1999.

[44] D.-s. Chang. Design of optimal control for a regression problem. *The Annals of Statistics*, 7(5):1078–1085, 1979.

[45] G. Chavent. Identification of functional parameters in partial differential equations. In R. E. Goodson and M. Polis, editors, *Identification of Parameters in Distributed Systems*, pages 31–48. The American Society of Mechanical Engineers, New York, 1974.

[46] G. Chavent. Distributed parameter systems: Identification. In M. G. Singh, editor, *Systems & Control Encyclopedia. Theory, Technology, Applications*, volume 2, pages 1193–1197. Pergamon Press, Oxford, 1987.

[47] G. Chavent. Identifiability of parameters in the output least square formulation. In É. Walter, editor, *Identifiability of Parametric Models*, pages 67–74. Pergamon Press, Oxford, 1987.

[48] G. Chavent. On the theory and practice of non-linear least-squares. *Adv. Water Resources*, 14(2):55–63, 1991.

[49] W. H. Chen and J. H. Seinfeld. Optimal location of process measurements. *International Journal of Control*, 21(6):1003–1014, 1975.

[50] X. Chen, H. Qi, L. Qi, and K.-l. Teo. Smooth convex approximation to the maximum eigenvalue function. *Journal of Global Optimization*, 2001. (Submitted for publication).

[51] L. H. Chiang, E. L. Russell, and R. D. Braatz. *Fault Detection and Diagnosis in Industrial Systems*. Springer-Verlag, London, 2001.

[52] C.-B. Chung and C. Kravaris. Identification of spatially discontinuous parameters in second-order parabolic systems by piecewise regularisation. *Inverse Problems*, 4:973–994, 1988.

[53] COMSOL AB. *Partial Differential Equation Toolbox for Use with Matlab. User's Guide*. The MathWorks, Inc., Natick, MA, 1995.

[54] D. Cook and V. Fedorov. Constrained optimization of experimental design. *Statistics*, 26:129–178, 1995.

[55] R. W. Cottle. Manifestations of the Schur complement. *Linear Algebra and Its Applications*, 8:189–211, 1974.

[56] D. R. Cox and N. Reid. *The Theory of the Design of Experiments*. Chapman and Hall, Boca Raton, FL, 2000.

[57] N. A. C. Cressie. *Statistics for Spatial Data*. John Wiley & Sons, New York, revised edition, 1993.

[58] R. F. Curtain and A. J. Pritchard. *Functional Analysis in Modern Applied Mathematics*. Academic Press, London, 1977.

[59] R. F. Curtain and H. Zwart. *An Introduction to Infinite-Dimensional Linear Systems Theory*. Texts in Applied Mathematics. Springer-Verlag, New York, 1995.

[60] R. Daley. *Atmospheric Data Analysis*. Cambridge University Press, Cambridge, 1991.

[61] J. Darlington, C. C. Pantelides, B. Rustem, and B. A. Tanyi. An algorithm for constrained nonlinear optimization under uncertainty. *Automatica*, 35:217–228, 1999.

[62] J. Darlington, C. C. Pantelides, B. Rustem, and B. A. Tanyi. Decreasing the sensitivity of open-loop optimal solutions in decision making under uncertainty. *European Journal of Operational Research*, 121:343–362, 2000.

[63] R. Dautray and J.-L. Lions. *Mathematical Analysis and Numerical Methods for Science and Technology*, volume 5: Evolution Problems I. Springer-Verlag, Berlin, 1992.

[64] N. de Nevers. *Air Pollution Control Engineering*. McGraw Hill, Boston, 2nd edition, 2000.

[65] M. A. Demetriou. Activation policy of smart controllers for flexible structures with multiple actuator/sensor pairs. In A. El Jai and M. Fliess, editors, *Proc. 14th Int. Symp. Mathematical Theory of Networks and Systems*, Perpignan, France, 19–23 June 2000, 2000. Published on CD-ROM.

[66] M. A. Demetriou, A. Paskaleva, O. Vayena, and H. Doumanidis. Scanning actuator guidance scheme in a 1-D thermal manufacturing process. *IEEE Transactions on Control Systems Technology*, 11(5):757–764, 2003.

[67] M. A. Demetriou and I. G. Rosen. On the persistence of excitation in the adaptive estimation of distributed parameter systems. Technical Report 93-4, Center for Applied Mathematical Sciences, University of Southern California, Los Angeles, Mar. 1993.

[68] A. Y. Dubovitskii and A. A. Milyutin. Extremum problems in the presence of restrictions. *U.S.S.R. Computational Mathematics and Mathematical Physics*, 5(3):395–453, 1965. (In Russian).

[69] G. E. Dullerud and F. Paganini. *A Course in Robust Control Theory. A Convex Approach.* Springer-Verlag, New York, 2000.

[70] A. Dvoretzky. Stochastic approximation revisited. *Advances in Applied Mathematics*, 7:220–227, 1986.

[71] G. N. Dyubin and V. G. Suzdal. *Introduction to Applied Game Theory.* Nauka, Moscow, 1981. (In Russian).

[72] A. El Jai. Distributed systems analysis via sensors and actuators. *Sensors and Actuators A*, 29:1–11, 1991.

[73] A. El Jai and M. Amouroux, editors. *Proceedings of the First International Workshop on Sensors and Actuators in Distributed Parameter Systems*, Perpignan, France, 16–18 December 1987, Perpignan, 1987. IFAC.

[74] A. El Jai and M. Amouroux. *Automatique des systèmes distribués.* Hermès, Paris, 1990.

[75] A. El Jai and A. J. Pritchard. *Sensors and Controls in the Analysis of Distributed Systems.* John Wiley & Sons, New York, 1988.

[76] Z. Emirsajłow. *The Linear Quadratic Control Problem for Infinite Dimensional Systems with Terminal Targets.* Technical University Publishers, Szczecin, 1991.

[77] S. M. Ermakov, editor. *Mathematical Theory of Experimental Design.* Nauka, Moscow, 1983. (In Russian).

[78] S. M. Ermakov and A. A. Zhigljavsky. *Mathematical Theory of Optimal Experiments.* Nauka, Moscow, 1987. (In Russian).

[79] M. Eslami. *Theory of Sensitivity in Dynamic Systems. An Introduction.* Springer-Verlag, Berlin, 1994.

[80] R. E. Ewing and J. H. George. Identification and control for distributed parameters in porous media flow. In F. Kappel, K. Kunisch, and W. Schappacher, editors, *Proc. 2nd Int. Conf. Distributed Parameter Systems*, Vorau, Austria, 1984, Lecture Notes in Control and Information Sciences, pages 145–161, Berlin, 1984. Springer-Verlag.

[81] R. R. Fedorenko. *Approximate Solution of Optimal Control Problems.* Nauka, Moscow, 1978. (In Russian).

[82] V. V. Fedorov. *Theory of Optimal Experiments.* Academic Press, New York, 1972.

[83] V. V. Fedorov. Optimal design with bounded density: Optimization algorithms of the exchange type. *Journal of Statistical Planning and Inference*, 22:1–13, 1989.

[84] V. V. Fedorov. Design of spatial experiments: Model fitting and prediction. Technical Report TM-13152, Oak Ridge National Laboratory, Oak Ridge, TN, 1996.

[85] V. V. Fedorov and P. Hackl. *Model-Oriented Design of Experiments*. Lecture Notes in Statistics. Springer-Verlag, New York, 1997.

[86] V. V. Fedorov and V. Khabarov. Duality of optimal design for model discrimination and parameter estimation. *Biometrika*, 73(1):183–190, 1986.

[87] K. Felsenstein. Optimal Bayesian design for discriminating among rival models. *Computational Statistics & Data Analysis*, 14:427–436, 1992.

[88] B. G. Fitzpatrick. Bayesian analysis in inverse problems. *Inverse Problems*, 7:675–702, 1991.

[89] B. G. Fitzpatrick. Large sample behavior in Bayesian analysis of nonlinear regression models. *Journal of Mathematical Analysis and Applications*, 192:607–626, 1995.

[90] B. G. Fitzpatrick and G. Yin. Empirical distributions in least squares estimation for distributed parameter systems. *Journal of Mathematical Systems, Estimation, and Control*, 5(1):37–57, 1995.

[91] W. H. Fleming and R. W. Rishel. *Deterministic and Stochastic Optimal Control*. Applications of Mathematics. Springer-Verlag, New York, 1975.

[92] I. Ford, D. M. Titterington, and C. P. Kitsos. Recent advances in nonlinear experimental design. *Technometrics*, 31(1):49–60, 1989.

[93] M. Galicki. The planning of robotic optimal motions in the presence of obstacles. *The International Journal of Robotics Research*, 17(3):248–259, Mar. 1998.

[94] M. Galicki and D. Uciński. Optimal control of robots under constraints on state variables. *Studia z Automatyki i Informatyki*, 23:95–111, 1998. (In Polish).

[95] M. Galicki and D. Uciński. Optimal trajectory planning for many redundant manipulators. In *Proc. 6th Nat. Conf. Robotics*, Świeradów-Zdrój, 9–12 September 1998, volume 1, pages 171–178, Wrocław, 1998. Technical University Press. (In Polish).

[96] M. Galicki and D. Uciński. Time-optimal motions of robotic manipulators. *Robotica*, 18:659–667, 2000.

[97] W. Gander and J. Hřebíček. *Solving Problems in Scientific Computing Using MAPLE and MATLAB*. Springer-Verlag, Berlin, 1993.

[98] J. Gilbert and L. Gilbert. *Elements of Modern Algebra*. Prindle, Weber & Schmidt, Boston, 1984.

[99] M. X. Goemans. Semidefinite programming in combinatorial optimization. *Mathematical Programming*, 79:143–162, 1997.

[100] L. Goldstein. Minimizing noisy functionals in Hilbert space: An extension of the Kiefer-Wolfowitz procedure. *Journal of Theoretical Probability*, 1(2):189–204, 1988.

[101] G. C. Goodwin and R. L. Payne. *Dynamic System Identification. Experiment Design and Data Analysis*. Mathematics in Science and Engineering. Academic Press, New York, 1977.

[102] V. Gopal and L. T. Biegler. Large scale inequality constrained optimization and control. *IEEE Control Systems*, 18(6):59–68, 1998.

[103] P. Grabowski. *Lecture Notes on Optimal Control Systems*. University of Mining and Metallurgy Publishers, Cracow, 1999.

[104] W. H. Greene. *Econometric Analysis*. Prentice Hall, Upper Saddle River, NJ, 5th edition, 2003.

[105] W. A. Gruver and E. Sachs. *Algorithmic Methods in Optimal Control*. Pitman Publishing Limited, London, 1980.

[106] W. W. Hager. Updating the inverse of a matrix. *SIAM Review*, 31(2):221–239, 1989.

[107] L. M. Haines. Optimal design for nonlinear regression models. *Communications in Statistics — Theory and Methods*, 22(6):1613–1627, 1993.

[108] P. W. Hammer. Parameter identification in parabolic partial differential equations using quasilinearization. *Journal of Mathematical Systems, Estimation, and Control*, 6(3):251–295, 1996.

[109] R. F. Hartl, S. P. Sethi, and R. G. Vickson. A survey of the maximum principles for optimal control problems with state constraints. *SIAM Review*, 37(2):181–218, June 1995.

[110] E. J. Haug, K. K. Choi, and V. Komkov. *Design Sensitivity Analysis of Structural Systems*. Mathematics in Science and Engineering. Academic Press, Orlando, FL, 1986.

[111] A. Heck. *Introduction to Maple*. Springer-Verlag, New York, 2nd edition, 1996.

[112] J. W. Helton and O. Merino. *Classical Control Using $\mathcal{H}^\infty$ Methods. Theory, Optimization and Design*. Society for Industrial and Applied Mathematics, Philadelphia, 1998.

[113] R. Hettich and K. O. Kortanek. Semi-infinite programming: Theory, methods and applications. *SIAM Review*, 35(3):380–429, 1993.

[114] N. G. Hogg. Oceanographic data for parameter estimation. In P. Malanotte-Rizzoli, editor, *Modern Approaches to Data Assimilation in Ocean Modeling*, Elsevier Oceanography, pages 57–76. Elsevier, Amsterdam, 1996.

[115] P. Holnicki, A. Kałuszko, M. Kurowski, R. Ostrowski, and A. Żochowski. An urban-scale computer model for short-term prediction of air pollution. *Archiwum Automatyki i Telemechaniki*, XXXI(1–2):51–71, 1986.

[116] D. Holtz and J. S. Arora. An efficient implementation of adjoint sensitivity analysis for optimal control problems. *Structural Optimization*, 13:223–229, 1997.

[117] A. D. Ioffe and V. M. Tikhomirov. *Theory of Extremal Problems*. Nauka, Moscow, 1974. (In Russian).

[118] V. Isakov. *Inverse Problems for Partial Differential Equations*. Applied Mathematical Sciences. Springer-Verlag, New York, 1998.

[119] M. Z. Jacobson. *Fundamentals of Atmospheric Modeling*. Cambridge University Press, Cambridge, 1999.

[120] J. Jakubowski and R. Sztencel. *An Introduction to Probability Theory*. SCRIPT, Warsaw, 2000. (In Polish).

[121] R. I. Jennrich. Asymptotic properties of non-linear least squares estimators. *The Annals of Mathematical Statistics*, 40(2):633–643, 1969.

[122] C. Johnson. *Numerical Solution of Partial Differential Equations by the Finite Element Method*. Cambridge University Press, Cambridge, 1987.

[123] R. Kalaba and K. Spingarn. *Control, Identification and Input Optimization*. Plenum Press, New York, 1982.

[124] D. C. Kammer. Sensor placement for on-orbit modal identification and correlation of large space structures. In *Proc. American Control Conf.*, San Diego, California, 23–25 May 1990, volume 3, pages 2984–2990, 1990.

[125] D. C. Kammer. Effects of noise on sensor placement for on-orbit modal identification of large space structures. *Transactions of the ASME*, 114:436–443, Sept. 1992.

[126] L. V. Kantorovich and G. P. Akilov. *Functional Analysis*. Nauka, Moscow, 1984. (In Russian).

[127] S. Karlin and W. J. Studden. *Tchebycheff Systems: With Applications in Analysis and Statistics*. John Wiley & Sons, New York, 1966.

[128] P. Kazimierczyk. Optimal experiment design; Vibrating beam under random loading. Technical Report 7/1989, Institute of Fundamental Technological Research, Polish Academy of Sciences, Warsaw, 1989.

[129] L. G. Khachiyan. A polynomial algorithm for linear programming. *Soviet Mathematics Doklady*, 20:191–194, 1979.

[130] A. Y. Khapalov. Optimal measurement trajectories for distributed parameter systems. *Systems & Control Letters*, 18(6):467–477, June 1992.

[131] A. I. Khuri and J. A. Cornell. *Response Surfaces. Design and Analyses.* Statistics: Textbooks and Monographs. Marcel Dekker, New York, 2nd edition, 1996.

[132] J. Kiefer and J. Wolfowitz. Optimum designs in regression problems. *The Annals of Mathematical Statistics*, 30:271–294, 1959.

[133] J. Klamka. *Controllability of Dynamical Systems.* Mathematics and Its Applications. Kluwer Academic Publishers, Dordrecht, The Netherlands, 1991.

[134] W. Kołodziej. *Mathematical Analysis.* Polish Scientific Publishers, Warsaw, 1979. (In Polish).

[135] J. Korbicz and D. Uciński. Sensors allocation for state and parameter estimation of distributed systems. In W. Gutkowski and J. Bauer, editors, *Proc. IUTAM Symposium*, Zakopane, Poland, 31 August–3 September 1993, pages 178–189, Berlin, 1994. Springer-Verlag.

[136] J. Korbicz, D. Uciński, A. Pieczyński, and G. Marczewska. Knowledge-based fault detection and isolation system for power plant. *Applied Mathematics and Computer Science*, 3(3):613–630, 1993.

[137] J. Korbicz, M. Z. Zgurovsky, and A. N. Novikov. Suboptimal sensors location in the state estimation problem for stochastic non-linear distributed parameter systems. *International Journal of Systems Science*, 19(9):1871–1882, 1988.

[138] J. Korbicz and M. Z. Zgurowski. *Estimation and Control of Stochastic Distributed-Parameter Systems.* Państwowe Wydawnictwo Naukowe, Warsaw, 1991. (In Polish).

[139] K. Kovarik. *Numerical Models in Grounwater Pollution.* Springer-Verlag, Berlin, 2000.

[140] A. Kowalewski. *Optimal Control of Infinite Dimensional Distributed Parameter Systems with Delays.* University of Mining and Metallurgy Publishers, Cracow, 2001. (In Polish).

[141] C. Kravaris and J. H. Seinfeld. Identification of parameters in distributed parameter systems by regularization. *SIAM Journal on Control and Optimization*, 23(2):217–241, Mar. 1985.

[142] A. Królikowski and P. Eykhoff. Input signals design for system identification: A comparative analysis. In *Prep. 7-th IFAC/IFORS Symp. Identification and System Parameter Estimation, York*, pages 915–920, 1985.

[143] M. Krzyśko. *Lectures on Probability Theory*. Wydawnictwa Naukowo-Techniczne, Warsaw, 2000. (In Polish).

[144] C. S. Kubrusly. Distributed parameter system identification: A survey. *International Journal of Control*, 26(4):509–535, 1977.

[145] C. S. Kubrusly and H. Malebranche. Sensors and controllers location in distributed systems — A survey. *Automatica*, 21(2):117–128, 1985.

[146] K. Kunisch. A review of some recent results on the output least squares formulation of parameter estimation problems. *Automatica*, 24(4):531–539, 1988.

[147] H. J. Kushner and A. Shwartz. Stochastic approximation in Hilbert space: Identification and optimization of linear continuous parameter systems. *SIAM Journal on Control and Optimization*, 23(5):774–793, Sept. 1985.

[148] H. J. Kushner and G. G. Yin. *Stochastic Approximation Algorithms and Applications*. Applications of Mathematics. Springer-Verlag, New York, 1997.

[149] P. K. Lamm. Estimation of discontinuous coefficients in parabolic systems: Applications to reservoir simulation. *SIAM Journal on Control and Optimization*, 25(1):18–37, Jan. 1987.

[150] K. Lange. *Numerical Analysis for Statisticians*. Springer-Verlag, New York, 1999.

[151] I. Lasiecka. Active noise control in an acoustic chamber: Mathematical theory. In S. Domek, R. Kaszyński, and L. Tarasiejski, editors, *Proc. 5th Int. Symp. Methods and Models in Automation and Robotics*, Międzyzdroje, Poland, 25–29 August 1998, volume 1, pages 13–22, Szczecin, 1998. Wyd. Uczelniane Polit. Szczecińskiej.

[152] I. Lasiecka and R. Triggiani. *Control Theory for Partial Differential Equations: Continuous and Approximation Theories*, volume I and II of *Encyclopedia of Mathematics and Its Applications*. Cambridge University Press, Cambridge, 2000.

[153] H. W. J. Lee, K. L. Teo, and A. E. B. Lim. Sensor scheduling in continuous time. *Automatica*, 37:2017–2023, 2001.

[154] H. W. J. Lee, K. L. Teo, V. Rehbock, and L. S. Jennings. Control parametrization enhancing technique for optimal discrete-valued control problems. *Automatica*, 35:1401–1407, 1999.

[155] A. S. Lewis. Derivatives of spectral functions. *Mathematics of Operations Research*, 21(3):576–588, 1996.

[156] A. S. Lewis. Nonsmooth analysis of eigenvalues. *Mathematical Programming*, 84:1–24, 1999.

[157] A. S. Lewis. The mathematics of eigenvalue optimization. *Mathematical Programming*, Ser. B 97:155–176, 2003.

[158] A. S. Lewis and M. L. Overton. Eigenvalue optimization. *Acta Numerica*, 5:149–190, 1996.

[159] LF 95. *Lahey/Fujitsu Fortran 95 Language Reference. Revision F.* Lahey Computer Systems, Inc., Incline Village, NV, 2000.

[160] J.-L. Lions. *Contrôle optimal de systèmes gouvernés par des équations aux dérivées partielles.* Études mathématiques. Dunod, Paris, 1968.

[161] Y. Lou and P. D. Christofides. Optimal actuator/sensor placement for nonlinear control of the Kuramoto-Sivashinsky equation. *IEEE Transactions on Control Systems Technology*, 11(5):737–745, 2003.

[162] B. Lucquin and O. Pironneau. *Introduction to Scientific Computing.* John Wiley & Sons, Chichester, 1998.

[163] D. G. Luenberger. *Optimization by Vector Space Methods.* John Wiley & Sons, New York, 1969.

[164] K. C. P. Machielsen. *Numerical Solution of Optimal Control Problems with State Constraints by Sequential Quadratic Programming in Function Space.* PhD thesis, Technische Universiteit Eindhoven, Eindhoven, The Netherlands, Mar. 1987.

[165] E. B. Magrab. *An Engineer's Guide to Matlab.* Prentice Hall, Upper Saddle River, NJ, 2000.

[166] P. Malanotte-Rizzoli, editor. *Modern Approaches to Data Assimilation in Ocean Modeling.* Elsevier Oceanography. Elsevier, Amsterdam, 1996.

[167] K. Malanowski, Z. Nahorski, and M. Peszyńska, editors. *Modelling and Optimization of Distributed Parameter Systems.* International Federation for Information Processing. Kluwer Academic Publishers, Boston, 1996.

[168] H. Malebranche. Simultaneous state and parameter estimation and location of sensors for distributed systems. *International Journal of Systems Science*, 19(8):1387–1405, 1988.

[169] K. Malinowski. Issues of hierarchical computations. In R. Wyrzykowski, B. Mochnacki, H. Piech, and J. Szopa, editors, *Proc. 3rd Int. Conf. Parallel Processing & Applied Mathematics*, Kazimierz Dolny, Poland, 14–17 September 1999, pages 110–120, Czestochowa, 1999.

[170] G. I. Marchuk. *Methods of Computational Mathematics.* Nauka, Moscow, 1989. (In Russian).

[171] H. Marcinkowska. *Distributions, Sobolev Spaces, Differential Equations.* Polish Scientific Publishers, Warsaw, 1993. (In Polish).

[172] MathWorks. *Using MATLAB. Version 6.* The MathWorks, Inc., Natick, MA, 2002.

[173] K. Maurin. *Analysis. Vol. I: Elements.* Mathematical Library. Polish Scientific Publishers, Warsaw, 5th edition, 1991. (In Polish).

[174] R. K. Mehra. Optimal input signals for parameter estimation in dynamic systems-survey and new results. *IEEE Transactions on Automatic Control,* 19(6):753–768, Dec. 1974.

[175] R. K. Mehra. Optimization of measurement schedules and sensor designs for linear dynamic systems. *IEEE Transactions on Automatic Control,* 21(1):55–64, 1976.

[176] L. P. Meissner. *Essential Fortran 90 & 95. Common Subset Edition.* Unicomp, 1997.

[177] V. P. Mikhailov. *Partial Differential Equations.* Nauka, Moscow, 1976. (In Russian).

[178] A. J. Miller. Non-linear programming using DONLP2: Fortran 90 version. Code obtainable from http://www.ozemail.com.au/~milleraj/donlp2.html, 1998.

[179] W. Mitkowski. *Stabilization of Dynamic Systems.* Wydawnictwa Naukowo-Techniczne, Warsaw, 1991. (In Polish).

[180] K. W. Morton. *Numerical Solution of Convection-Diffusion Problems.* Chapman & Hall, London, 1996.

[181] W. G. Müller. *Collecting Spatial Data. Optimum Design of Experiments for Random Fields.* Contributions to Statistics. Physica-Verlag, Heidelberg, 2nd revised edition, 2001.

[182] W. G. Müller and A. Pázman. Design measures and approximate information matrices for experiments without replications. *Journal of Statistical Planning and Inference,* 71:349–362, 1998.

[183] W. G. Müller and A. Pázman. Measures for designs in experiments with correlated errors. *Biometrika,* 90(2):423–434, 2003.

[184] W. G. Müller and A. C. Ponce de Leon. Discriminating between two binary data models: Sequentially designed experiments. *Journal of Statistical Computation and Simulation,* 55:87–100, 1996.

[185] A. Munack. Optimal sensor allocation for identification of unknown parameters in a bubble-column loop bioreactor. In A. V. Balakrishnan

and M. Thoma, editors, *Analysis and Optimization of Systems, Part 2,* Lecture Notes in Control and Information Sciences, volume 63, pages 415–433. Springer-Verlag, Berlin, 1984.

[186] R. E. Munn. *The Design of Air Quality Monitoring Networks.* Macmillan Publishers Ltd, London, 1981.

[187] K. Nakano and S. Sagara. Optimal measurement problem for a stochastic distributed parameter system with movable sensors. *International Journal of Systems Science,* 12(12):1429–1445, 1981.

[188] K. Nakano and S. Sagara. Optimal scanning measurement problem for a stochastic distributed-parameter system. *International Journal of Systems Science,* 19(7):1069–1083, 1988.

[189] W. Näther. *Effective Observation of Random Fields.* BSB B. G. Teubner Verlagsgesellschaft, Leipzig, 1985.

[190] I. M. Navon. Practical and theoretical aspects of adjoint parameter estimation and identifiability in meteorology and oceanography. available at http://www.scri.fsu.edu/~navon/pubs/index.html.

[191] P. Neittaanmäki and D. Tiba. *Optimal Control of Nonlinear Parabolic Systems. Theory, Algorithms, and Applications.* Monographs and Textbooks in Pure and Applied Mathematics. Marcel Dekker, Inc., New York, 1994.

[192] R. Nixdorf. An invariance principle for a finite dimensional stochastic approximation method in a Hilbert space. *Journal of Multivariate Analysis,* 15:252–260, 1984.

[193] M. Nørgaard, O. Ravn, N. K. Poulsen, and L. K. Hansen. *Neural Networks for Modelling and Control of Dynamic Systems. A Practitioner's Handbook.* Advanced Textbooks in Control and Signal Processing. Springer-Verlag, London, 2000.

[194] D. Nychka, W. W. Piegorsch, and L. H. Cox, editors. *Case Studies in Environmental Statistics.* Lecture Notes in Statistics, volume 132. Springer-Verlag, New York, 1998.

[195] D. Nychka and N. Saltzman. Design of air-quality monitoring networks. In D. Nychka, W. W. Piegorsch, and L. H. Cox, editors, *Case Studies in Environmental Statistics,* Lecture Notes in Statistics, volume 132, pages 51–76. Springer-Verlag, New York, 1998.

[196] T. E. O'Brien and J. O. Rawlings. A nonsequential design procedure for parameter estimation and model discrimination in nonlinear regression models. *Journal of Statistical Planning and Inference,* 55:77–93, 1996.

[197] D. Y. Oh and H. C. No. Determination of the minimal number and optimal sensor location in a nuclear system with fixed incore detectors. *Nuclear Engineering and Design,* 152:197–212, 1994.

[198] S. Omatu and K. Matumoto. Distributed parameter identification by regularization and its application to prediction of air pollution. *International Journal of Systems Science*, 22(10):2001–2012, 1991.

[199] S. Omatu and K. Matumoto. Parameter identification for distributed systems and its application to air pollution estimation. *International Journal of Systems Science*, 22(10):1993–2000, 1991.

[200] S. Omatu and J. H. Seinfeld. *Distributed Parameter Systems: Theory and Applications*. Oxford Mathematical Monographs. Oxford University Press, New York, 1989.

[201] M. L. Overton. Large-scale optimization of eigenvalues. *SIAM Journal on Control and Optimization*, 2(1):88–120, 1992.

[202] P. Y. Papalambros and D. J. Wilde. *Principles of Optimal Design. Modeling and Computation*. Cambridge University Press, Cambridge, 2nd edition, 2000.

[203] A. Papoulis. *Probability, Random Variables, and Stochastic Processes*. McGraw Hill, New York, 3rd edition, 1991.

[204] E. Parzen. An approach to time series analysis. *The Annals of Mathematical Statistics*, 32(4):951–989, 1961.

[205] E. Parzen. *Stochastic Processes*. Holden-Day, San Francisco, 1962.

[206] M. Patan. *Optimal Observation Strategies for Parameter Estimation of Distributed Systems*. PhD thesis, University of Zielona Góra, Zielona Góra, Poland, 2004.

[207] V. V. Patel, G. Deodhare, and T. Viswanath. Some applications of randomized algorithms for control system design. *Automatica*, 38:2085–2092, 2002.

[208] A. Pázman. *Foundations of Optimum Experimental Design*. Mathematics and Its Applications. D. Reidel Publishing Company, Dordrecht, The Netherlands, 1986.

[209] A. Pázman and W. G. Müller. Optimal design of experiments subject to correlated errors. *Statistics & Probability Letters*, 52:29–34, 2001.

[210] G. C. Pflug. *Optimization of Stochastic Models. The Interface Between Simulation and Optimization*. Engineering and Computer Science: Discrete Event Dynamic Systems. Kluwer Academic Publishers, Boston, 1996.

[211] G. A. Phillipson. *Identification of Distributed Systems*. Modern Analytic and Computational Methods in Science and Mathematics. Elsevier, New York, 1971.

[212] D. A. Pierre. *Optimization Theory with Applications.* Series in Decision and Control. John Wiley & Sons, New York, 1969.

[213] N. Point, A. Vande Wouwer, and M. Remy. Practical issues in distributed parameter estimation: Gradient computation and optimal experiment design. *Control Engineering Practice,* 4(11):1553–1562, 1996.

[214] E. Polak. On the mathematical foundations of nondifferentiable optimization in engineering design. *SIAM Review,* 29(1):21–89, 1987.

[215] E. Polak. *Optimization. Algorithms and Consistent Approximations.* Applied Mathematical Sciences. Springer-Verlag, New York, 1997.

[216] M. P. Polis. The distributed system parameter identification problem: A survey of recent results. In *Proc. 3rd IFAC Symp. Control of Distributed Parameter Systems,* Toulouse, France, pages 45–58, 1982.

[217] A. C. Ponce de Leon and A. C. Atkinson. Optimum experimental design for discriminating between two rival models in the presence of prior information. *Biometrika,* 78(3):601–608, 1991.

[218] W. H. Press, S. A. Teukolsky, W. T. Vetterling, and B. P. Flannery. *Numerical Recipes in FORTRAN. The Art of Parallel Scientific Computing.* Cambridge University Press, Cambridge, 2nd edition, 1992.

[219] W. H. Press, S. A. Teukolsky, W. T. Vetterling, and B. P. Flannery. *Numerical Recipes in Fortran 90. The Art of Parallel Scientific Computing.* Cambridge University Press, Cambridge, 2nd edition, 1996.

[220] L. Pronzato. Removing non-optimal support points in D-optimum design algorithms. *Statistics & Probability Letters,* 63:223–228, 2003.

[221] L. Pronzato and É. Walter. Robust experiment design via stochastic approximation. *Mathematical Biosciences,* 75:103–120, 1985.

[222] L. Pronzato and É. Walter. Robust experiment design via maximin optimization. *Mathematical Biosciences,* 89:161–176, 1988.

[223] B. N. Pshenichnyi. *Necessary Conditions for an Extremum.* Marcel Dekker, Inc., New York, 1971.

[224] F. Pukelsheim. *Optimal Design of Experiments.* Probability and Mathematical Statistics. John Wiley & Sons, New York, 1993.

[225] F. Pukelsheim and S. Rieder. Efficient rounding of approximate designs. *Biometrika,* 79(4):763–770, 1992.

[226] F. Pukelsheim and A. Wilhelm. A contribution to the discussion of the paper "Constrained optimization of experimental design" by D. Cook and V. Fedorov. *Statistics,* 26:168–172, 1995.

[227] R. Pytlak. *Numerical Methods for Optimal Control Problems with State Constraints.* Springer-Verlag, Berlin, 1999.

[228] R. Pytlak and R. B. Vinter. A feasible directions algorithm for optimal control problems with state and control constraints: Convergence analysis. *SIAM Journal on Control and Optimization*, 36:1999–2019, 1998.

[229] R. Pytlak and R. B. Vinter. Feasible direction algorithm for optimal control problems with state and control constraints: Implementation. *Journal of Optimization Theory and Applications*, 101(3):623–649, June 1999.

[230] A. Quarteroni, R. Sacco, and F. Saleri. *Numerical Mathematics*. Springer-Verlag, New York, 2000.

[231] Z. H. Quereshi, T. S. Ng, and G. C. Goodwin. Optimum experimental design for identification of distributed parameter systems. *International Journal of Control*, 31(1):21–29, 1980.

[232] E. Rafajłowicz. Design of experiments for parameter identification of the static distributed systems. *Systems Science*, 4(4):349–361, 1978.

[233] E. Rafajłowicz. Design of experiments for eigenvalue identification in distributed-parameter systems. *International Journal of Control*, 34(6):1079–1094, 1981.

[234] E. Rafajłowicz. Optimal experiment design for identification of linear distributed-parameter systems: Frequency domain approach. *IEEE Transactions on Automatic Control*, 28(7):806–808, July 1983.

[235] E. Rafajłowicz. Optimization of actuators for distributed parameter systems identification. *Problems of Control and Information Theory*, 13(1):39–51, 1984.

[236] E. Rafajłowicz. Adaptive input sequence design for linear distributed-parameter systems identification. *Large Scale Systems*, 11:43–58, 1986.

[237] E. Rafajłowicz. *Choice of Optimum Input Signals in Linear Distributed-Parameter Systems Identification*. Monographs. Technical University Press, Wrocław, 1986. (In Polish).

[238] E. Rafajłowicz. Optimum choice of moving sensor trajectories for distributed parameter system identification. *International Journal of Control*, 43(5):1441–1451, 1986.

[239] E. Rafajłowicz. Information equivalence of sensors-controllers configurations in identification of homogeneous static distributed systems. In Y. Sunahara, S. G. Tzafestas, and T. Futagami, editors, *Proc. IMACS/IFAC Int. Symp. Modelling and Simulation of Distributed Parameter Systems*, Hiroshima, Japan, 6–9 October 1987, pages 553–557, 1987.

[240] E. Rafajłowicz. Distributed parameter systems analysis by tracking stationary points of their states. *International Journal of Systems Science*, 19(4):613–620, 1988.

[241] E. Rafajłowicz. Reduction of distributed system identification complexity using intelligent sensors. *International Journal of Control*, 50(5):1571–1576, 1989.

[242] E. Rafajłowicz. Time-domain optimization of input signals for distributed-parameter systems identification. *Journal of Optimization Theory and Applications*, 60(1):67–79, 1989.

[243] E. Rafajłowicz. Optimum input signals for parameter estimation in systems described by linear integral equations. *Computational Statistics & Data Analysis*, 9:11–19, 1990.

[244] E. Rafajłowicz. Optimal measurement schedules and sensors' location for estimating intensities of random sources. In S. Bańka, S. Domek, and Z. Emirsajłow, editors, *Proc. 2nd Int. Symp. Methods and Models in Automation and Robotics*, Międzyzdroje, Poland, 30 August–2 September 1995, volume 1, pages 169–174, Szczecin, 1995. Wyd. Uczelniane Polit. Szczecińskiej.

[245] E. Rafajłowicz. *Algorithms of Experimental Design with Implementations in MATHEMATICA*. Akademicka Oficyna Wydawnicza PLJ, Warsaw, 1996. (In Polish).

[246] E. Rafajłowicz. Selective random search for optimal experiment designs. In A. C. Atkinson, L. Pronzato, and H. P. Wynn, editors, *MODA 5 — Advances in Model-Oriented Data Analysis and Experimental Design: Proc. 5th Int. Workshop in Marseilles*, France, June 22–26 1998, Contributions to Statistics, chapter 9, pages 75–83. Physica-Verlag, Heidelberg, 1998.

[247] E. Rafajłowicz and W. Myszka. Computational algorithm for input-signal optimization in distributed-parameter systems identification. *International Journal of Systems Science*, 17(6):911–924, 1986.

[248] M. M. Rao. *Measure Theory and Integration*. John Wiley & Sons, New York, 1987.

[249] P. A. Raviart and J.-M. Thomas. *Introduction à l'analyse numérique des équations aux dérivées partielles*. Mathématiques appliquées pour la maîtrise. Masson, Paris, 1992.

[250] R. Reemtsen and S. Görner. Numerical methods for semi-infinite programming: A survey. In R. Reemtsen and J.-J. Rückmann, editors, *Semi-Infinite Programming*, pages 195–275. Kluwer Academic Publishers, Boston, 1998.

[251] H. Reinhard. *Équations aux dérivées partielles. Introduction.* Dunod, Paris, 1991.

[252] M. Renardy and R. C. Rogers. *An Introduction to Partial Differential Equations.* Texts in Applied Mathematics. Springer-Verlag, New York, 1993.

[253] P. Révész. Robbins-Monro procedure in a Hilbert space and its application in the theory of learning processes I. *Studia Scientiarum Mathematicarum Hungarica,* 8:391–398, 1973.

[254] P. Révész. Robbins-Monro procedure in a Hilbert space II. *Studia Scientiarum Mathematicarum Hungarica,* 8:469–472, 1973.

[255] R. T. Rockafellar. *Convex Analysis.* Princeton University Press, Princeton, 1970.

[256] S. Rolewicz. *Functional Analysis and Control Theory.* Polish Scientific Publishers, Warsaw, 1974. (In Polish).

[257] E. W. Sacks. Semi-infinite programming in control. In R. Reemtsen and J.-J. Rückmann, editors, *Semi-Infinite Programming,* pages 389–411. Kluwer Academic Publishers, Boston, 1998.

[258] J. Sacks and D. Ylvisaker. Designs for regression problems with correlated errors. *The Annals of Mathematical Statistics,* 37(1):66–89, 1966.

[259] J. Sacks and D. Ylvisaker. Designs for regression problems with correlated errors: Many parameters. *The Annals of Mathematical Statistics,* 39(1):49–69, 1968.

[260] J. Sacks and D. Ylvisaker. Designs for regression problems with correlated errors III. *The Annals of Mathematical Statistics,* 41(6):2057–2074, 1970.

[261] M. Sahm and R. Schwabe. A note on optimal bounded designs. In A. Atkinson, B. Bogacka, and A. Zhigljavsky, editors, *Optimum Design 2000,* chapter 13, pages 131–140. Kluwer Academic Publishers, Dordrecht, The Netherlands, 2001.

[262] K. Schittkowski. *Numerical Data Fitting in Dynamical Systems — A Practical Introduction with Applications and Software.* Kluwer Academic Publishers, Dordrecht, The Netherlands, 2002.

[263] R. Schwabe. *Optimum Designs for Multi-Factor Models.* Springer-Verlag, New York, 1996.

[264] A. L. Schwartz. *Theory and Implementation of Numerical Methods Based on Runge-Kutta Integration for Solving Optimal Control Problems.* PhD thesis, University of California, Berkeley, 1996.

[265] G. A. F. Seber and C. J. Wild. *Nonlinear Regression.* John Wiley & Sons, New York, 1989.

[266] K. Shimizu and E. Aiyoshi. Necessary conditions for min-max problems and algorithms by a relaxation procedure. *IEEE Transactions on Automatic Control,* AC-25(1):62–66, 1980.

[267] A. Shwartz and N. Berman. Abstract stochastic approximations and applications. *Stochastic Processes and their Applications,* 31:133–149, 1989.

[268] J. Sikora. *Numerical Approaches to the Impedance and Eddy-Current Tomographies.* Technical University Press, Warsaw, 2000. (In Polish).

[269] S. D. Silvey. *Optimal Design. An Introduction to the Theory for Parameter Estimation.* Chapman and Hall, London, 1980.

[270] S. D. Silvey, D. M. Titterington, and B. Torsney. An algorithm for optimal designs on a finite design space. *Communications in Statistics — Theory and Methods,* 14:1379–1389, 1978.

[271] S. Simons. Minimax theorems. In C. A. Floudas and P. M. Pardalos, editors, *Encyclopedia of Optimization,* volume 3, pages 284–289. Kluwer Academic Publishers, Dordrecht, The Netherlands, 2001.

[272] E. Skubalska-Rafajłowicz and E. Rafajłowicz. Searching for optimal experimental designs using space-filling curves. *Applied Mathematics and Computer Science,* 8(3):647–656, 1998.

[273] J. Sokołowski and J.-P. Zolesio. *Introduction to Shape Optimization: Shape Sensitivity Analysis.* Computational Mathematics. Springer-Verlag, Berlin, 1992.

[274] J. C. Spall. *Introduction to Stochastic Search and Optimization. Estimation, Simulation, and Control.* John Wiley & Sons, Hoboken, NJ, 2003.

[275] P. Spellucci. A new technique for inconsistent QP problems in the SQP method. *Mathematical Methods of Operations Research,* 47:355–400, 1998.

[276] P. Spellucci. An SQP method for general nonlinear programs using only equality constrained subproblems. *Mathematical Programming,* 82:413–448, 1998.

[277] M. C. Spruill and W. J. Studden. A Kiefer-Wolfowitz theorem in a stochastic process setting. *The Annals of Statistics,* 7(6):1329–1332, 1979.

[278] R. C. St. John and N. R. Draper. D-optimality for regression designs: A review. *Technometrics,* 17(1):15–23, 1975.

[279] W. E. Stewart, Y. Shon, and G. E. P. Box. Discrimination and goodness of fit of multiresponse mechanistic models. *AIChE Journal*, 44(6):1404–1412, 1998.

[280] J. F. Sturm. *Primal-Dual Interior Point Approach to Semidefinite Programming*. PhD thesis, Erasmus University, Rotterdam, 1997. Tinbergen Institute Series 156.

[281] P. J. Sturm, R. A. Almbauer, and R. Kunz. Air quality study for the city of Graz, Austria. In H. Power, N. Moussiopoulos, and C. A. Brebbia, editors, *Urban Air Pollution, Vol. 1*, chapter 2, pages 43–100. Computational Mechanics Publications, Southampton, 1994.

[282] N.-Z. Sun. *Inverse Problems in Groundwater Modeling*. Theory and Applications of Transport in Porous Media. Kluwer Academic Publishers, Dordrecht, The Netherlands, 1994.

[283] N.-Z. Sun. *Mathematical Modeling of Groundwater Pollution*. Springer-Verlag, New York, 1996.

[284] A. Sydow, T. Lux, P. Mieth, M. Schmidt, and S. Unger. The DYMOS model system for the analysis and simulation of regional air pollution. In R. Grützner, editor, *Modellierung und Simulation im Umweltbereich*, pages 209–219. Vieweg-Verlag, Wiesbaden, 1997.

[285] A. Sydow, T. Lux, H. Rosé, W. Rufeger, and B. Walter. Conceptual design of the branch-oriented simulation system DYMOS (dynamic models for smog analysis). *Transactions of the Society for Computer Simulation International*, 15(3):95–100, 1998.

[286] T. Szántai. Approximation of multivariate probability integrals. In C. A. Floudas and P. M. Pardalos, editors, *Encyclopedia of Optimization*, volume 1, pages 53–59. Kluwer Academic Publishers, Dordrecht, The Netherlands, 2001.

[287] J. C. Taylor. *An Introduction to Measure and Probability*. Springer-Verlag, New York, 1997.

[288] R. Tempo, E. W. Bai, and F. Dabbene. Probabilistic robustness analysis: Explicit bounds for the minimum number of samples. *Systems & Control Letters*, 30:237–242, 1997.

[289] R. Tempo and F. Dabbene. Probabilistic robustness analysis and design of uncertain systems. In G. Picci and D. S. Gilliam, editors, *Dynamical Systems, Control, Coding, Computer Vision. New Trends, Interfaces, and Interplay*, Progress in Systems and Control Theory, pages 263–282. Birkhäuser, Basel, 1999.

[290] K. L. Teo and Z. S. Wu. *Computational Methods for Optimizing Distributed Systems*. Academic Press, Orlando, FL, 1984.

[291] A. N. Tikhonov and V. Y. Arsenin. *Solutions of Ill-Posed Problems.* John Wiley & Sons, New York, 1977.

[292] D. M. Titterington. Aspects of optimal design in dynamic systems. *Technometrics*, 22(3):287–299, 1980.

[293] C. Tocci and S. Adams. *Applied Maple for Engineers and Scientists.* Artech House, Boston, 1996.

[294] B. Torsney. Computing optimising distributions with applications in design, estimation and image processing. In Y. Dodge, V. V. Fedorov, and H. P. Wynn, editors, *Optimal Design and Analysis of Experiments*, pages 316–370. Elsevier, Amsterdam, 1988.

[295] B. Torsney and S. Mandal. Construction of constrained optimal designs. In A. Atkinson, B. Bogacka, and A. Zhigljavsky, editors, *Optimum Design 2000*, chapter 14, pages 141–152. Kluwer Academic Publishers, Dordrecht, The Netherlands, 2001.

[296] D. A. Tortorelli and P. Michaleris. Design sensitivity analysis: Overview and review. *Inverse Problems in Engineering*, 1(1):71–105, 1994.

[297] H. J. H. Tuenter. The minimum L_2-distance projection onto the canonical simplex: A simple algorithm. *Algo Research Quarterly*, pages 53–56, Dec. 2001.

[298] D. Uciński. Optimal sensor location for parameter identification of distributed systems. *Applied Mathematics and Computer Science*, 2(1):119–134, 1992.

[299] D. Uciński. Optimal design of moving sensors trajectories for identification of distributed parameter systems. In *Proc. 1st Int. Symp. Methods and Models in Automation and Robotics*, Międzyzdroje, Poland, 1–3 September 1994, volume 1, pages 304–309, Szczecin, 1994. Wyd. Uczelniane Polit. Szczecińskiej.

[300] D. Uciński. Optimal selection of measurement locations for identification of parameters in distributed systems. In S. Bańka, S. Domek, and Z. Emirsajłow, editors, *Proc. 2nd Int. Symp. Methods and Models in Automation and Robotics*, Międzyzdroje, Poland, 30 August–2 September 1995, volume 1, pages 175–180, Szczecin, 1995. Wyd. Uczelniane Polit. Szczecińskiej.

[301] D. Uciński. Optimal planning of moving sensor trajectories in parameter estimation of distributed systems. In *Proc. 6th Nat. Conf. Robotics, Świeradów-Zdrój, 24–26 September 1996*, volume 1, pages 225–233, Wrocław, 1996. Technical University Press. (In Polish).

[302] D. Uciński. Optimal planning of sensor movements along given paths for distributed parameter systems identification. In *Proc. 3rd Int.*

Symp. Methods and Models in Automation and Robotics, Międzyzdroje, Poland, 10–13 September 1996, volume 1, pages 187–192, Szczecin, 1996. Wyd. Uczelniane Polit. Szczecińskiej.

[303] D. Uciński. Measurement optimization with moving sensors for parameter estimation of distributed systems. In A. Sydow, editor, *Proc. 15th IMACS World Congress on Scientific Computation, Modelling and Applied Mathematics,* Berlin, Germany, August 1997, volume 5, pages 191–196, Berlin, 1997. Wissenschaft & Technik Verlag.

[304] D. Uciński. Optimal location of scanning sensors for parameter identification of distributed systems from noisy experimental data. In P. S. Szczepaniak, editor, *Proc. 9th Int. Symp. System-Modelling-Control,* Zakopane, Poland, 27 April–1 May 1998, 1998. Published on CD-ROM.

[305] D. Uciński. A robust approach to the design of optimal trajectories of moving sensors for distributed-parameter systems identification. In A. Beghi, L. Finesso, and G. Picci, editors, *Proc. 13th Int. Symp. Mathematical Theory of Networks and Systems,* Padova, Italy, 6–10 July 1998, pages 551–554, Padova, 1998. Il Poligrafo.

[306] D. Uciński. Towards a robust-design approach to optimal location of moving sensors in parameter identification of DPS. In S. Domek, R. Kaszyński, and L. Tarasiejski, editors, *Proc. 5th Int. Symp. Methods and Models in Automation and Robotics,* Międzyzdroje, Poland, 25–29 August 1998, volume 1, pages 85–90, Szczecin, 1998. Wyd. Uczelniane Polit. Szczecińskiej.

[307] D. Uciński. Optimum design of sensor locations in parameter estimation of distributed systems. *Studia z Automatyki i Informatyki,* 24:151–167, 1999. (In Polish).

[308] D. Uciński. Robust design of sensor locations in parameter estimation of distributed systems. In Z. Bubnicki and J. Józefczyk, editors, *Proc. 13th Nat. Conf. Automatics,* Opole, Poland, 21–24 September 1999, volume 1, pages 289–292, 1999. (In Polish).

[309] D. Uciński. A technique of robust sensor allocation for parameter estimation in distributed systems. In P. M. Frank, editor, *Proc. 5th European Control Conf.,* Karlsruhe, Germany, 31 August–3 September 1999. EUCA, 1999. Published on CD-ROM.

[310] D. Uciński. Optimal sensor location for parameter estimation of distributed processes. *International Journal of Control,* 73(13):1235–1248, 2000.

[311] D. Uciński. Optimization of sensors' allocation strategies for parameter estimation in distributed systems. *Systems Analysis — Modelling — Simulation,* 37:243–260, 2000.

[312] D. Uciński. Sensor motion planning with design criteria in output space. In A. C. Atkinson, P. Hackl, and W. Müller, editors, mODa 6, *Proc. 6th Int. Workshop on Model-Oriented Data Analysis*, Puchberg/Schneeberg, Austria, 2001, pages 235–242, Heidelberg, 2001. Physica-Verlag.

[313] D. Uciński and A. C. Atkinson. Experimental design for time-dependent models with correlated observations. *Studies in Nonlinear Dynamics & Econometrics*, 8(2), 2004. Article No. 13.

[314] D. Uciński and B. Bogacka. T-optimum design for discrimination between two multivariate dynamic models. *Journal of the Royal Statistical Society: Series B (Statistical Methodology)*, 2004. (Accepted for publication).

[315] D. Uciński and B. Bogacka. T-optimum designs for multiresponse dynamic heteroscedastic models. In A. Di Bucchianico, H. Läuter, and H. P. Wynn, editors, mODa 7, *Proc. 7th Int. Workshop on Model-Oriented Data Analysis*, Heeze, The Netherlands, 2004, pages 191–199, Heidelberg, 2004. Physica-Verlag.

[316] D. Uciński and M. A. Demetriou. An approach to the optimal scanning measurement problem using optimum experimental design. In *Proc. American Control Conference*, Boston, MA, 2004. Published on CD-ROM.

[317] D. Uciński and A. El Jai. On weak spreadability of distributed-parameter systems and its achievement via linear-quadratic control techniques. *IMA Journal of Mathematical Control and Information*, 14:153–174, 1997.

[318] D. Uciński and S. El Yacoubi. Modelling and simulation of an ecological problem by means of cellular automata. In S. Domek, R. Kaszyński, and L. Tarasiejski, editors, *Proc. 5th Int. Symp. Methods and Models in Automation and Robotics*, Międzyzdroje, Poland, 25–29 August 1998, volume 1, pages 289–293, Szczecin, 1998. Wyd. Uczelniane Polit. Szczecińskiej.

[319] D. Uciński and S. El Yacoubi. Parameter estimation of cellular automata models. In R. Wyrzykowski, B. Mochnacki, H. Piech, and J. Szopa, editors, *Proc. 3rd Int. Conf. Parallel Processing & Applied Mathematics*, Kazimierz Dolny, Poland, 14–17 September 1999, pages 168–176, Częstochowa, 1999.

[320] D. Uciński and J. Korbicz. Parameter identification of two-dimensional distributed systems. *International Journal of Systems Science*, 21(2):2441–2456, 1990.

[321] D. Uciński and J. Korbicz. An algorithm for the computation of optimal measurement trajectories in distributed parameter systems iden-

tification. In A. Isidori, S. Bittanti, E. Mosca, A. D. Luca, M. D. Di Benedetto, and G. Oriolo, editors, *Proc. 3rd European Control Conf.*, Roma, Italy, 5–8 September 1995, volume 2, pages 1267–1272. EUCA, 1995.

[322] D. Uciński and J. Korbicz. Optimal location of movable sensors for distributed-parameter system identification. In P. Borne, M. Staroswiecki, J. P. Cassar, and S. El Khattabi, editors, *Proc. Symp. Control, Optimization and Supervision of 1st IMACS Int. Multiconf. CESA'96*, Lille, France, 9–12 July 1996, volume 2, pages 1004–1009, 1996.

[323] D. Uciński and J. Korbicz. Measurement optimization with minimax criteria for parameter estimation in distributed systems. In G. Bastin and M. Gevers, editors, *Proc. 4rd European Control Conf.*, Brussels, Belgium, 1–4 July 1997. EUCA, 1997. Published on CD-ROM.

[324] D. Uciński and J. Korbicz. Optimization of sensors' allocation strategies for parameter estimation in distributed systems. In P. Borne, M. Ksouri, and A. El Kamel, editors, *Proc. Symp. Modeling, Analysis and Control of 2nd IMACS Int. Multiconf. CESA'98*, Nabeul-Hammamet, Tunisia, 1–4 April 1998, volume 1, pages 128–133, 1998.

[325] D. Uciński and J. Korbicz. On robust design of sensor trajectories for parameter estimation of distributed systems. In *Proc. 14th IFAC World Congress*, Beijing, China, 5–9 July 1999, volume H: Modeling, Identification, Signal Processing I, pages 73–78, 1999.

[326] D. Uciński and J. Korbicz. Path planning for moving sensors in parameter estimation of distributed systems. In *Proc. 1st Workshop on Robot Motion and Control RoMoCo'99*, Kiekrz, Poland, 28–29 June, 1999, pages 273–278, 1999.

[327] D. Uciński and J. Korbicz. Optimal sensor allocation for parameter estimation in distributed systems. *Journal of Inverse and Ill-Posed Problems*, 9(3):301–317, 2001.

[328] D. Uciński, J. Korbicz, and M. B. Zaremba. On optimization of sensors motions in parameter identification of two-dimensional distributed systems. In J. W. Neuwenhuis, C. Praagman, and H. L. Trentelman, editors, *Proc. 2nd European Control Conf.*, Groningen, The Netherlands, 28 June–1 July 1993, volume 3, pages 1359–1364. ECCA, 1993.

[329] D. Uciński and M. Patan. Optimal location of discrete scanning sensors for parameter estimation of distributed systems. In *Proc. 15th IFAC World Congress*, Barcelona, Spain, 22–26 July 2002, 2002. Published on CD-ROM.

[330] M. van de Wal and B. de Jager. A review of methods for input/output selection. *Automatica*, 37:487–510, 2001.

[331] M. van Loon. Numerical smog prediction, I: The physical and chemical model. Technical Report NM-R9411, Centrum voor Wiskunde en Informatica, Amsterdam, 1994.

[332] M. van Loon. Numerical smog prediction, II: Grid refinement and its application to the Dutch smog prediction model. Technical Report NM-R9523, Centrum voor Wiskunde en Informatica, Amsterdam, 1995.

[333] A. Vande Wouwer, N. Point, S. Porteman, and M. Remy. On a practical criterion for optimal sensor configuration — Application to a fixed-bed reactor. In *Proc. 14th IFAC World Congress,* Beijing, China, 5–9 July, 1999, volume I: Modeling, Identification, Signal Processing II, Adaptive Control, pages 37–42, 1999.

[334] L. Vandenberghe and S. Boyd. Semidefinite programming. *SIAM Review*, 38(1):49–95, 1996.

[335] L. Vandenberghe and S. Boyd. Connections between semi-infinite and semidefinite programming. In R. Reemtsen and J.-J. Rückmann, editors, *Semi-Infinite Programming*, pages 277–294. Kluwer Academic Publishers, Boston, 1998.

[336] L. Vandenberghe and S. Boyd. Applications of semidefinite programming. *Applied Numerical Mathematics*, 29:283–299, 1999.

[337] L. Vandenberghe, S. Boyd, and S.-P. Wu. Determinant maximization with linear matrix inequality constraints. *SIAM Journal on Matrix Analysis and Applications*, 19(2):499–533, 1998.

[338] A. Venot, L. Pronzato, É. Walter, and J.-F. Lebruchec. A distribution-free criterion for robust identification, with applications in system modelling and image processing. *Automatica*, 22(1):105–109, 1986.

[339] A. D. Ventzel. *A Course on the Theory of Stochastic Processes.* Nauka, Moscow, 1975. (In Russian).

[340] M. Vidyasagar. *A Theory of Learning and Generalization with Applications to Neural Networks and Control Systems.* Springer-Verlag, London, 1997.

[341] M. Vidyasagar. An introduction to some statistical aspects of PAC learning theory. *Systems & Control Letters*, 34:115–124, 1998.

[342] M. Vidyasagar. Statistical learning theory and randomized algorithms for control. *IEEE Control Systems*, 18(6):69–85, Dec. 1998.

[343] M. Vidyasagar. Randomized algorithms for robust controller synthesis using statistical learning theory. *Automatica*, 37:1515–1528, 2001.

[344] G. Wahba. On the regression design problem of Sacks and Ylvisaker. *The Annals of Mathematical Statistics*, 42(3):1035–1053, 1971.

[345] G. Wahba. Parameter estimation in linear dynamic systems. *IEEE Transactions on Automatic Control*, AC-25(2):235–238, 1980.

[346] H. Walk and L. Zsidó. Convergence of the Robbins-Monro method for linear problems in a Banach space. *Journal of Mathematical Analysis and Applications*, 139:152–177, 1989.

[347] É. Walter and L. Pronzato. Robust experiment design: Between qualitative and quantitative identifiabilities. In É. Walter, editor, *Identifiability of Parametric Models*, pages 104–113. Pergamon Press, Oxford, 1987.

[348] É. Walter and L. Pronzato. Qualitative and quantitative experiment design for phenomenological models — A survey. *Automatica*, 26(2):195–213, 1990.

[349] É. Walter and L. Pronzato. *Identification of Parametric Models from Experimental Data*. Communications and Control Engineering. Springer-Verlag, Berlin, 1997.

[350] J. Warga. *Optimal Control of Differential and Functional Equations*. Academic Press, New York, 1972.

[351] D. Williams. *Weighing the Odds. A Course in Probability and Statistics*. Cambridge University Press, Cambridge, 2001.

[352] W.-K. Wong. A unified approach to the construction of minimax designs. *Biometrika*, 79(3):611–619, 1992.

[353] G. Yin and B. G. Fitzpatrick. On invariance principles for distributed parameter identification algorithms. *Informatica*, 3(1):98–118, 1992.

[354] G. Yin and Y. M. Zhu. On H-valued Robbins-Monro processes. *Journal of Multivariate Analysis*, 34:116–140, 1990.

[355] M. B. Zarrop. *Optimal Experiment Design for Dynamic System Identification*. Springer-Verlag, New York, 1979.

[356] M. B. Zarrop and G. C. Goodwin. Comments on "Optimal inputs for system identification". *IEEE Transactions on Automatic Control*, AC-20(2):299–300, Apr. 1975.

[357] H. Zwart and J. Bontsema. An application-driven guide through infinite-dimensional systems theory. In G. Bastin and M. Gevers, editors, *European Control Conference 1997: Plenaries and Mini-Courses*, pages 289–328. CIACO, Ottignies/Louvain-la-Neuve, 1997.

Index

Milton Keynes UK
Ingram Content Group UK Ltd.
UKHW020319111024
449327UK00040B/1398